Biostimulants for Sustainable Agriculture and Crop Production

Biostimulants for Sustainable Agriculture and Crop Production

Edited by Thelma Bosso

Larsen & Keller
www.larsen-keller.com

Biostimulants for Sustainable Agriculture and Crop Production
Edited by Thelma Bosso
ISBN: 979-8-88836-076-7 (Hardback)

Larsen & Keller
Published by Larsen and Keller Education,
5 Penn Plaza,
19th Floor,
New York, NY 10001, USA

Cataloging-in-Publication Data

Biostimulants for sustainable agriculture and crop production / edited by Thelma Bosso.
p. cm.
Includes bibliographical references and index.
ISBN 979-8-88836-076-7
1. Plant growth promoting substances. 2. Sustainable agriculture. 3. Growth (Plants). 4. Agriculture. I. Bosso, Thelma.
SB128 .B56 2023
631.8--dc23

For more information regarding Larsen and Keller Education and its products, please visit the publisher's website www.larsen-keller.com

Table of Contents

Preface

This book aims to highlight the current researches and provides a platform to further the scope of innovations in this area. This book is a product of the combined efforts of many researchers and scientists, after going through thorough studies and analysis from different parts of the world. The objective of this book is to provide the readers with the latest information of the field.

Biostimulants are synthetic or natural substances that can be used in soil, seeds and plants, and which influence tolerance towards abiotic stress, uptake of nutrients and nutrient efficiency. Seaweed extracts, helpful fungi and bacteria, humic acids, and liquid manure composting are some of the several types of biostimulants. They are increasingly being used in sustainable agricultural practices. These practices strive to strike a balance between the need to produce food and preserving the ecological system. The three primary goals of sustainable agriculture are maintaining environmental health, ensuring economic profitability, and guaranteeing social and economic equity. Biostimulants are compatible with the most progressive farming practices utilized in integrated crop management. They act as soil conditioners and assist the plants in securing nutrients, increasing crop productivity, and repelling pests. This book aims to shed light on the use of biostimulants in sustainable agriculture and crop production. It presents researches and studies performed by experts across the globe. This book will prove immensely beneficial to professionals and students involved in this area of agriculture science.

I would like to express my sincere thanks to the authors for their dedicated efforts in the completion of this book. I acknowledge the efforts of the publisher for providing constant support. Lastly, I would like to thank my family for their support in all academic endeavors.

Editor

1

The Application of a Plant Biostimulant Based on Seaweed and Yeast Extract Improved Tomato Fruit Development and Quality

Giuseppe Mannino [1], Cristina Campobenedetto [1,2], Ivano Vigliante [2], Valeria Contartese [2], Carla Gentile [3] and Cinzia M. Bertea [1,*]

1 Department of Life Sciences and Systems Biology, Innovation Centre, Plant Physiology Unit, University of Turin, 10135 Turin, Italy; giuseppe.mannino@unito.it (G.M.); cristina.campobenedetto@unito.it (C.C.)
2 Green Has Italia S.p.A, 12043 Canale (CN), Italy; c.campobenedetto@greenhasitalia.it (C.C.); i.vigliante@greenhasitalia.com (I.V.); v.contartese@greenhasitalia.it (V.C.)
3 Department of Biological, Chemical and Pharmaceutical Sciences and Technologies (STEBICEF), University of Palermo, 90128 Palermo, Italy; carla.gentile@unipa.it
* Correspondence: cinzia.bertea@unito.it

Abstract: Plant biostimulants are under investigation as innovative products to improve plant production and fruit quality, without resulting in environmental and food contaminations. Here, the effects of the application of Expando, a biostimulant based on seaweed and yeast extracts, on plant productivity, fruit ripening times, and fruit quality of *Solanum lycopersicum* var. Micro-Tom were evaluated. After biostimulant treatment, a two-week reduction of ripening times and a concomitant enhancement of the production percentage during the earliest ripening times, in terms of both fruit yield (+110%) and size (+85%), were observed. Concerning fruit quality, proximate analysis showed that tomatoes treated with the biostimulant had better nutritional composition compared to untreated samples, since both the quality of unsatured fatty acids (C16:3ω3: +328%; C18:2ω6: −23%) and micronutrients essential for human health (Fe: +14%; Cu: +21%; Zn: +24%) were increased. From a nutraceutical point of view, despite strong changes in bioactive compound profile not being observed, an increase of the antioxidant properties was recorded in fruits harvested by plants treated with the biostimulant (2,2′-azino-bis(3-ethylbenzothiazoline-6-sulphonic acid (ABTS): +38%; 2,2-diphenyl-1-picrylhydrazyl (DPPH): +11%). In conclusion, the biostimulant application was able to reduce the ripening times and fruit size, while slightly increasing nutritional and nutraceutical values, leading to more marketable tomato fruits.

Keywords: *Solanum lycopersicum*; polyphenols; lycopene; tocopherols; carotenoids; mineral content; fruit quality; ripening time; fruit size; fruit quality

---

1. Introduction

In the last decades, the world population has been constantly growing, and it is estimated to reach 10.6 billion in 2050 [1]. Feeding the current world population is considered already today a hard challenge. To date, waste reduction, changing diet habits, and enhancing the potential production of crops is applied as a strategy to monitor this global warning [2]. On the other hand, considering global climate changes, salinization, desertification, natural erosion, and fertility reduction of soils, finding new areas to use as agricultural land is not an easy task to achieve [1]. Consequently, the improvement of plant resilience to adverse conditions and the improvement of crop production by using agrochemicals and organic fertilizers has largely been employed in the agronomic practices [2,3]. However, in the last century, the use of these products massively increased worldwide, resulting not only in a well-awaited intensification of food production, but also in serious environmental contaminations [2,4]. In addition, scientific evidence suggests that the use of agrochemicals may affect crop quality, in terms of sensorial, nutritional, and nutraceutical traits [2–4], and is also a potential risk for human health [5,6].

Food safety is nowadays a primary concern for the vast majority of consumers [2,4], and in particular they are worried about chemical hazards, especially those perceived to be invisible and having long term effects or serious health implications [5,6]. For this reason, together with consumer awareness toward the benefits of food intake with high nutritional and nutraceutical properties [7,8], consumers' preference for organic foods grown in an eco-sustainable way is also rising because they are perceived to be safe, healthy, and environmentally friendly [9].

In order to satisfy both agronomic and consumer needs, sustainable agriculture should be a valid alternative to conventional techniques, providing food sufficiently for all, while reducing environmental risks and allowing the production of high quality foods [10]. Moreover, changes in current unsustainable agricultural practices into more sustainable procedures may not only contribute to food security but also to mitigating climate change [11]. In this context, the use of plant wastes as natural biostimulants has aimed to increase sustainable productivity, and quality of fruit has turned out to be one of the most attractive frontiers in the agronomic field [12–16]. Despite both the differences in the raw materials employed and the large variety of phytochemicals present in the various biostimulant formulations [17], their application in very small amounts seems to be enough to promote plant growth. In particular, scientific evidence shows that treatment with biostimulants can affect plant growth processes by enhancing water and/or nutrient uptake [18], root and shoot growth [12,14,19], activity of key enzymes involved in primary or secondary metabolism [17,20], and plant antioxidant defense system [21]. Moreover, treatments with biostimulants seem to be able to positively affect fruit quality in terms of nutritional values and nutraceutical properties [13–16]. However, with the exception of few cases, the effects of biostimulant treatment on crops was mainly investigated at the seedling stage, or measuring plant physiological parameters under abiotic stresses [18,21–23]. Moreover, despite the literature presenting a few scientific papers relating to the effects derived from the application of commercial biostimulants [24,25], little knowledge is available on the effect of biostimulant application on fruit yield, proximate composition, and nutraceutical properties in the absence of environmental stress.

Here, we evaluated the potential effect derived from the application of a commercial biostimulant based on seaweed and selected yeast extracts in terms of productivity, pomological traits, proximate composition, and nutraceutical values of tomatoes (*Solanum lycopersicum* L. var. Micro-Tom) produced by plants grown in standard conditions. Seaweed and yeast extracts are receiving a greater acceptance in horticulture as plant biostimulants due to several beneficial effects against abiotic stresses [17,26–28]. However, the effects deriving from the application of biostimulants based on the combination of seaweed and yeast extracts on tomato fruit quality grown under standard conditions are scarcely available in the literature.

In this work, in order to understand if the biostimulant could affect plant productivity, fruit ripening times, and fruit quality, we monitored tomato plants after the application of the biostimulant under controlled conditions. In particular, concerning the evaluation of the tomato quality, we evaluated both nutritional (macro- and micro-nutrients) and nutraceutical (total polyphenol content, total carotenoid content, and antioxidant activities).

2. Materials and Methods

2.1. Plant Material and Treatments with the Biostimulant

Tomato (*Solanum lycopersicum* L. var. Micro-Tom) seeds were sowed in greenhouse under controlled temperature (day: 25 ± 3 °C; night: 18 ± 4 °C) and humidity (63% ± 6%) conditions. At the first true leaf appearance, tomato plants were transplanted in pots containing soil (50% unfertilized peat + 50%

expanded clay) and grown until fruit production stage. For the experimentation, treatments were performed using a commercial biostimulant (Expando®, Green Has Italia S.P.A., Canale, Italy) applied via foliar application. The label of the product claims to contain 3% (*w*/*w*) organic nitrogen, 4% (*w*/*w*) phosphoric anhydride, 6% (*w*/*w*) potassium oxide, 0.02% (*w*/*w*) boron, 0.1% (*w*/*w*) molybdenum, 0.02% (*w*/*w*) manganese, and 12% (*w*/*w*) organic carbon. The pH (in 1% (*w*/*w*) water solution) and electrical conductivity (in water solution 1g L^{-1}) were respectively 6.50 ± 0.50 and 350 µS cm^{-1}. Concerning the treatments, the plants were divided into 3 different groups. The first one included 20 plants that were treated 3 times (during flowering, fruit-set, and fruit growing stages) with a biostimulant dosage equal to 1.5 mL/L^{-1} (triple treatment). The second groups included plants that received an additional biostimulant application during ripening phase at the same dosage (quadruple treatment). The last group was composed by plants treated using the same experimental protocol but with replacing the biostimulant with water. All plants received the same maintenance, and Hoagland solution was applied to the plants twice a week. Treatment application was performed using a hand sprayer until the complete wetting of all foliar surfaces (run-off condition), and without the use of surfactants.

2.2. *Evaluation of Productivity and Pomological Analysis*

Sixty days after the sowing, tomato fruits were harvested upon reaching a full and uniform red color. Fruits that met ripening criteria were collected every 4 days over a 30-day period. For each harvest time (HT), the fruits were counted and weighted using a digital scale (Cubis Sartorius, Gottingen, Germany). Yield per plant was determined by collecting all fruits from each plant. Moreover, after the last HT, the fruits of each tomato plant were grouped and photographed using a digital camera (Canon EOS M100, Tokyo, Japan). The resulting images were analyzed using ImageJ software and the external diameter (ED) was automatically calculated. On the basis of this parameter, we divided the tomatoes into 16 different fruit size classes, as shown in Figure 1.

fruit caliber	1\2	3\4	5\6	7\8	9\10	11\12	13\14	15\16
from (cm)	< 0.450	0.525	0.675	0.825	0.975	1.125	1.275	> 1.425
to (cm)	0.525	0.675	0.825	0.975	1.125	1.275	1.425	

Figure 1. Tomato caliber classes and their relative external diameter (cm). After the last harvesting time, all fully colored fruits were grouped and photographed. In order to divide the fruits into the different size classes, we analyzed the resulting images by ImageJ.

2.3. *Proximate Composition*

2.3.1. Sugar Content

The total content of sugars was determined as previously described [8]. Briefly, the fresh tomatoes were pulped and homogenized, and the resulting juices were vortexed (4000× *g*, 4 °C). Two grams of the centrifuged juice were mixed with 1 mL of 5% (*v*/*v*) phenol (Sigma Aldrich, Waltham, MA, USA) in a glass tube, and subsequently 5 mL of H_2SO_4 was rapidly added to the mixture. During the 10 min of incubation time, carbohydrates contained in the juice produced furfural derivatives that were spectrophotometrically detected at 490 nm. Reference solutions were prepared in the same manner but replacing the juice with distilled water. The measure was repeated twice. Quantification was performed using an external calibration curve of glucose, and the results were expressed as milligram per 100 g of fresh weight (FW).

2.3.2. Protein Content

In order to quantify the protein content in tomatoes, we employed the Kjeldahl method as previously described [8,29]. Briefly, 1 gram of dried tomato sample was accurately weighed and then digested at 420 °C for 1 h with 10 mL of H_2SO_4. Then, 15 mL of 30% (*w*/*v*) H_2O_2 was slowly added in the presence of 0.2 g of CuO (8S Heating Digester, Velp Scientifica, Usmate Velate, Italy). Digested samples were then distilled using UDK Semi-Automatic Kjeldahl Distillation Unit (Velp Scientifica, Usmate Velate, Italy). In order to produce an alkaline distillation environment, we employed 32% (*w*/*v*) NaOH and distilled ammonia. Finally, each sample was separately collected in a volumetric flask containing 25 mL of 4% (*w*/*v*) H_3BO_3. The titrations were performed with standardized 0.25 N HCl. The total protein content was indirectly calculated using a nitrogen conversion factor of 6.25 [30].

2.3.3. Lipid Content and Fatty Acid Profile

Fifty grams of dried tomato was refluxed in a Soxhlet apparatus using 250 mL of petroleum benzene in a weighed glass flask, as previously described [31]. The oils were recovered by distilling the organic solvent in a rotary evaporator at 50 °C and until a constant weight was measured. Fatty acid profile was obtained via trans-esterification of 500 mg of tomato sample with 1 mL of 10% (*w*/*v*) boron tri-fluoride solubilized in methanol [31,32]. Fifty micrograms of heptadecanoic acid (C17:0) were added in each sample as internal standard. The obtained fatty acid methyl esters (FAMEs) were purified by the serial addition of 1 mL water and 1 mL hexane. Purification was repeated twice, and the organic phases were pooled together. After centrifugation (8000× *g*, 5 min, 4 °C) and dehydration by using anhydrous $MgSO_4$, the phases containing FAMEs were employed for gas chromatographic (GC) analysis. FAME identification was performed by Gas Chromatography (GC) coupled with Mass Spectrometry (MS) (GCMS-TQ8040, Shimadzu, Kyoto, Japan) and quantified by Gas Chromatography with Flame Ionization Detection (FID) (GC-2010 Plus, Shimadzu, Kyoto, Japan). For the quantification, an external curve of heptadecanoic-methyl ester (Limit of Detection: 0.2 μg/mL; Limit of Quantification: 0.7 μg/mL) was employed. Helium was used as GC carrier during the analysis with a constant flux of 1 mL min^{-1}. The chromatographic separation was obtained in ZB5-MS (30 m length, 250 μm diameter and stationary phase thickness of 0.25 μm, 5% phenyl-arylene, and 95% poly-dimethyl siloxane) column (Phenomenex, Torrance, CA, USA). Chromatographic temperature conditions employed for the separation consisted of 60 °C held for 1 min, and then raised 10 °C per minute until 180 °C. Temperature was then brought to 230 °C in 40 min and to 320 °C in 20 min. This temperature was held for an additional 5 min. The same column and chromatographic conditions were used for both GC–MS analyses. Concerning the mass spectrometer, the ionization energy of the ion source was set to 70 eV and the acquisition mode to 50–350 *m*/*z*. Compounds were identified through comparison of mass fragmentation spectra with reference NIST 98 spectra or by comparison of Kovats indexes and internal standard co-injection of pure standards (Sigma-Aldrich, Waltham, MA, USA).

2.3.4. Moisture, Ashes, and Mineral Content

Water content was evaluated by drying the homogenized tomato fruits in an oven (Nuve, Ankara, Turkey) at 90 °C for at least 2 hours and until the dried weight remained constant. Through the weight loss, the percentage of water contained in the samples was calculated. The total mineral content was calculated by incubating the samples in a muffle at 400 °C for 24 h. For the determination of the mineral content, we digested tomato samples using a MARS 6 microwave digestion system (CEM, Matthews, NC, USA), as previously described [8]. Briefly, approximately 500 mg of dried tomato sample was weighed directly into a 100 mL polytetrafluoroethylene (PTFE) digestion vial, and then 9 mL of (65% *w*/*w*) HNO_3 and 1 mL of (30% *w*/*v*) H_2O_2 were added. The temperature was gradually increased to 200 °C over 20 min and held for further 15 min to ensure complete digestion. Once cooled, the digested samples were diluted to a final volume of 50 mL with distilled H_2O. All measurements were performed using a microwave plasma-atomic emission spectrometer (Agilent 4200 MP-AES) fitted with a double-pass cyclonic spray chamber and OneNeb nebulizer. N_2 gas was supplied from a tank. All wavelengths were selected in the software library according to the sensitivity that was required. The calibration standards were prepared by diluting a 1000 mg/L multi-element standard solution (Sigma Aldrich, Waltham, MA, USA and Scharlab S.L., Barcelona, Spain) in 1% (*v*/*v*) HNO_3.

2.4. Ascorbic Acid Content

Ascorbic acid content (AAC) was determined as previously described by Bajaj [33], with some modifications. Briefly, the fresh fruits were crashed and homogenized, and then extracted 1:20 (*w*/*v*) using a mixture containing 0.05 M oxalic acid and 0.2 mM ethylenediaminetetraacetic acid (EDTA) for 10 min. Consequently, the extracts were filtered through a filter-paper and centrifuged (5000× *g*, 10 min, 4 °C). To 5 mL of the surfactant, we added 0.5 mL of a mixture containing 0.9 M HPO_3 and 40% (*v*/*v*) acetic acid. Finally, 1 mL of 5% (*v*/*v*) H_2SO_4 and 2 mL of 5% (*w*/*v*) ammonium molybdate were added. After 15 min, the absorbance was measured at 760 nm against a blank prepared replacing the sample with 5 mL of the mixture of EDTA/oxalic acid. Quantification was performed using an external calibration curve of pure ascorbic acid (AA), and the results were expressed as milligram of AA per 100 g of FW.

2.5. Total Tocopherol Content

Total tocopherol content (TTC) was determined as previously reported [34]. Briefly, the fresh fruits were pulped and homogenized, and then placed in a Soxhlet apparatus with the aim of performing the extraction using hexane/ethyl acetate mixture at 85:15 (*v*/*v*). The extraction was performed in the dark for 8 h. Finally, samples were centrifuged (6000× *g*, 24 °C) for 10 min, and then 1 mL of the surfactant was added to 3 mL of a mixture containing 0.6 M H_2SO_4, 28 mM Na_2PO_4, and 4 mM $MoNH_4$. After reduction of molybdenum during the incubation at 37 °C for 90 min, the absorbance of each sample was measured at 695 nm against a blank prepared replacing the sample with water. Quantification was performed using an external calibration curve of pure α-tocopherol, and the results were expressed as milligram of α-tocopherol per 100 g of FW.

2.6. Total Carotenoid Content

The total carotenoid content (TCC) was evaluated as previously reported [23], with some modifications [7]. Briefly, the homogenized tomato samples were extracted with 50:25:25 (*v*/*v*/*v*) hexane/acetone/ethanol mixture, using a 1:10 (*w*/*v*) ratio. After vortexing and centrifugation (10 min at 5000× *g*, 25 °C), the upper phase containing carotenoids was removed using a glass pipette and collected in a glass test tube. Extraction was repeated twice, and the different organic phases were pooled together. The combined organic fraction was firstly washed with 10 mL of saturated aqueous NaCl and then with 5 mL of 10% (*w*/*v*) K_2CO_3. Finally, the organic layer was dried with $CaCl_2$. The excess of solvent was allowed to evaporate at room temperature for a few minutes in the dark. The entire extraction procedure was repeated twice with the aim to obtain 3 different technical replicates. One hundred milligrams of the obtained dried extract was mixed with 10 mL of 2:3 (*v*/*v*) acetone/hexane mixture for 1 min and then filtered on a paper filter. The absorbance was recorded at 3 different wavelengths (453, 505, and 663 nm) using a UV–VIS spectrophotometer (Cary 50, Agilent Technologies, Santa Clara, CA, USA), and TCC was calculated using the following formula:

$$TCC\left(\frac{mg}{100}mL\right) = 0.216 \times Abs_{663} - 0.304 \times Abs_{505} + 0.452 \times Abs_{453}$$

The obtained data were adjusted for both the extraction volume and sample weight, and expressed as micrograms of β-carotene per 100 g of FW.

2.7. *Lycopene Content*

The amount of lycopene was spectrophotometrically determined as previously described by Rayhan and colleagues [35], but with some modifications. Briefly, the fresh fruits were pulped and homogenized, and the obtained juices were extracted repeatedly using 1:1 (*v*/*v*) acetone/petroleum ether mixture until the complete color loss of the samples. Lycopene was collected in the petroleum ether fraction after the addition of an equal volume of deionized water. In order to completely remove lycopene from the acetone/water fraction, after the separation of the petroleum ether phase from the acetone/aqueous phase, we again extracted the acetone fraction with an equal amount of petroleum ether. The different petroleum ether fractions were combined, and the residual water content was removed by the adding Na_2SO_4 until the formation of a precipitate. The extraction procedure was repeated twice for each sample in order to obtain 3 different technical replicates. After an appropriate dilution, the color of the petroleum ether fraction was measured spectrophotometrically (Cary 50, Agilent Technologies, Santa Clara, CA, USA) in a quartz cuvette at 503 nm, against a blank consisting of petroleum ether. Lycopene content of the samples was calculated by using an external calibration curve of pure lycopene, and data were expressed as milligram of lycopene per 100 g of FW.

2.8. *Total Polyphenol Content*

Polyphenols were extracted from the fruits following the protocol previously described [34], with some modifications [36]. Briefly, the homogenized tomato samples were weighted and extracted with 90% (*v*/*v*) MetOH using a 1:10 (*w*/*v*) ratio. Samples were sonicated at room temperature for 15 min and then macerated under constant agitation for 24 h. After maceration, they were centrifuged (15 min at 6000× *g*, 4 °C) and the resulting supernatants were filtered. Methanolic extracts were stored at −20 °C until total polyphenol content (TPC) was determined via Folin–Ciocalteu assay, as previously reported [36]. Quantification was performed using an external calibration curve with gallic acid (GA). Analyses were performed in triplicate, and data were expressed as millimole GA equivalents (GAE) per 100 g of FW.

2.9. *Evaluation of Antioxidant Properties*

The same methanolic extracts used for TPC quantification were employed for the evaluation of the fruit antioxidant properties. The determination of the antioxidant properties included both the evaluation of radical scavenging activity via 2,2′-azino-bis(3-ethylbenzothiazoline-6-sulphonic acid (ABTS) and 2,2-diphenyl-1-picrylhydrazyl (DPPH) assay, and the evaluation of the reducing power via ferric reducing antioxidant power (FRAP) assay [34]. The inhibition percentage for each assay was measured using the following equation:

$$AA\% = \frac{A_{blank} - A_{sample}}{A_{blank}} \times 100$$

where AA% is the percentage of color reduction of the reagent, A_{blank} is the absorbance of blank, and A_{sample} is the absorbance of the sample read at the specific wavelength of each assay. 6-Hydroxy-2,5,7,8-tetramethylchroman-2-carboxylic acid (Trolox) was employed as reference standard, and the antioxidant activity of each assay was expressed as millimole of Trolox equivalent (TE) per 100 g of FW.

2.9.1. ABTS assay

In order to form the radical 2,2′-azino-bis(3-ethylbenzothiazoline-6-sulphonic acid (ABTS), we prepared 7 mM ABTS in distilled water and incubated it for 16 h with 2.45 mM $K_2S_2O_8$ at room temperature. The radical $ABTS^{\cdot+}$ solution was diluted in methanol until reaching a final absorbance equal to 0.70 at 734 nm. For the assay, 1 mL of the ABTS mixture was incubated for 5 min with different dilutions of the previously prepared tomato methanolic extracts. The decay of the radical

$ABTS^{\cdot+}$ resulting from the incubation with the different dilutions of tomato methanolic extracts was monitored by reading the color decrease at 734 nm.

2.9.2. DPPH assay

At 1 mL of 0.1 mM 2,2-diphenyl-1-picrylhydrazyl ($DPPH^{\cdot}$), we added different concentrations of the previously prepared tomato methanolic extracts. After vigorous shaking, the mixture was incubated for 30 min in the dark and at 25 °C. The reduction of the radical DPPH resulting from the incubation with the different dilutions of tomato methanolic extracts was monitored by reading the color decrease at 517 nm.

2.9.3. FRAP assay

To 300 mM CH_3COONa, we added 36% (*v*/*v*) HCl until a pH value equal to 3.6 was measured. Consequently, 10 mM 2,4,6-Tris(2-pyridyl)-s-triazine (TPTZ) and 20 mM $FeCl_3$ were added to the acetate mixture at an 8:1:1 (*v*/*v*/*v*) ratio. The composed mixture was then incubated at 37 °C and for 10 min with different concentrations of the previously prepared tomato methanolic extracts. The decay of the absorbance was monitored at 595 nm.

2.10. Statistycal Analysis

Differences in the pomological attributes at different harvest times were analyzed using the one-way analysis of variance (ANOVA; general linear model); meanwhile, those between the treatments at the same harvest time were tested with Tukey's high significance difference (HSD) test at $p \leq 0.05$. At least 5 different replicates were used for chemical analyses, and statistical differences were analyzed using the one-way analysis of variance (ANOVA; general linear model). All statistical analyses were performed using SPSS ver. 24 software (IMB, Armonk, NY, USA).

3. Results and Discussion

The improvement of the pomological traits is one of the main focuses of modern agriculture practices [37]. In particular, one of the main challenges is to find new methodologies that are able to reduce the fruit ripening time with the concomitant enhancement of pomological traits, such as fruit appearance, number, weight, and size [37]. However, in addition to the agronomic needs to maximize the production yield and to improve the pomological aspects of crop products, the consumers' demand for products with high nutritional and nutraceutical values is also increasingly rising [22,38]. In this context, seaweed and yeast extracts are receiving a greater acceptance in horticulture as plant biostimulants, due to several beneficial effects against abiotic stresses [18,24–26]. However, the effects deriving from the application of biostimulants based on the combination of seaweed and yeast extracts on tomato fruit quality grown under standard conditions are scarcely available in the literature. For these reasons, in this work we evaluated whether the treatments with a commercial biostimulant based on seaweed and yeast extract could positively affect not only the agronomical aspects of tomato plants, but also the proximate composition, nutritional values, and nutraceutical properties of their fruits.

3.1. Biostimulant Treatments Reduced the Ripening Time and Increased Yield in Early Harvest Stages

Fruit maturation is one of the most important plant stages for crops, and it seriously affects the production times in agriculture. Among the possible strategies aimed to increase plant productivity, decreasing the fruit ripening times and increasing the production yield are the most popular solutions among farmers [17,37]. Consequently, "grow faster, grow more, but stronger" was for a long time the main objective of modern plant biotechnologies. As an alternative to plant engineering, specific products, including plant biostimulants, are widely used for this purpose [11,17,39]. Here, we evaluated the possible effect of the commercial plant biostimulant based on seaweed and yeast extract on tomato fruit maturation and yield.

In our experimental conditions, both untreated and biostimulant-treated plants started to produce ripe fruits after about 75 days from transplanting. However, differences in the fruit number, weight, yield, and ripening time were observed within 30 days. In general, the untreated plants produced 22.08 ± 5.89 fruits per plant, of which about 80% during the medium/later harvest time (13–21 days after the appearance of the first ripe fruits). In particular, 17%, 21%, and 38% of the total fruit amount was collected between the fourth and sixth sampling times (Figure 2A).

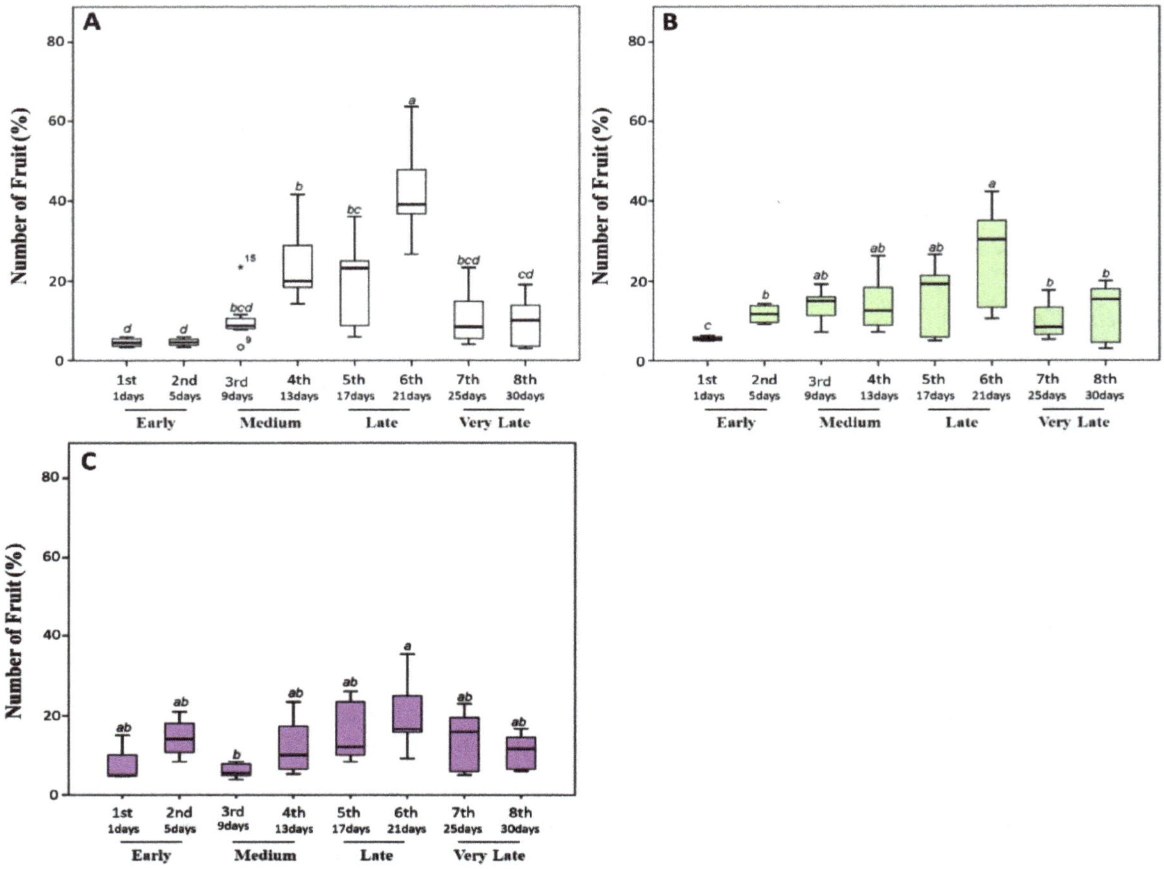

Figure 2. Fruit percentage (%) produced by tomato plants during the harvesting time (30 days). After the appearance of the first ripe fruits, the fully red-colored fruits were harvested every 4 days. (**A**) The percentage of fruit number harvested from untreated plants. (**B**,**C**) Those harvested from plants treated with a triple or a quadruple dosage of biostimulant, respectively. Within the same panel, the different lowercase letters on the top of each boxplot indicate significant differences at $p \leq 0.05$, as measured by Tukey's multiple range test. The letter "a" denotes the highest value. For additional statistical analysis information, see Table S1.

On the other hand, tomato plants treated with a triple or quadruple dosage of biostimulant produced a non-significantly lower amount of fruit. In particular, 18.92 ± 3.63 and 19.83 ± 2.52 fruits were respectively produced by plants treated with the biostimulant three or four times. However, a statistically significant ($p \leq 0.05$) decrease in the fruit percentage during the latest stages and an increase during the earliest harvest times was observed when tomato plants were treated with the biostimulant (Figure 2B,C). In particular, the percentage of the total fruit amount collected during the earliest stages increased from 8.18% ± 1.95% (untreated plants) to 17.21% ± 3.05% and to 21.75% ± 7.32% for plants treated three or four times, respectively. Consequently, the amount of fruit produced at the latest stages significantly ($p \leq 0.05$) decreased from 42.51% ± 8.06% to 25.95% ± 6.74% and to 20.13% ± 8.12%, respectively. These data are indicative of a reduction in fruit ripening times after the application of the biostimulant. Our results are consistent with the findings of Soppelsa and colleagues, who treated strawberry plants with different products, including aminoacid-based biostimulant [12]. In particular, their results showed that, although the number of fruits did not significantly increase with the application of the biostimulant, the treated plants had an overall earlier ripening and a larger share of fruits picked during the earlier stages.

Finally, after the last harvesting time, the unripe fruits were also picked and counted. The untreated plants produced 8.35 ± 2.10 unripe fruits; meanwhile, 15.74 ± 3.59 and 17.23 ± 2.22 were harvested from plants treated with a triple or quadruple dosage of biostimulant, respectively. These data, although excluded from further analyses, suggested that plants treated with the biostimulant were still able to produce fruits compared to untreated plants.

Concerning weight yield data, we recorded a production of 44.89 ± 6.62 g of tomato per plants in untreated plants, of which about 80% was harvested during the late and very late stages (sixth to eighth harvest times) (Figure 3A). In particular, 51%, 17%, and 21% of the total fruit production was collected starting from the 21st day. Similar data were also previously documented for untreated Micro-Tom grown in greenhouse in similar experimental conditions [40–42].

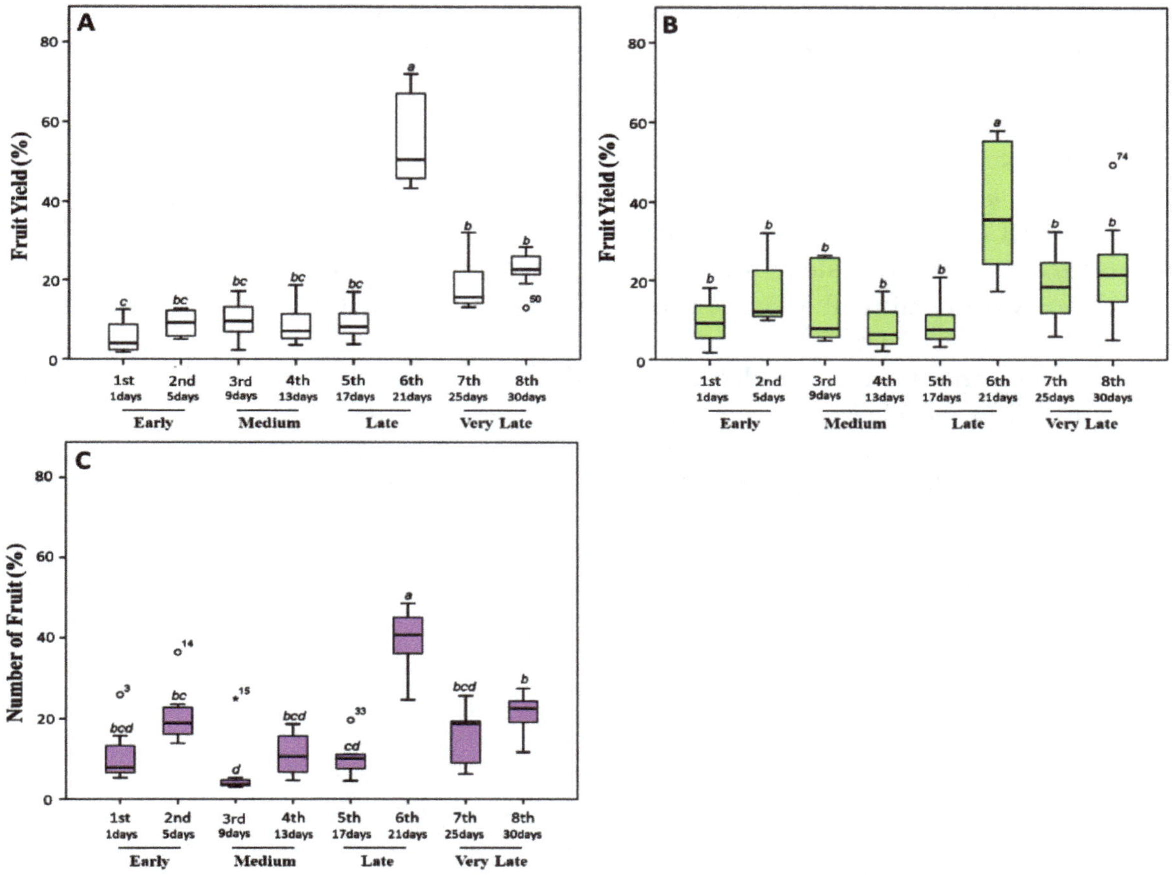

Figure 3. Fruit yield percentage (%) produced by tomato plants during the harvesting time (30 days). After the appearance of the first ripe fruits, the fully red-colored fruits were harvested every 4 days. (**A**) The fruit weight percentage harvested from untreated plants. (**B**,**C**) Those harvested from plants treated with a triple or a quadruple dosage of biostimulant, respectively. Within the same panel, the different lowercase letters on the top of each boxplot indicate significant differences at $p \leq 0.05$, as measured by Tukey's multiple range test. The letter "a" denotes the highest value. For additional statistical analysis information, see Table S1.

When tomato plants were treated with a triple (Figure 3B) or quadruple dosage (Figure 3C) of the biostimulant, a significant ($p \leq 0.05$) decrease in the yield percentage during the late and very late stages and an increase during early times were observed. In particular, the yield percentage in the earliest stages increased from 8.56% ± 1.70% grams per plant (untreated plants) to 20.98% ± 4.04% (triple dosage of the biostimulant) and to 20.89% ± 3.38% (quadruple dosage of the biostimulant). Moreover, an increment in the total weight (from 44.89 ± 6.62 g per plant to 59.83 ± 5.58 and 54.50 ± 3.78 g, respectively, for triple and quadruple treated plants) was observed. The increase in yield production of Micro-Tom testing the action of biostimulants under greenhouse conditions has been reported in limited research studies [43]. Normally, the effects induced by the application of biostimulants on plant growth and crop productivity seem to be different from the nutritional effect observed after the application of commercial fertilizers. In particular, Colla et al. [44] and Ali et al. [45] showed that the improved productivity of tomato plants could be linked to signaling molecules, including polysaccharides, soluble peptides, oligopeptides, and free amino acids, which represent about the 30–40% of the total content of commercial seaweed extracts. Indeed, these molecules act in plants as signaling compounds regulating plant growth and development, and promoting endogenous phytohormonal biosynthesis [46].

3.2. Biostimulant Treatments Increased the Size of Tomato Fruits

Fruit size is an important factor directly correlated to the perception of high-quality fruits, and is of critical consideration not only from a consumer point of view, but also in terms of commercial value [7]. However, for most fruits produced by commercial cultivars, the size is too small to satisfy market demand. Consequently, farmers prefer to increase the fruit size at the expense of the total amount of fruits [47].

Micro-Tom is one of the crops that produces the smallest fruits within the genus *Solanum lycopersicum* [40]. Indeed, their external diameter ranges between a few hundred millimeters up to one and half centimeters, and only rarely does the fruit diameter exceed 1 centimeter [40]. For this reason, although the taste of Micro-Tom is completely comparable to other tomato varieties, this cultivar finds various difficulties in marketing due its bad production yield and fruit size; however, is largely employed as a laboratory model.

In this work, in order to understand if the biostimulant application could affect fruit size and uniformity, we classified the fully red-colored tomatoes harvested from untreated or biostimulant-treated plants on the basis of their external diameters into 16 different size classes (Figure 1, Table S2). Data concerning the classification and the dimension analysis are displayed in Figure 4.

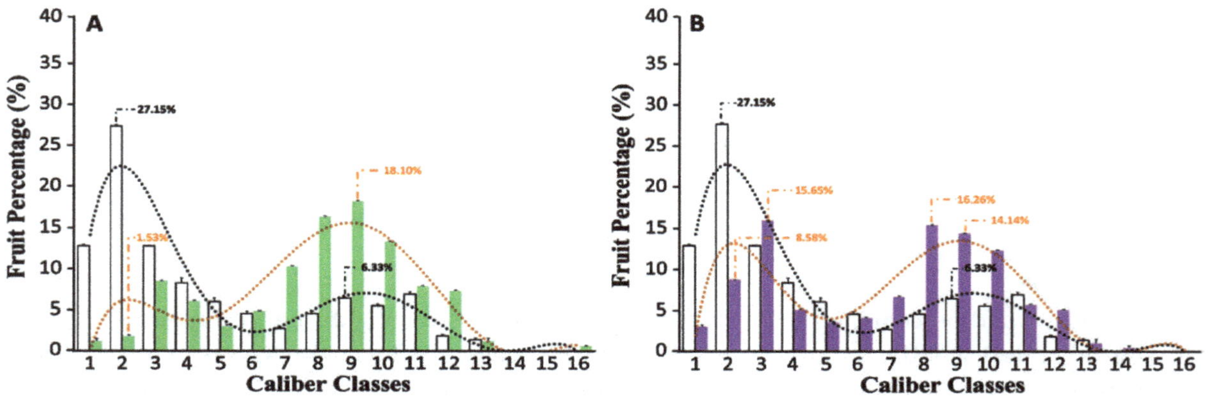

Figure 4. Size of the fruits produced by untreated tomato plants, or treated with a triple or quadruple dosage of biostimulant. (**A**) The fruit caliber of plants that were non-treated (grey) versus plants treated with a triple dosage (green) of biostimulant. (**B**) The fruit caliber of non-treated plants (grey) versus plants treated with a quadruple dosage (violet) of biostimulant. For each panel, the dotted line shows the distribution trend of fruit size (untreated = black; treated with the biostimulant = orange). The *x*-axis reports the caliber classes in which the fruits have been grouped according to their external diameter. The caliber classes and the external diameter of each group is reported in Table S2.

In our experimental conditions, untreated plants produced very small fruits, and more than 50% of the total production had an external diameter (ED) ranging between 0.350 and 0.600 cm (Figure 4). In particular, 13% of the fruits belonged to the first caliber class (ED less than 0.450 cm), 27% to the second (ED comprising between 0.500 and 0.525 cm), and about 13% to the third (ED comprising between 0.525 and 0.600 cm). On the other hand, plants treated with a triple or quadruple dosage of the biostimulant significantly ($p \leq 0.05$) produced fruits larger than untreated plants (Figure 4). Specifically, the best results were obtained when tomato plants were subjected to a triple dosage of the biostimulant treatment (Figure 4A). In this case, about 60% of the total production was concentrated between the 7th and 10th caliber classes (ED ranged between 0.825 and 1.125 cm). Similar results were also obtained from plants treated with a quadruple dosage of the biostimulant (Figure 4B). However, in this case, a large percentage of fruits with a lower ED was also counted.

Actually, the mechanism behind the increase of fruit yield and size after the treatment with the biostimulants is unclear and unknown. However, several authors correlated these effects to the potential intensification of plant enzymatic systems to the biostimulant chelating metal activity, or to their auxin- and gibberellin-like activity [12,19]. In particular, it was recently reported in the literature that the application of seaweed-based biostimulants on crop plants could lead not only to enhanced root development [27], biomass accumulation [19,26], and triggering in flowering and fruit set [48], but also to improved fruit yield and large sized fruits with superior quality [19,49].

3.3. *Biostimulant Treatments Affected the Proximate Composition of Tomato Fruits*

In general, biostimulants help to increase plant production by improving their fitness [17]. Their mode of action is not well understood yet, but several pieces of evidence support the beneficial effects of biostimulation also on food composition [15]. However, despite numerous studies focused their attention on plant growth and yield following biostimulant treatments, there are only few studies reporting the potential influence on the fruit quality [17,50]. These aspects are of special interest, particularly for those plants commonly used for the consumption or for the preparation of cosmetic, nutraceutical, and pharmaceutical products.

In our experimental conditions, both untreated and biostimulant-treated plants produced fruits with a moisture content of about 90%. Even if small significant ($p < 0.05$) differences were observed in the water content, important variations of the total amount of proteins, lipids, and sugars were recorded (Table 1). In particular, the most significant effect derived from the application of this biostimulant on tomato plants was observed on fruit lipid content and fatty acid profile. Indeed, tomato fruits harvested from plants treated with a triple dosage of the biostimulant had an enrichment of 20% in the total fat content, ranging from 0.528 ± 0.016 to 0.631 ± 0.018 mg per gram of FW. Accordingly with data previously reported in the literature [51], our GC–MS and GC–FID analyses allowed for the identification and the quantification of eight different fatty acids (Table 2). Unsaturated fatty acids (UFA) were the main fatty acids in our tomato samples, reaching about 75% of the total fatty acid content, both in untreated and biostimulant-treated samples. The main fatty acid detected in our samples was linoleic acid (C18:2ω6), followed by oleic acid (C18:1ω9), and palmitic acid (C16:0). However, after the treatment with the biostimulant, strong changes in their percentage content were observed. In particular, the percentage content of roughanic acid (C16:3ω3), palmitoleic acid (C16:1ϖ7), oleic acid (C18:1ϖ9), elaidic acid (C18:1ϖ9), and sapienic acid (C16:1ϖ-10) increased in parallel to the decrease of linoleic acid (C18:2ϖ6). On the other hand, saturated fatty acids did not vary significantly (Table 2).

Table 1. Nutritional, mineral, and bioactive compound composition of fruits harvested from untreated plants, or treated with a triple or quadruple dosage of the biostimulant. Data are expressed as mean ± SD. Within the same row, different lowercase letters indicate statistical differences among the samples, as measured by one-way ANOVA analysis followed by Tukey's test ($p \leq 0.05$). Letter "a" denotes the highest content. For additional statistical analysis information, see Table S4.

	Untreated	Triple Dosage	Quadruple Dosage
	Nutritional Content *(g per 100 g of FW)*		
Protein content	14.569 ± 0.201 ^{b}	15.038 ± 0.138 ^{a}	14.712 ± 0.194 ^{ab}
Fat content	0.528 ± 0.016 ^{b}	0.631 ± 0.018 ^{a}	0.531 ± 0.016 ^{b}
Sugar content	2.285 ± 0.034 ^{b}	2.411 ± 0.078 ^{a}	2.364 ± 0.051 ^{ab}
TSS (°Brix)	6.832 ± 0.321 ^{a}	7.001 ± 0.175 ^{a}	7.122 ± 0.489 ^{a}
Moisture content	90.438 ± 0.154 ^{a}	90.252 ± 0.143 ^{a}	89.088 ± 0.015 ^{b}
Ashes	0.761 ± 0.021 ^{a}	0.783 ± 0.012 ^{a}	0.758 ± 0.009 ^{a}
Available energy *(kcal)*	72.168 ± 1.084 ^{c}	75.475 ± 1.026 ^{a}	73.083 ± 1.124 ^{b}
	Mineral Content *(mg per 100 g of FW)*		
K	541.671 ± 0.772 ^{ab}	549.724 ± 4.953 ^{a}	529.188 ± 9.972 ^{b}
Na	10.874 ± 0.283 ^{c}	13.535 ± 0.471 ^{a}	12.799 ± 0.271 ^{b}
Ca	27.653 ± 0.469 ^{a}	23.612 ± 0.425 ^{b}	23.248 ± 0.139 ^{b}
Mg	20.574 ± 0.053 ^{a}	21.027 ± 0.375 ^{a}	20.977 ± 0.424 ^{a}
P	113.791 ± 2.546 ^{b}	126.579 ± 2.487 ^{a}	126.637 ± 0.996 ^{a}
Cl	2.753 ± 0.033 ^{b}	2.394 ± 0.080 ^{c}	2.985 ± 0.047 ^{a}
Fe	0.505 ± 0.022 ^{b}	0.577 ± 0.014 ^{a}	0.563 ± 0.02 ^{a}
Cu	0.096 ± 0.003 ^{b}	0.117 ± 0.008 ^{a}	0.111 ± 0.007 ^{a}
Zn	0.089 ± 0.005 ^{c}	0.111 ± 0.006 ^{a}	0.103 ± 0.002 ^{b}
Mn	0.145 ± 0.002 ^{b}	0.163 ± 0.004 ^{a}	0.165 ± 0.006 ^{a}
Si	0.506 ± 0.013 ^{b}	0.625 ± 0.021 ^{a}	0.386 ± 0.003 ^{c}
B	0.046 ± 0.005 ^{a}	0.052 ± 0.003 ^{a}	0.049 ± 0.004 ^{a}
Mo	0.013 ± 0.001 ^{b}	0.016 ± 0.002 ^{a}	0.014 ± 0.002 ^{ab}
	Bioactive Compound Content *(mg per 100 g of FW)*		
Lycopene	7.182 ± 0.173 ^{a}	6.972 ± 0.312 ^{a}	7.013 ± 0.392 ^{a}
Carotenoids	13.524 ± 1.235 ^{a}	14.091 ± 0.924 ^{a}	13.892 ± 0.783 ^{a}
Tocopherols	0.398 ± 0.057 ^{b}	0.424 ± 0.036 ^{ab}	0.445 ± 0.074 ^{a}
Ascorbic acid	15.925 ± 1.267 ^{a}	14.180 ± 2.202 ^{a}	14.781 ± 1.892 ^{a}
Polyphenols	35.652 ± 1.723 ^{a}	33.420 ± 2.279 ^{a}	35.318 ± 2.323 ^{a}

Table 2. Fatty acid percentage composition of untreated and biostimulant-treated tomato fruits. Data are expressed as mean ± SD. Within the same row, different lowercase letters indicate statistical differences among the samples, as measured by one-way ANOVA analysis followed by Tukey's test ($p \leq 0.05$). Letter "a" denotes the highest content. For additional statistical analysis information, see Table S5.

	Untreated	Triple Dosage	Quadruple Dosage
C16:3ω3	0.202 ± 0.026 ^{b}	0.866 ± 0.009 ^{a}	0.236 ± 0.007 ^{b}
C16:1ω7	0.431 ± 0.044 ^{b}	0.655 ± 0.102 ^{a}	0.404 ± 0.015 ^{b}
C16:1ω10	0.089 ± 0.017 ^{a}	0.155 ± 0.012 ^{a}	0.079 ± 0.054 ^{a}
C16:0	19.734 ± 0.935 ^{a}	19.604 ± 0.150 ^{a}	19.997 ± 1.925 ^{a}
C18:2ω6	48.984 ± 2.988 ^{a}	37.964 ± 0.013 ^{b}	49.220 ± 3.007 ^{a}
C18:1ω9	24.441 ± 1.262 ^{a}	27.154 ± 2.273 ^{a}	23.848 ± 0.359 ^{a}
C18:1ω9	1.410 ± 0.122 ^{b}	2.352 ± 0.039 ^{a}	1.332 ± 0.045 ^{b}
C18:0	4.128 ± 0.181 ^{a}	4.060 ± 0.045 ^{a}	4.516 ± 0.336 ^{a}
SFA	23.863 ± 1.117 ^{b}	29.664 ± 0.195 ^{a}	24.514 ± 2.261 ^{b}
UFA	75.559 ± 4.459 ^{a}	69.149 ± 2.449 ^{a}	75.122 ± 3.490 ^{a}
MUFA	26.372 ± 1.445 ^{b}	30.318 ± 2.427 ^{a}	25.664 ± 0.475 ^{b}
PUFA	49.187 ± 3.014 ^{a}	38.831 ± 0.022 ^{b}	49.457 ± 3.015 ^{a}
SFA/UFA	0.315 ± 0.013 ^{b}	0.428 ± 0.014 ^{a}	0.326 ± 0.009 ^{b}

Fatty acid profile is now a nutritional hot topic, and the presence of these molecules in foods is very attractive for the consumer due to the different benefits derived from their intake [31,32]. Among the variations in the fatty acid profile evaluated in our experiments, the increment of C16:3ω3 and the concomitant reduction in C18:2ω6 are the most interesting effects observed. Indeed, even if C16:3ω3 and C18:2ω6 are both essential fatty acids that cannot be synthesized in the human body and must therefore be absorbed through the diet, recent studies have shown that dietary imbalance of C18:2ω6/C16:3ω3 ratio can affect human health [52]. In particular, the concomitant intake of a large amount of ω6 and low amount of ω3 are associated to an increased production of pro-inflammatory cytokines such as tumor necrosis factor-alpha (TNF-α), interleukin-1 (IL-1), and interleukin-6 (IL-6), and thus excessively increase inflammation [52].

In the present study, we also evaluated the mineral composition and the content of some vitamins in tomato fruits harvested from both untreated and biostimulant-treated plants. This quantification is reported in Table 1. In our experimental conditions, independently from the treatments, K was the most abundant in all the analyzed fruits, ranging between 73 and 75% of the total mineral content; meanwhile, Na was only about 1.5%. When tomato plants were treated with the biostimulant, MP-AES analysis revealed changes in mineral profile (Table 1). In particular, the fruits harvested from biostimulant-treated plants were enriched in K, Na, P, Fe, Cu, Zn, Mn, and Mo, recording the highest values in plants treated with three biostimulant dosages.

Micronutrients are used in several plant physiological processes, and in a relatively small amount, constituting less than 0.1% of dry plant tissue [53]. The microelement uptake by plants depends both on the content originally present in the soil, as well as from the content that becomes available after plant watering or after plant treatment with chemicals, fertilizers, or biostimulants [54,55]. Plants increase micronutrient uptake in specific conditions, with the aim of increasing the binding of solar energy [55], improving the conversion of light into chemical energy [55], intensifying primary or secondary metabolism [54], or enhancing the enzymatic activity of a large number of enzymes [54]. Microelements are stored in different plant districts, including edible organs. The consumption of microelement-enriched fruits may have an important impact on human health due to their involvement in several biochemical processes. Indeed, insufficient dietary intakes of microelements impair several functions of the central nervous system, reproductive system, enzyme activities, and energy metabolism, thus leading to serious illnesses [53].

In our experimental conditions, we observed an average increase of about 15% of Fe, Cu, Zn, and Mn in tomato fruits produced by plants treated with the biostimulant (Table 1). The most significant effect was observed in tomato fruits harvested from plants treated with a triple dosage of the biostimulant. In this case, an increment of 24% and 21% was recorded for Zn and Cu, respectively.

Tomato fruits are also an excellent source of ascorbic acid, tocopherols, and retinol compared to other regularly consumed vegetables [56]. They are very important antioxidant vitamins that are involved in different human physiological pathways. For example, ascorbic acid is required for the correct functioning of several enzymes and contributes to the regular immune system function [57]. Tocopherols are a class of four different methylated phenols largely studied for their functional bioactivity, playing a key role in age-related macular degeneration and chronic diseases, such as cardiovascular disease, cancer, and Alzheimer's disease. Finally, retinol is an essential vitamin resulting from the metabolism of β-carotene, whose deficiency results in serious illnesses [58].

Our spectrophotometric analyses were aimed at the direct quantification of the total content of ascorbic acid and tocopherols, and at the indirect estimation of retinol through the evaluation of the total carotenoid content (Table 1). Our data suggest that the application of the biostimulant did not drastically change the vitamin content, except for tocopherols. However, a strong limitation of spectrophotometric assays is the impossibility of quantifying each specific form of tocopherol or carotenoid. Consequently, we cannot exclude that the application of the biostimulant does not effectively affect the qualitative composition of carotenoids, even in cases in which quantitative changes were not recorded.

3.4. Biostimulant Treatments Did Not Affect the Phytochemical Composition of Tomato Fruits, but Increased Their Antioxidant Properties

Tomatoes are the second most important commercial crop worldwide, and they are a very important component of the Mediterranean diet [56,59]. The importance of tomato consumption is linked to its particular phytochemical composition as it is a rich source of bioactive and antioxidant compounds, including lycopene, carotenoids, and polyphenols [59].

Recent findings showed that biostimulant applications may be able to modify the plant primary and secondary metabolism, stimulating the production of bioactive compounds in roots, stems, leaves, and fruits [44,60]. For example, Kocira and colleagues, studying the effects resulting from the application of three different biostimulants from *Ulva fasciata*, *Sargassum ilicifolium*, and *Gracilaria corticata* on beans, reported not only an increase in yield, but also in the production of polyphenols and carotenoids [60]. Similar results were also described after the application of biostimulants on other plants, including *Prunus armeniaca* [14], *Phaseolus vulgaris* [61], *Malus pumila* [13], and *Solanum lycopersicum* [15].

However, the treatment of crops with unconventional products may negatively affect the fruit quality by reducing their nutraceutical components [21,62–64]. For example, Grabowska et al. showed that treatments with commercial plant biostimulants on two different tomato cultivars increased the lycopene content in the fruits but also caused a strong reduction of other qualitative traits, including carotenes, ascorbic acid, soluble sugars, and total antioxidant activity [64]. Consequently, it is important not only for improving the agronomic aspects such as yield and size, but also to verify and guarantee that the fruit nutraceutical properties are not compromised. Consequently, another aim of our work was the evaluation of the total content of polyphenols (Figure 5), carotenoids, and lycopene (Table 1) both in untreated and biostimulant-treated fruits.

Plant polyphenols are highly desired components in the human diet due to their beneficial antioxidant properties and their preventive role in different oxidative stress-related illnesses, including cancer, cardiovascular diseases, and neurodegenerative diseases [38]. In our experimental conditions, TPC of Micro-Tom fruits ranged between 32.42 ± 2.27 mg GAE per 100 g of FW and 35.31 ± 2.32 mg GAE per 100 g of FW. Our determinations are in accordance with data reported in the literature for Micro-Tom by other authors [56,59]. When tomato plants were treated with the biostimulant, changes in TPC were not observed. While unusual, these results are closely comparable with data from other studies. For example, Stefania Sut and colleagues showed a positive impact of the application of six different biostimulants on growth and production of *Acmella oleracea* plants, with limited effects on the biosynthesis of secondary metabolites [65]. Despite TPC not being affected by the biostimulant treatments, variations in antioxidant properties were observed (Figure 5). In particular, we recorded an increase in the scavenging activity after the biostimulant application. Indeed, when tomato plants were treated with a triple dosage, DPPH and ABTS values were, respectively, 27% and 8% higher compared to untreated plants; meanwhile, they were 38% and 11% higher, respectively, in tomato plants treated four times. Changes in radical scavenging activity independently from TPC variations may be linked to specific qualitative changes in the polyphenol composition after the treatments [22] as well as to the presence of other non-polyphenolic antioxidant compounds, such as carotenoids, alkaloids, lycopene, and vitamins [38,66]. On the other hand, when the reducing activity was measured by FRAP assay, we did not register any statistically significant ($p \leq 0.05$) difference between fruits harvested from untreated or biostimulant-treated plants. The antioxidant differences observed between ABTS/DPPH and FRAP assays may instead have been related to the variability in pH, hydrophilicity, or action mechanism of the three different assays [22 36 66].

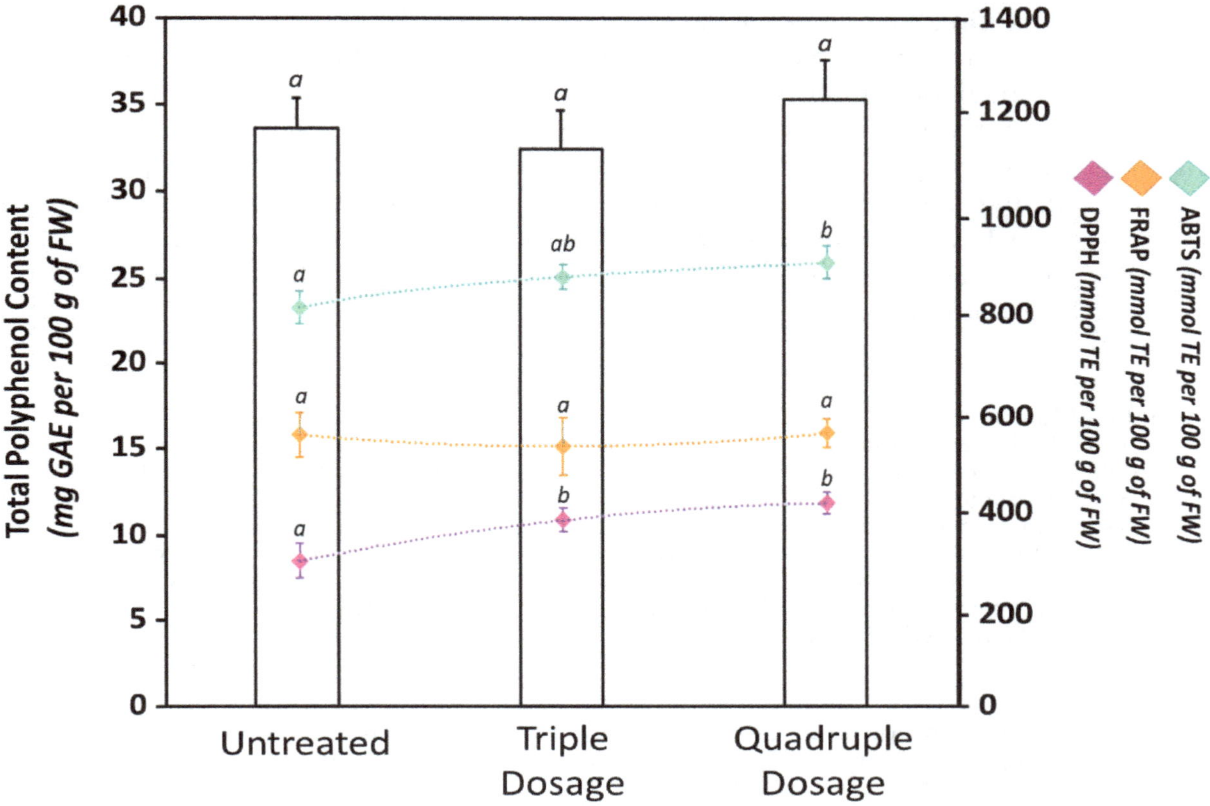

Figure 5. Total polyphenol content (TPC), radical scavenging activity (2,2′-azino-bis(3-ethylbenzothiazoline-6-sulphonic acid (ABTS) and 2,2-diphenyl-1-picrylhydrazyl (DPPH)), and ferric reducing antioxidant power (FRAP) of untreated and biostimulant-treated tomato fruits. The bars show the TPC measured by Folin–Ciocalteu assay; meanwhile, the points indicate the antioxidant properties measured by ABTS (light-blue), DPPH (violet), or FRAP (orange). Values are expressed as mean ± standard deviation (SD) of three experiments. Within the same dataset, different lowercase letters indicate significant differences at $p \leq 0.05$ as measured by Tukey's multiple range test. The letter "a" denotes the lowest content. For additional statistical information, see Table S6.

Color changes in ripening tomato fruits are correlated to different events, including the transition of chloroplasts to chromoplasts, the degradation of chlorophyll, and the synthesis of carotenoids and lycopene [56]. Due to the consumer preference for fully colored fruits, researchers have assessed new methodologies to enhance their production [50]. In particular, it has been demonstrated that the application of fertilizers, chemicals, or biostimulants may affect their synthesis [25,49,50], following not only in early maturation, but also in bioactive component-enriched foods [15,50,67]. However, at the end of our experimentations, although the fruits harvested from biostimulant-treated plants gained a fully red color before the corresponding untreated plants, we did not observe significant ($p \leq 0.05$) enrichment of the lycopene and β-carotene contents (Table 1). These data suggest that, although the biostimulant is capable of accelerating the normal maturation process promoting lycopene and carotenoid biosynthesis, it is not able to modify its typical phytochemical composition.

4. Conclusions

In the present study, we investigated the effect of the application of plant biostimulant based on seaweed and yeast extracts on tomato yield, maturation, and fruit quality. Our data show how the application of the biostimulant was able not only to drastically reduce fruit ripening times, but also to enhance the production yield in terms of both fruit amount and size, leading to more marketable products. Moreover, nutritional analysis revealed changes in proximate composition and mineral content, as well as increased level of unsatured fatty acids and micronutrients essential for human health. Finally, regarding the number of bioactive compounds, the obtained results indicate limited influence of the biostimulant treatments on the biosynthesis of secondary metabolites at the fruit level. Indeed, the total content of polyphenols, carotenoids, and lycopene were not significantly improved compared to fruits harvested from control plants. Despite this, an increase in antioxidant properties, in terms of radical scavenging activity, was observed. These data suggest that despite quantitative changes not occurring, possible and punctual changes in the qualitative composition of polyphenols and carotenoids may have occurred. In conclusion, the biostimulant application was able to reduce the ripening times and fruit size, while slightly increasing nutritional and nutraceutical values, leading to more marketable tomato fruits.

Supplementary Materials:
Table S1: Tukey's high significance difference (HSD) post hoc differences in fruit number percentage and fruit weight percentage produced during the experimentation time (30 days) and collected during the specific harvest time. Table S2: Percentage of fruits harvested according to their external diameter. Table S3: Tukey's HSD post hoc differences in fruit size of ripe fruits produced during the experimentation time (30 days) and collected during the specific harvest time. Table S4: Tukey's HSD post hoc differences in proximate composition, mineral content, and bioactive compounds determined in untreated and biostimulant-treated tomato fruits. Table S5: Tukey's HSD post hoc differences in fatty acid composition. Table S6: Tukey's HSD post hoc differences in radical scavenging activity (DPPH and ABTS), reducing activity (FRAP), and total polyphenol content (TPC).

Author Contributions: Conceptualization, G.M., V.C., and C.M.B.; data curation, G.M. and C.G.; formal analysis, G.M.; funding acquisition, C.M.B.; investigation, G.M. and C.G.; methodology, G.M. and C.G.; project administration, G.M. and C.M.B.; software, G.M. and I.V.; supervision, C.C. and C.M.B.; validation, G.M. and C.M.B.; visualization, G.M. and I.V.; writing—original draft, G.M., C.G., and C.M.B.; writing—review and editing, G.M., C.C., I.V., V.C., C.G., and C.M.B. All authors have read and agreed to the published version of the manuscript.

Acknowledgments: The authors are grateful to Regione Piemonte for supporting the activity of the PhD Program in Apprenticeship to C.C. Moreover, the University of Turin is thankful to Green Has Italia S.p.A for providing the biostimulant free of charge.

References

1. Tripathi, A.D.; Mishra, R.; Maurya, K.K.; Singh, R.B.; Wilson, D.W. Estimates for world population and global food availability for global health. In *The Role of Functional Food Security in Global Health*; Elsevier: Amsterdam, The Netherlands, 2019; pp. 3–24.
2. Carvalho, F.P. Agriculture, pesticides, food security and food safety. *Environ. Sci. Policy* **2006**, *9*, 685–692. [CrossRef]
3. Szentpétery, Z.; Hegedűs, Z.; Jolánkai, M. Impact of agrochemicals on yield quality and pesticide residues of winter wheat varieties. *Cereal Res. Commun.* **2005**, *33*, 635–640. [CrossRef]
4. Ramankutty, N.; Mehrabi, Z.; Waha, K.; Jarvis, L.; Kremen, C.; Herrero, M.; Rieseberg, L.H. Trends in global agricultural land use: Implications for environmental health and food security. *Annu. Rev. Plant Biol.* **2018**, *69*, 789–815. [CrossRef] [PubMed]
5. Pandey, S.; Joshi, N.; Kumar, M. Agrochemicals and human well-being: A review in context of Indian agriculture. *Int. J. Cosmet. Sci.* **2020**, *8*, 1539–1543. [CrossRef]
6. Elahi, E.; Weijun, C.; Zhang, H.; Nazeer, M. Agricultural intensification and damages to human health in relation to agrochemicals: Application of artificial intelligence. *Land use policy* **2019**, *83*, 461–474. [CrossRef]
7. Gentile, C.; Di Gregorio, E.; Di Stefano, V.; Mannino, G.; Perrone, A.; Avellone, G.; Sortino, G.; Inglese, P.; Farina, V. Food quality and nutraceutical value of nine cultivars of mango (*Mangifera indica* L.) fruits grown in Mediterranean subtropical environment. *Food Chem.* **2019**, *277*, 471–479. [CrossRef]
8. Farina, V.; Tinebra, I.; Perrone, A.; Sortino, G.; Palazzolo, E.; Mannino, G.; Gentile, C. Physicochemical, Nutraceutical and Sensory Traits of Six Papaya (*Carica papaya* L.) Cultivars Grown in Greenhouse Conditions in the Mediterranean Climate. *Agronomy* **2020**, *10*, 501. [CrossRef]

9. Annunziata, A.; Vecchio, R. Organic farming and sustainability in food choices: An analysis of consumer preference in Southern Italy. *Agric. Agric. Sci. Procedia* **2016**, *8*, 193–200. [CrossRef]
10. Sahota, A.; Haumann, B.; Givens, H.; Baldwin, C.J. Ecolabeling and consumer interest in sustainable products. *Sustain. Food Ind.* **2009**, 159–181.
11. Calvo, P.; Nelson, L.; Kloepper, J.W. Agricultural uses of plant biostimulants. *Plant Soil* **2014**, *383*, 3–41. [CrossRef
12. Soppelsa, S.; Kelderer, M.; Casera, C.; Bassi, M.; Robatscher, P.; Matteazzi, A.; Andreotti, C. Foliar Applications of Biostimulants Promote Growth, Yield and Fruit Quality of Strawberry Plants Grown under Nutrient Limitation. *Agronomy* **2019**, *9*, 483. [CrossRef]]
13. Graziani, G.; Ritieni, A.; Cirillo, A.; Cice, D.; Di Vaio, C. Effects of Biostimulants on Annurca Fruit Quality and Potential Nutraceutical Compounds at Harvest and during Storage. *Plants* **2020**, *9*, 775. [CrossRef] [PubMed
14. Tarantino, A.; Lops, F.; Disciglio, G.; Lopriore, G. Effects of plant biostimulants on fruit set, growth, yield and fruit quality attributes of 'Orange rubis®'apricot (*Prunus armeniaca* L.) cultivar in two consecutive years. *Sci. Hortic.* **2018**, *239*, 26–34. [CrossRef]]
15. Francesca, S.; Arena, C.; Hay Mele, B.; Schettini, C.; Ambrosino, P.; Barone, A.; Rigano, M.M. The Use of a Plant-Based Biostimulant Improves Plant Performances and Fruit Quality in Tomato Plants Grown at Elevated Temperatures. *Agronomy* **2020**, *10*, 363. [CrossRef]
16. Abubakar, A.R.; Ashraf, N.; Ashraf, M. Effect of plant biostimulants on fruit cracking and quality attributes of pomegranate cv. Kandhari kabuli. *Sci. Res. Essays* **2013**, *8*, 2171–2175.
17. Rouphael, Y.; Colla, G. Synergistic biostimulatory action: Designing the next generation of plant biostimulants for sustainable agriculture. *Front. Plant Sci.* **2018**, *871*. [CrossRef]
18. Bargiacchi, E.; Miele, S.; Romani, A.; Campo, M. Biostimulant activity of hydrolyzable tannins from sweet chestnut (Castanea sativa Mill.). *Acta Hortic.* **2012**, *1009*, 111–116. [CrossRef]
19. Khan, W.; Rayirath, U.P.; Subramanian, S.; Jithesh, M.N.; Rayorath, P.; Hodges, D.M.; Critchley, A.T.; Craigie, J.S.; Norrie, J.; Prithiviraj, B. Seaweed extracts as biostimulants of plant growth and development. *J. Plant Growth Regul.* **2009**, *28*, 386–399. [CrossRef]
20. Campobenedetto, C.; Mannino, G.; Agliassa, C.; Acquadro, A.; Contartese, V.; Garabello, C.; Bertea, C.M. Transcriptome Analyses and Antioxidant Activity Profiling Reveal the Role of a Lignin-Derived Biostimulant Seed Treatment in Enhancing Heat Stress Tolerance in Soybean. *Plants* **2020**, *9*, 1308. [CrossRef]
21. Campobenedetto, C.; Grange, E.; Mannino, G.; Van Arkel, J.; Beekwilder, J.; Karlova, R.; Garabello, C.; Contartese, V.; Bertea, C.M. A Biostimulant Seed Treatment Improved Heat Stress Tolerance During Cucumber Seed Germination by Acting on the Antioxidant System and Glyoxylate Cycle. *Front. Plant Sci.* **2020**, [CrossRef]
22. Mannino, G.; Di Stefano, V.; Lauria, A.; Pitonzo, R.; Gentile, C. Vaccinium macrocarpon (Cranberry)-Based Dietary Supplements: Variation in Mass Uniformity, Proanthocyanidin Dosage and Anthocyanin Profile. Demonstrates Quality Control Standard Needed. *Nutrients* **2020**, *12*, 992. [CrossRef] [PubMed]
23. Mannino, G.; Nerva, L.; Gritli, T.; Novero, M.; Fiorilli, V.; Bacem, M.; Bertea, C.M.; Lumini, E.; Chitarra, W.; Balestrini, R. Effects of Different Microbial Inocula on Tomato Tolerance to Water Deficit. *Agronomy* **2020** *10*, 170. [CrossRef]
24. Deng, Q.; Xia, H.; Lin, L.; Wang, J.; Yuan, L.; Li, K.; Zhang, J.; Lv, X.; Liang, D. SUNRED, a natural extract-based biostimulant, application stimulates anthocyanin production in the skins of grapes. *Sci. Rep.* **2019**, *9*, 2590, [CrossRef] [PubMed]
25. Barrajón-Catalán, E.; Álvarez-Martínez, F.J.; Borrás, F.; Pérez, D.; Herrero, N.; Ruiz, J.J.; Micol, V. Metabolomic analysis of the effects of a commercial complex biostimulant on pepper crops. *Food Chem.* **2020**, *310*, 125818. [CrossRef] [PubMed]
26. Battacharyya, D.; Babgohari, M.Z.; Rathor, P.; Prithiviraj, B. Seaweed extracts as biostimulants in horticulture. *Sci. Hortic.* **2015**, *196*, 39–48. [CrossRef]
27. Kumar, R.; Trivedi, K.; Anand, K.G.V.; Ghosh, A. Science behind biostimulant action of seaweed extract on growth and crop yield: Insights into transcriptional changes in roots of maize treated with Kappaphycus alvarezii seaweed extract under soil moisture stressed conditions. *J. Appl. Phycol.* **2020**, *32*, 599–613. [CrossRef]
28. Contartese, V.; Garabello, C.; Occhipinti, A.; Barbero, F.; Bertea, C.M. Effects of a new biostimulant on gene expression and metabolic responses of tomato plants. In Proceedings of the II World Congress on the Use of Biostimulants in Agriculture 1148, Florence, Italy, 16–19 November 2015; pp. 35–42.

29. Bradstreet, R.B. Kjeldahl method for organic nitrogen. *Anal. Chem.* **1954**, *26*, 185–187. [CrossRef]
30. Jung, S.; Rickert, D.A.; Deak, N.A.; Aldin, E.D.; Recknor, J.; Johnson, L.A.; Murphy, P.A. Comparison of Kjeldahl and Dumas methods for determining protein contents of soybean products. *J. Am. Oil Chem. Soc.* **2003**, *80*, 1169. [CrossRef]
31. Mannino, G.; Gentile, C.; Maffei, M.E. Chemical partitioning and DNA fingerprinting of some pistachio (*Pistacia vera* L.) varieties of different geographical origin. *Phytochemistry* **2019**, *160*, 40–47. [CrossRef]
32. Vigliante, I.; Mannino, G.; Maffei, M.E. Chemical Characterization and DNA Fingerprinting of Griffonia simplicifolia Baill. *Molecules* **2019**, *24*, 1032. [CrossRef]
33. Bajaj, K.L.; Kaur, G. Spectrophotometric determination of L-ascorbic acid in vegetables and fruits. *Analyst* **1981**, *106*, 117–120. [CrossRef] [PubMed]
34. Parimelazhagan, T. *Pharmacological Assays of Plant-Based Natural Products;* Springer: Berlin/Heidelberg, Germany, 2015; Volume 71, ISBN 3319268112.
35. Rayhan, M.; Mumtaz, B.; Motalab, M.; Zubair, M.A.; Haque, M.Z.; Saha, B.K. Extraction and Quantification of Lycopene, β-Carotene and Total Phenolic Contents from Papaya (Carica papaya) and Formulation of Lycopene Enriched Fruit Drinks. *Am. J. Food Nutr.* **2019**, *7*, 55–63.
36. Vigliante, I.; Mannino, G.; Maffei, M.E. OxiCyan®, a phytocomplex of bilberry (Vaccinium myrtillus) and spirulina (Spirulina platensis), exerts both direct antioxidant activity and modulation of ARE/Nrf2 pathway in HepG2 cells. *J. Funct. Foods* **2019**, *61*, 103508. [CrossRef]
37. Altieri, M.A.; Nicholls, C.I.; Montalba, R. Technological approaches to sustainable agriculture at a crossroads: An agroecological perspective. *Sustainability* **2017**, *9*, 349. [CrossRef]
38. Mannino, G.; Gentile, C.; Porcu, A.; Agliassa, C.; Caradonna, F.; Bertea, C.M. Chemical Profile and Biological Activity of Cherimoya (Annona cherimola Mill.) and Atemoya (Annona atemoya) Leaves. *Molecules* **2020**, *25*, 2612. [CrossRef] [PubMed]
39. Rouphael, Y.; Colla, G. Biostimulants in Agriculture. *Front. Plant Sci.* **2020**, *11*. [CrossRef]
40. Huther, C.M.; Martinazzo, E.G.; Schock, A.A.; Rombaldi, C.V.; Bacarin, M.A. Production components in transformed and untransformed 'Micro-Tom'tomato plants. *Rev. Ciência Agronômica* **2018**, *49*, 85–92.
41. Martí, E.; Gisbert, C.; Bishop, G.J.; Dixon, M.S.; García-Martínez, J.L. Genetic and physiological characterization of tomato cv. Micro-Tom. *J. Exp. Bot.* **2006**, *57*, 2037–2047. [CrossRef]
42. Gratão, P.L.; Monteiro, C.C.; Antunes, A.M.; Peres, L.E.P.; Azevedo, R.A. Acquired tolerance of tomato (Lycopersicon esculentum cv. Micro-Tom) plants to cadmium-induced stress. *Ann. Appl. Biol.* **2008**, *153*, 321–333. [CrossRef]
43. Rothan, C.; Just, D.; Fernandez, L.; Atienza, I.; Ballias, P.; Lemaire-Chamley, M. Culture of the tomato Micro-Tom cultivar in greenhouse. In *Plant Signal Transduction*; Springer: Berlin/Heidelberg, Germany, 2016; pp. 57–64.
44. Colla, G.; Cardarelli, M.; Bonini, P.; Rouphael, Y. Foliar applications of protein hydrolysate, plant and seaweed extracts increase yield but differentially modulate fruit quality of greenhouse tomato. *Hort. Sci.* **2017**, *52*, 1214–1220. [CrossRef]
45. Ali, N.; Farrell, A.; Ramsubhag, A.; Jayaraman, J. The effect of Ascophyllum nodosum extract on the growth, yield and fruit quality of tomato grown under tropical conditions. *J. Appl. Phycol.* **2016**, *28*, 1353–1362. [CrossRef]
46. Cerdán, M.; Sánchez-Sánchez, A.; Oliver, M.; Juárez, M.; Sánchez-Andreu, J.J. Effect of foliar and root applications of amino acids on iron uptake by tomato plants. In Proceedings of the IV Balkan Symposium on Vegetables and Potatoes 830, Plovdiv, Bulgaria, 9–12 September 2008; pp. 481–488.
47. Guardiola, J.L.; García-Luis, A. Increasing fruit size in Citrus. Thinning and stimulation of fruit growth. *Plant Growth Regul.* **2000**, *31*, 121–132. [CrossRef]
48. Kavipriya, R.; Dhanalakshmi, P.K.; Jayashree, S.; Thangaraju, N. Seaweed extract as a biostimulant for legume crop, green gram. *J. Ecobiotechnol.* **2011**, *3*, 16–19.
49. Rouphael, Y.; Colla, G.; Giordano, M.; El-Nakhel, C.; Kyriacou, M.C.; De Pascale, S. Foliar applications of a legume-derived protein hydrolysate elicit dose-dependent increases of growth, leaf mineral composition, yield and fruit quality in two greenhouse tomato cultivars. *Sci. Hortic.* **2017**, *226*, 353–360. [CrossRef]
50. Caruso, G.; De Pascale, S.; Cozzolino, E.; Cuciniello, A.; Cenvinzo, V.; Bonini, P.; Colla, G.; Rouphael, Y. Yield and nutritional quality of Vesuvian Piennolo tomato PDO as affected by farming system and biostimulant application. *Agronomy* **2019**, *9*, 505. [CrossRef]

51. Benítez, J.J.; Castillo, P.M.; Del Río, J.C.; León-Camacho, M.; Domínguez, E.; Heredia, A.; Guzmán-Puyol, S.; Athanassiou, A.; Heredia-Guerrero, J.A. Valorization of tomato processing by-products: Fatty acid extraction and production of bio-based materials. *Materials* **2018**, *11*, 2211.
52. Simopoulos, A.P. Omega-6/omega-3 essential fatty acid ratio and chronic diseases. *Food Rev. Int.* **2004**, *20*, 77–90. [CrossRef]
53. Rattan, R.K.; Patel, K.P.; Manjaiah, K.M.; Datta, S.P. Micronutrients in soil, plant, animal and human health. *J. Indian Soc. Soil Sci.* **2009**, *57*, 546–558.
54. Bhat, B.A.; Islam, S.T.; Ali, A.; Sheikh, B.A.; Tariq, L.; Islam, S.U.; Dar, T.U.H. Role of Micronutrients in Secondary Metabolism of Plants. In *Plant Micronutrients*; Springer: Berlin/Heidelberg, Germany, 2020; pp. 311–329.
55. Römheld, V.; Marschner, H. Function of micronutrients in plants. *Micronutr. Agric.* **1991**, *4*, 297–328.
56. Gomez, P.; Ferrer, M.Á.; Fernández-Trujillo, J.P.; Calderon, A.; Artes, F.; Egea-Cortines, M.; Weiss, J. Structural changes, chemical composition and antioxidant activity of cherry tomato fruits (cv. Micro-Tom) stored under optimal and chilling conditions. *J. Sci. Food Agric.* **2009**, *89*, 1543–1551. [CrossRef]
57. Gupta, U.; Gupta, S. Role of vitamins in human health and nutrition: Sources and morbidity. *Curr. Nutr. Food Sci.* **2015**, *11*, 105–115. [CrossRef]
58. Chudow, M.; Adams, B. ABC's of Vitamin Supplementation in Critical Illness. *J. Pharm. Pract.* **2020**, 0897190020958234. [CrossRef] [PubMed]
59. Frusciante, L.; Carli, P.; Ercolano, M.R.; Pernice, R.; Di Matteo, A.; Fogliano, V.; Pellegrini, N. Antioxidant nutritional quality of tomato. *Mol. Nutr. Food Res.* **2007**, *51*, 609–617. [CrossRef] [PubMed]
60. Kocira, A.; Świeca, M.; Kocira, S.; Złotek, U.; Jakubczyk, A. Enhancement of yield, nutritional and nutraceutical properties of two common bean cultivars following the application of seaweed extract (Ecklonia maxima). *Saudi J. Biol. Sci.* **2018**, *25*, 563–571. [CrossRef]
61. Kocira, A.; Kocira, S.; Świeca, M.; Złotek, U.; Jakubczyk, A.; Kapela, K. Effect of foliar application of a nitrophenolate–based biostimulant on the yield and quality of two bean cultivars. *Sci. Hortic.* **2017**, *214*, 76–82. [CrossRef]
62. El-Garhy, H.A.S.; Abdel-Rahman, F.A.; Shams, A.S.; Osman, G.H.; Moustafa, M. Comparative Analyses of Four Chemicals Used to Control Black Mold Disease in Tomato and Its Effects on Defense Signaling Pathways, Productivity and Quality Traits. *Plants* **2020**, *9*, 808. [CrossRef]
63. Drobek, M.; Frąc, M.; Cybulska, J. Plant Biostimulants: Importance of the Quality and Yield of Horticultural Crops and the Improvement of Plant Tolerance to Abiotic Stress—A Review. *Agronomy* **2019**, *9*, 335. [CrossRef]
64. Grabowska, A.; Kunicki, E.; Jezdinsky, A.; Kalisz, A.; Sekara, A. The effect of biostimulants on the quality parameters of tomato grown for the processing industry. *Eff. Biostimulants Qual. Parameters Tomato Grown Process. Ind.* **2015**, 203–217.
65. Sut, S.; Ferrarese, I.; Shrestha, S.S.; Kumar, G.; Slaviero, A.; Sello, S.; Altissimo, A.; Pagni, L.; Gattesco, F.; Dall'Acqua, S. Comparison of Biostimulant Treatments in Acmella oleracea Cultivation for Alkylamides Production. *Plants* **2020**, *9*, 818. [CrossRef]
66. Mannino, G.; Perrone, A.; Campobenedetto, C.; Schittone, A.; Margherita Bertea, C.; Gentile, C. Phytochemical profile and antioxidative properties of Plinia trunciflora fruits: A new source of nutraceuticals. *Food Chem.* **2020**, *307*. [CrossRef]
67. Polo, J.; Mata, P. Evaluation of a biostimulant (Pepton) based in enzymatic hydrolyzed animal protein in comparison to seaweed extracts on root development, vegetative growth, flowering, and yield of gold cherry tomatoes grown under low stress ambient field conditions. *Front. Plant Sci.* **2018**, *8*, 2261. [CrossRef] [PubMed]

2

Exogenous Application of Amino Acids Improves the Growth and Yield of Lettuce by Enhancing Photosynthetic Assimilation and Nutrient Availability

Shumaila Khan [1,†], Hongjun Yu [1,†], Qiang Li [1,†], Yinan Gao [2], Basheer Noman Sallam [1], Heng Wang [1], Peng Liu [1] and Weijie Jiang [1,*]

1 Institute of Vegetables and Flowers, Chinese Academy of Agricultural Sciences (CAAS), No.12 Zhongguancun South Street, Haidian District, Beijing 100081, China; shumaila_khan@yahoo.com (S.K.); liqiang05@caas.cn (Q.L.); yuhongjun@caas.cn (H.Y.); basir_zen@yahoo.com (B.N.S.); 82101181122@caas.cn (H.W.); 82101179107@caas.cn (P.L.)

2 College of Horticulture and Landscape, Tianjin Agricultural University, Tianjin 300000, China; ilyzp1@gmail.com

* Correspondence: jiangweijie@caas.cn

† The authors equally contributed to this work.

Abstract: As natural plant growth stimulators, amino acids are widely used to improve the yield and quality of crops. Several studies have illustrated the effects of different amino acids on lettuce plant parts. However, the effects of applying single amino acids on root growth remain elusive. The objective of this study was to evaluate the effect of root application of L-methionine on the growth of lettuce. In this study, two successive experiments on butterhead lettuce were conducted under hydroponic conditions. Three amino acids, L-methionine (20 mg/L), L-glycine (210 mg/L), and L-tryptophan (220 mg/L), were applied separately. L-methionine significantly increased the growth performance by 23.60%, whereas growth using L-tryptophan and L-glycine decreased by 98.78% and 27.45%, respectively. Considering the results of the first experiment, a second experiment was established with different concentrations of L-methionine (2200 mg/L, 220 mg/L, 22 mg/L, 2.2 mg/L, 0.2 mg/L, and 0.02 mg/L). The plants were allowed to grow for four weeks. Leaf width, plant area, leaf area, chlorophyll contents, etc., were evaluated. The results show that plant growth significantly improved by applying L-methionine at the lowest concentrations of 0.2 mg/L and 0.02 mg/L, which can, therefore, improve hydroponic production of lettuce and, accordingly, human nutrition.

Keywords: L-methionine; L-tryptophan; L-glycine; lettuce; nitrogen

1. Introduction

Lettuce (*Lactuca sativa* L.) is one of the main vegetable crops widely cultivated in China and consumed as a salad throughout the world. Due to its high nutritional value provided by mineral elements, vitamins, and folate, which play significant roles in human nutrition and diet, lettuce has become the focal point of several studies [1–4].

Nitrogen is an essential element for lettuce plants, an integral component of protein, phospholipid, and chloroplast [5,6]. Nitrogen uptake, assimilation, and utilization play essential roles in plant growth and development [7,8]. The plants mainly take up nitrogen in the form of nitrate (NO_3^-) and ammonium (NH_4^+) or N_2 from the atmosphere through nitrogen-fixing bacteria [9–11]. The application of a large amount of chemical fertilizer to ensure high crop yield causes serious issues for agricultural products [12] and the environment [13,14]. Hence, there is a need to look for sustainable horticultural practices to counteract chemical-based agribusiness. In this respect, the application of amino acids, as a type of growth-promoting substance, supplies plant nutrients but also improves plant quality, which ultimately boosts the yield and commercial output of crops [15]. Therefore, it has become popular in sustainable agriculture [16–20]

Amino acids as biostimulants (substances that promote plant growth, improve nutrient availability, and enhance plant quality) [20,21] are not only getting popular for mitigating injuries caused by abiotic stresses [22] but also serve as hormone precursors [20,23–25]; signaling factors of different physiological progressions, such as glutamate receptors (GRLs) [24–26]; and regulators of nitrogen uptake [27], root development [25,28–30], and antioxidant metabolism [25,30–32]. Better root development supported by the addition of amino acids can boost nitrogen fixation, which induces an enhanced root surface for nutrient uptake [29,33].

Direct application of amino acids and their products could modulate N uptake and assimilation; this phenomenon is mediated by enzymes engaged with N assimilation [15,25,30]. In addition, it could be a follow-up to the flagging pathway that controls N securing amino acids in roots, which are mostly accessible as supplements [30]. Additionally, application of amino acids was also found to increase K^+ in plants both in the presence of salt stress and without salt application [25,30,32]

A recent study [31] showed that seed treatment or foliar application of amino acids had different effects on soybean crops. An amino acid applied individually acts as a signaling component, i.e., increases antioxidant enzyme activity and causes efficient nutrient uptake [25,34,35]. Different investigations have demonstrated a positive impact of foliar application of amino acid mixtures on plants, for example, increased production in *Solanum lycopersicum* L. [36] and accumulation of dry matter, chlorophyll [37], starch, and polysaccharides in *Vicia faba* L. [38].

Previous studies have indicated an association of L-methionine with the biosynthesis of growth regulating substances such as cytokinins, auxins, and brassinosteroids in plants [39,40]. L-methionine functions as a precursor of a significant number of essential biomolecules such as vitamins, polyamine, cofactors [41], and antioxidants such as glutathione, which are considered to be significant determinants of cellular redox homeostasis and many defense compounds [19]. All of these biomolecules contain sulfur moieties that act as functional groups and are derived from L-methionine. In plants, L-methionine biosynthesis plays a central role in fixing inorganic sulfur from the environment, providing the only metabolic sulfide donor for the generation of glutathione, phytochelatins, iron–sulfur clusters, vitamin cofactors, and multiple secondary metabolites [42,43].

However, there is little data on the impact of isolated amino acids, particularly in root application. Additionally, the majority of investigations have been carried out utilizing a mixture of amino acids and other methods of application, such as foliar application and seed treatment. Hence, this study is based on the hypothesis that the root application of individual amino acids can improve the uptake of nitrogen and other growth-related factors, which can lead to increased productivity of lettuce plants. Therefore, the objective of the present work was to evaluate the effect of the separate application of L-tryptophan, L-glycine, and L-methionine in nutrient solution on the growth, yield, and physiology of lettuce plants.

2. Materials and Methods

The research study was carried out at the Vegetables and Flowers Institute, Chinese Academy of Agricultural Sciences, Beijing, China, in 2017–2018 to determine the regulation of lettuce plant growth response under different amino acids and concentrations.

2.1. Plant Material and Growth Conditions

Sowing of butterhead lettuce seeds was done under controlled conditions using a mixture of peat moss with an average of 2–3 seeds per hole. All cultural practices were maintained in order to have a good plant stand. Average minimum and maximum monthly temperatures were set to 24 °C and 34 °C. Plants were provided with natural sunlight with a light intensity of approximately

900–1000 µmoles m^{-2}/s. At pre-emergence stages, the nutrient solution was applied once a week. Plants with at least 2 fully expanded leaves 30 days after sowing were transferred to a closed-loop hydroponic system (Figure 1).

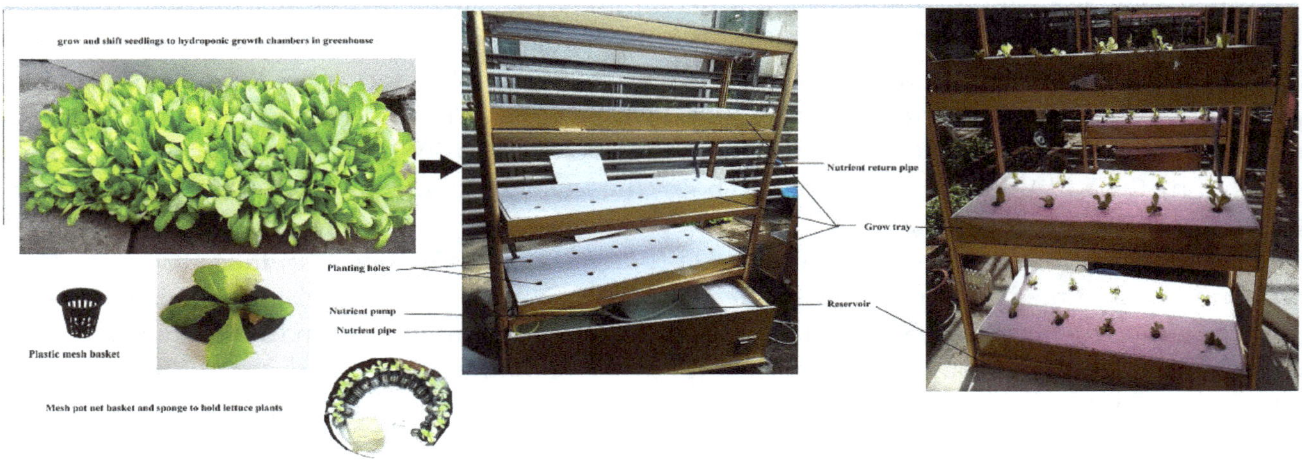

Figure 1. The hydroponic system, mesh basket, and other materials used to grow lettuce plants.

Briefly, the hydroponic system consisted of 3 growing nutrient trays (used as 3 replicates) with 10 holes each, with a distance of 21.2 × 20.5 cm between the holes. The height of the hydroponic growing stand was 160 cm from the ground, while the length and width of the growing tray were 102 cm and 38 cm, respectively. The capacity of the water reservoir was 80 L with 6 cm depth and recyclable. Plastic mesh used to cover the plants was 6 cm long and 2 cm wide.

A pH of 6.0–6.3 and electrical conductivity (EC) of 1.5–2.0 mS cm^{-1} of nutrient solution were maintained regularly for optimal plant growth.

Plantlets were grown with 75% strength Hoagland nutrient solution containing the following nutrients (in mg/L), as previously described [44]: $Ca(NO_3)_2$ = 1122; KNO_3 = 910; KH_2PO_4 = 272; NH_4NO_3 = 40; $MgSO_4$ = 247; EDTA (Ethylenediamine tetraacetic acid Ferric Sodium Salt) = 16.80; $ZnSO_4$ = 1.20; $Na_2B_4O_7$ = 0.28; Na_2MoO_4 = 0.20; $CuSO_4$ = 0.10; and $MnSO_4$ = 0.86.

2.2. Application of Three Amino Acids on Lettuce

The experiment was conducted from December 2017 to February 2018 with 3 replications. The concentration of 3 amino acids, L-methionine, L-tryptophan, and L-glycine, was kept at 20 mg/L, 210 mg/L, and 220 mg/L, respectively. This gave 4 treatment combinations (3 amino acids and 1 control treatment). The amino acid treatment was started 8 days after transplanting into the nutrient solution to prevent plants from undergoing nutrient shock. Data were recorded every week and the crop was harvested after 30 days of treatment.

2.3. Application of L-Methionine Concentrations on Lettuce

The second experiment was conducted from January to March 2018 with a single amino acid, L-methionine (selected from experiment 1), in 6 concentrations, as 3 treatments and 3 replications. Plants were transplanted to the nutrient solution and treated with the amino acid after 8 days. The concentration of L-methionine applied was 2200 mg/L, 220 mg/L, and 22 mg/L for the treatment and 2.2 mg/L, 0.2 mg/L, and 0.02 mg/L for the control. All other experimental conditions were the same as in the first experiment. The tanks of nutrient solution were refreshed weekly.

2.4. Data Collection and Analysis

Data were recorded for the following morphological and physiological parameters: root length, leaf length, leaf width, leaf area, plant area, chlorophyll content, and fresh and dry mass of root and shoot.

2.4.1. Vegetative Growth Parameters

The number of leaves, plant height, plant diameter, and leaf area were measured every 7 days following standard procedures as proposed in [45–47].

$$AF\left(cm^2\right) = 0.7 \times Length\ (cm) \times Width\ (cm) - 2.4$$

Leaf length and width were measured by using a measuring tape/scale.

Root length was measured by separating roots from plants and placing them on paper and blotting them, then using a measuring tape (recorded in cm). Fresh weight per plant was square root transformed to normalize the error distribution before the analysis, as described [48], using an electronic balance (S = 0.1 g) (Acculab V-1200). The harvested plants were rinsed with distilled water, then the roots were blotted on filter paper and dried completely in an oven at 60–65 °C to determine dry weight [49]. The following formulas were used to calculate the index of growth traits [45].

Relative growth rate (RGR) was calculated by the following formula:

$$\mathrm{RGR} = (\mathrm{lnW_2} - \mathrm{lnW_1})(\mathrm{t_2} - \mathrm{t_1})$$

where W_2 and W_1 denote the plant's dry mass (g) at time t_2 and t_1, respectively.

The net assimilation rate was calculated by the following formula:

$$\mathrm{NAR} = \mathrm{dW}/(\mathrm{A} \times \mathrm{dt})$$

where A is the area of assimilation organs (cm^2), dW is the dry mass increment (g), and dt is the time of cultivation (days). Root mass ratio (RMR; root mass per unit total plant mass) was calculated as described in [49].

2.4.2. Physiological Measurements

Total chlorophyll content was estimated by using a portable The Soil Plant Analysis Development (SPAD) chlorophyll apparatus (SPAD-502 Plus, Konica Minolta, Tokyo, Japan). The leaf net photosynthesis rate for 3 independent lettuce seedlings per experimental replicate was determined using a portable LI-6400 photosynthesis system (Li-Cor 6400-18, Lincoln, NE, USA) [40,50]. The set values used were as follows: photosynthetic photon flux density, 500 $\mu mol \cdot m^{-2} \cdot s^{-1}$; air flow rate inside the sample chamber, 400 $\mu mol \cdot s^{-1}$.

The nutrient contents were measured in dried leaves ground by an electric mortar (multipurpose high-speed disintegrator, Dingia), sieved, weighed out to 0.2 g, and digested by concentrated nitric acid (HNO_3, 5–6 mL) carefully under a laminar flow hood cabinet. All nutrients were analyzed using an optical emission spectrometer (Optima 5300 DV Spectrometer, Shelton, CT, USA).

The total N was determined by the Kjeldahl method [51].

2.5. Statistical Analysis

The recorded data were subjected to analysis of variance (ANOVA) and fixed-factor models [52], and Duncan's multiple range test was used to assess the significance of treatment differences by means of IBM SPSS Statistics for Windows (version 20.0, IBM Corp., Armonk, NY, USA).

3. Results

3.1. Application of Three Amino Acids on Lettuce

The effects of different amino acids were studied for vegetative growth. The plants grown in the modified nutrient solution (Hoagland and amino acids) showed varied responses for metric traits.

The vegetative indicators responded positively and significantly to all applied L-methionine concentrations (Figure 2A–F). Leaf length, width, and the number increased in response to L-methionine application, and decreased with L-glycine and L-tryptophan. The leaf length of L-methionine treated plants increased by 11.41% compared to control plants, while it decreased by 13.76% and 61.92% in L-glycine and L-tryptophan treated plants, respectively (Figure 2A). A significant increase in leaf width (17.46%) was also found with L-methionine, but there was an 18.25% and 63.49% decrease in response to L-glycine and L-tryptophan, respectively (Figure 2B). Similarly, leaf area and leaf numbers also increased under L-methionine treatment (31.41% and 50.4%, respectively), while leaf area decreased under L-glycine and L-tryptophan (29.67% and 86.25%, respectively) and leaf numbers decreased under L-tryptophan (50.36%) compared to control (Figure 2C,D).

Furthermore, plant height and area also had an encouraging response to L-methionine application (Figure 2E,F). The results revealed that there was an abrupt change in plant height and area, which were reduced by 82.91% and 90.78%, respectively, upon L-tryptophan application.

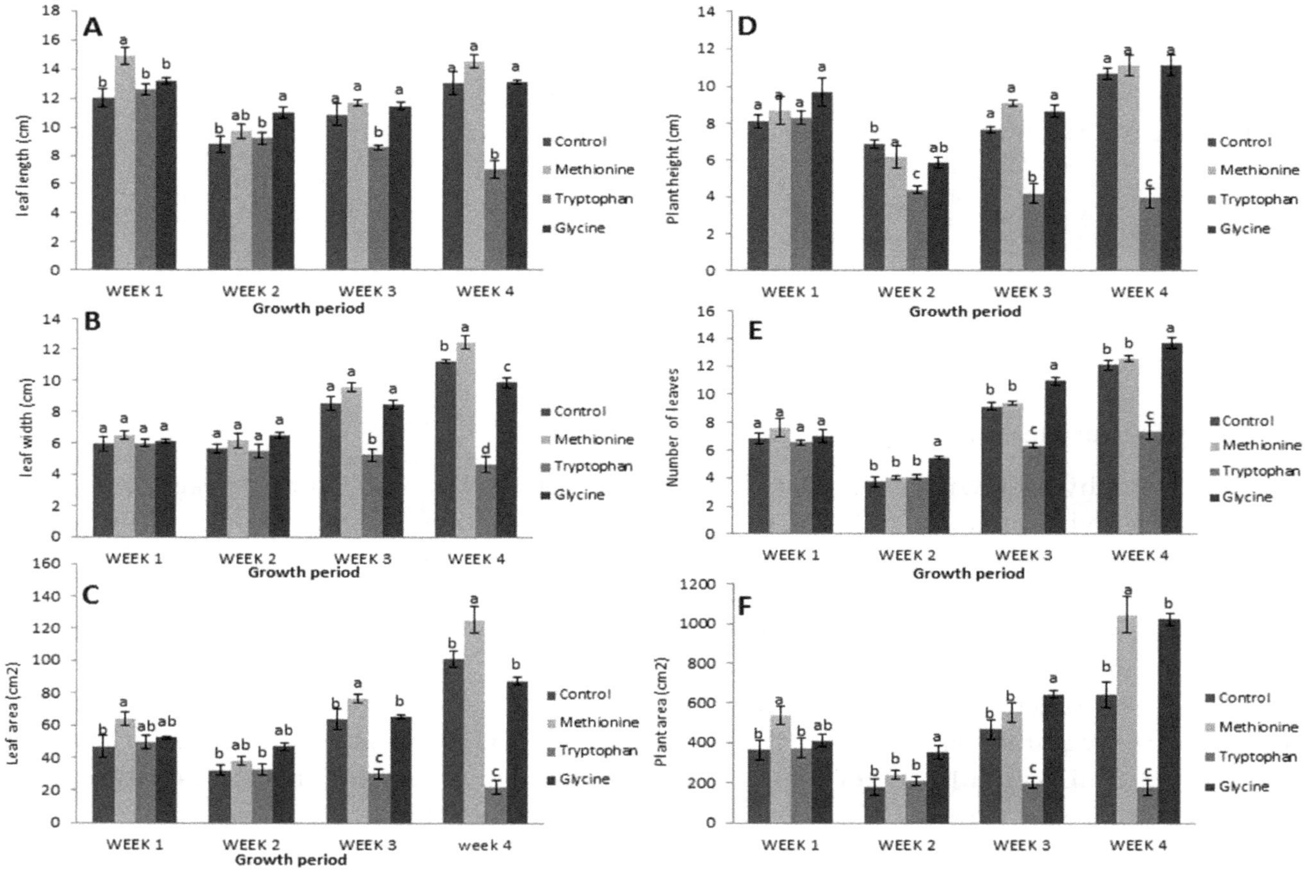

Figure 2. (**A**) Leaf length, (**B**) leaf width, (**C**) leaf area, (**D**) plant height, (**E**) leaf number, and (**F**) plant area with L-methionine (20 mg/L), L-glycine (220 mg/L), and L-tryptophan (210 mg/L). Means followed by the same lowercase letters do not differ significantly from each other in the comparison between amino acid treatments each week using Duncan's multiple range test ($p < 0.05$).

Likewise, root length, shoot-to-root ratio, relative water content, net assimilation rate, and fresh and dry biomass were positively affected by L-methionine treatment and negatively by L-tryptophan and L-glycine (Figure 3). The results revealed that lettuce plants showed a significant increase in root length (Figure 3A), relative water content (Figure 3D), and net assimilation rate (Figure 3C) in response to L-methionine application, and a decrease with the other two amino acids. Interestingly, the shoot-to-root ratio was found to be higher in response to L-tryptophan, which suggests that amino acids other than L-methionine also have an important role in plant growth. Moreover, a relative increase in fresh and dry biomass was observed with L-methionine application, by 20.88% and 15.71%, respectively, and a decrease with L-tryptophan and L-glycine treated plants (Figure 3D). Taken together, these results indicate that biostimulants, specifically the amino acid L-methionine, play a critical role in the growth and development of lettuce plants.

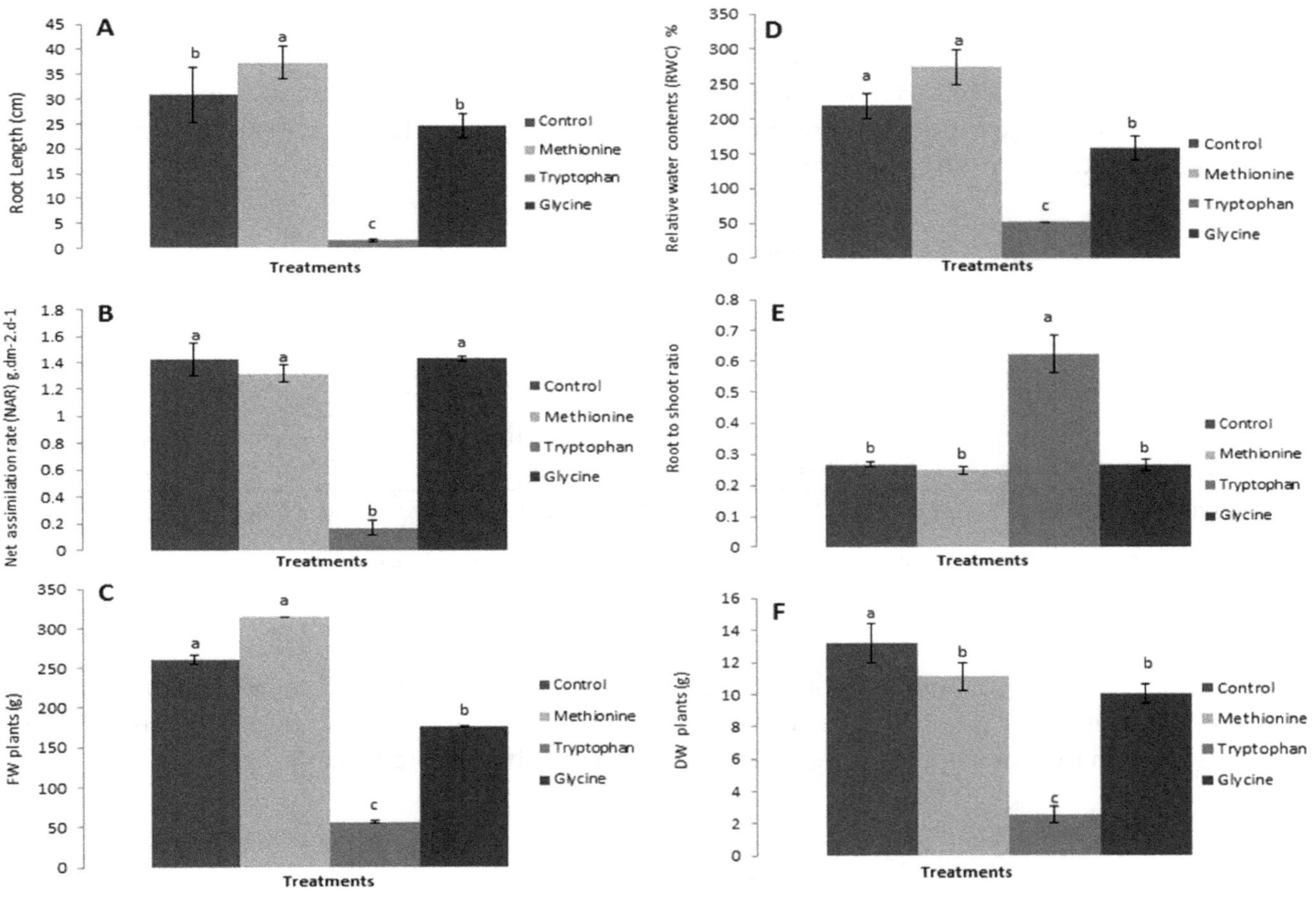

Figure 3. Effects of amino acids on (**A**) root length, (**B**) net assimilation rate (NAR), (**C**) plant fresh weight (FW), (**D**) relative water content (RWC), (**E**) shoot-to-root ratio (S:R), and (**F**) plant dry weight (DW). Means followed by the same lowercase letters do not differ significantly from each other according to Duncan's multiple range test ($p < 0.05$).

3.1.1. Photosynthetic Measurements

Photosynthetic measurements of lettuce leaves (Table 1) included amino acid application effect, rate of net photosynthesis, stomatal conductance, and transpiration rate. They were significantly affected by applied amino acids compared to control, except for total chlorophyll content. Transpiration rate and intracellular CO_2 were relatively higher among all amino acid treated plants. The high levels of photosynthesis and chlorophyll content suggest that amino acids are important regulators of photosynthesis in lettuce, ultimately leading to higher yield and biomass.

Table 1. Effects of amino acids on physiological indicators of lettuce plants.

Treatments	Net Photosynthesis Rate (Pn) (μmol CO_2 m^{-2} s^{-1})	Stomatal Conductance (mol m^{-2} s^{-1})	Ci (μmol/mol)	Transpiration Rate (Tr) (mol H_2O m^{-2} s^{-1})	Total Chlorophyll Content (SPAD Values)
Control	7.31 a	0.07 a	335.78 b	1.86 a	16.5 n.s.
L-methionine (Meth)	4.33 b	0.03 b	369.55 ab	0.97 bc	16.7 n.s.
L-tryptophan (Try)	3.08 c	0.03 b	426.67 a	0.83 c	15.6 n.s.
L-glycine (Gly)	4.63 b	0.04 b	390.34 ab	1.53 b	17.4 n.s.

L-Meth (20 mg/L), L-Try (220 mg/L), and L-Gly (210 mg/L). n.s., not significant. Means followed by the same letters are not statistically different from each other, according to Duncan's multiple range test ($p < 0.05$).

In addition, relative growth parameters including leaf area index, leaf dry matter content, root mass ratio, specific leaf area, and leaf area ratio were higher in response to L-tryptophan treatment (Table 2). The relative growth rate was found to be higher with L-glycine application. L-methionine and L-glycine treated plants had a nonsignificant association with all other traits compared to control plants. These results show that along with L-methionine, L-tryptophan is also an important player, especially for relative growth enhancement in lettuce.

Table 2. Effects of amino acids on growth indices of lettuce plants.

Treatments	LAI (cm^2 cm^{-2})	LDMC (g g^{-1})	RMR (g g^{-1})	SLA (cm^2 g^{-1})	LAR (cm^2 g^{-1})	RGR (g g^{-1} d^{-1})
Control	2.73 b	0.05 b	0.01 b	0.07 n.s.	22.67 bc	4.44 b
L-methionine	2.06 b	0.03 b	0.01 b	0.04 n.s.	6.38 c	1.18 c
L-tryptophan	26.18 a	0.58 a	0.15 a	0.10 n.s.	63.65 a	1.34 c
L-glycine	3.85 b	0.05 b	0.01 b	0.07 n.s.	35.78 b	7.31 a

L-Methionine (20 mg/L), L-Tryptophan (220 mg/L), and L-Glycine (210 mg/L). LAI, leaf area index; LDMC, leaf dry matter content; RMR, root mass ratio; SLA, specific leaf area; LAR, leaf area ratio; RGR, relative growth rate; n.s. not significant. Means followed by the same letters are not statistically different from each other, according to Duncan's multiple range test ($p < 0.05$).

However, vitamin C content was not significantly affected by the applied concentrations, although there was a tendency for it to be higher in L-methionine and L-glycine compared to control (Table 3). In contrast, dry matter percentages were higher in L-tryptophan treated plants.

Table 3. Effects of amino acids on vitamin C content and dry matter percentage of lettuce leaves.

Treatments	Vitamin C (mg 100 g^{-1})	Dry Matter (%)
Control	0.25 n.s.	13.4 bc
L-methionine	0.3 n.s.	10.1 c
L-tryptophan	0.2 n.s.	94.5 a
L-glycine	0.3 n.s.	24.7 b

L-Methionine (20 mg/L), L-Tryptophan (220 mg/L), and L-Glycine (210 mg/L). n.s., not significant. Means followed by the same letters are not statistically different from each other, according to Duncan's multiple range test ($p < 0.05$).

3.1.2. Nutrient Contents

Nutrient content analysis (Table 4) revealed a dynamic change in response to all amino acid concentrations. The content of essential elements such as nitrogen, phosphorus, and potassium varied among treatments.

Table 4. Effects of amino acids on essential elements of lettuce.

Treatments	N	P	K	S	Ca	Mg	Fe	Cu	Na	Zn	Al
	(% DW)	(mg/g)									
Control	1.4 b	12.3 d	208.1 b	9.8 b	93.7 a	22.9 a	6.8 a	0.1 c	18.3 a	0.2 b	13.1 a
L-methionine	4.3 a	32.2 b	420.7 a	10.1 b	66.3 c	12.6 b	2.1 b	0.06 d	4.48 b	0.4 a	7.1 b
L-tryptophan	3.7 a	36.8 a	421.4 a	13.1 a	81.5 b	13.4 b	2.8 b	0.14 b	3.95 b	0.4 a	9.0 ab
L-glycine	4.3 a	21.7 c	230.8 b	13.3 a	44.6 d	12.6 b	6.9 a	0.23 a	3.40 b	0.3 a	9.6 ab

L-methionine (20 mg/L), L-tryptophan (220 mg/L), and L-glycine (210 mg/L); Macronutrients: N, nitrogen; P, phosphorus; K, potassium; S, sulfur; Ca, calcium; Mg, magnesium. Micronutrients: Fe, iron; Cu, copper; Mo, molybdenum; Na, sodium; Zn, zinc; Al, aluminum. Means followed by the same letters are not statistically different from each other, according to Duncan's multiple range test ($p < 0.05$).

3.2. *Application of Different L-Methionine Concentrations on Lettuce*

In the second trial, L-methionine at a higher concentration had a reduced effect on plant growth and physiology. The results show that lower levels of L-methionine significantly contributed to enhancing the number of leaves, plant height, and leaf length and width (Figure 4). Remarkably, 0.22 mg/L concentration of L-methionine resulted in a gradual increase in vegetative growth compared to control plants during all weeks. In contrast, higher levels were negatively associated with the corresponding measurements. Lettuce plants had short stature and fewer leaves in response to 2200 mg/L and 220 mg/L of L-methionine (Figure 4B,D). Concomitantly, more than 80% and 50% decreases in these two traits were found with increasing amino acid levels.

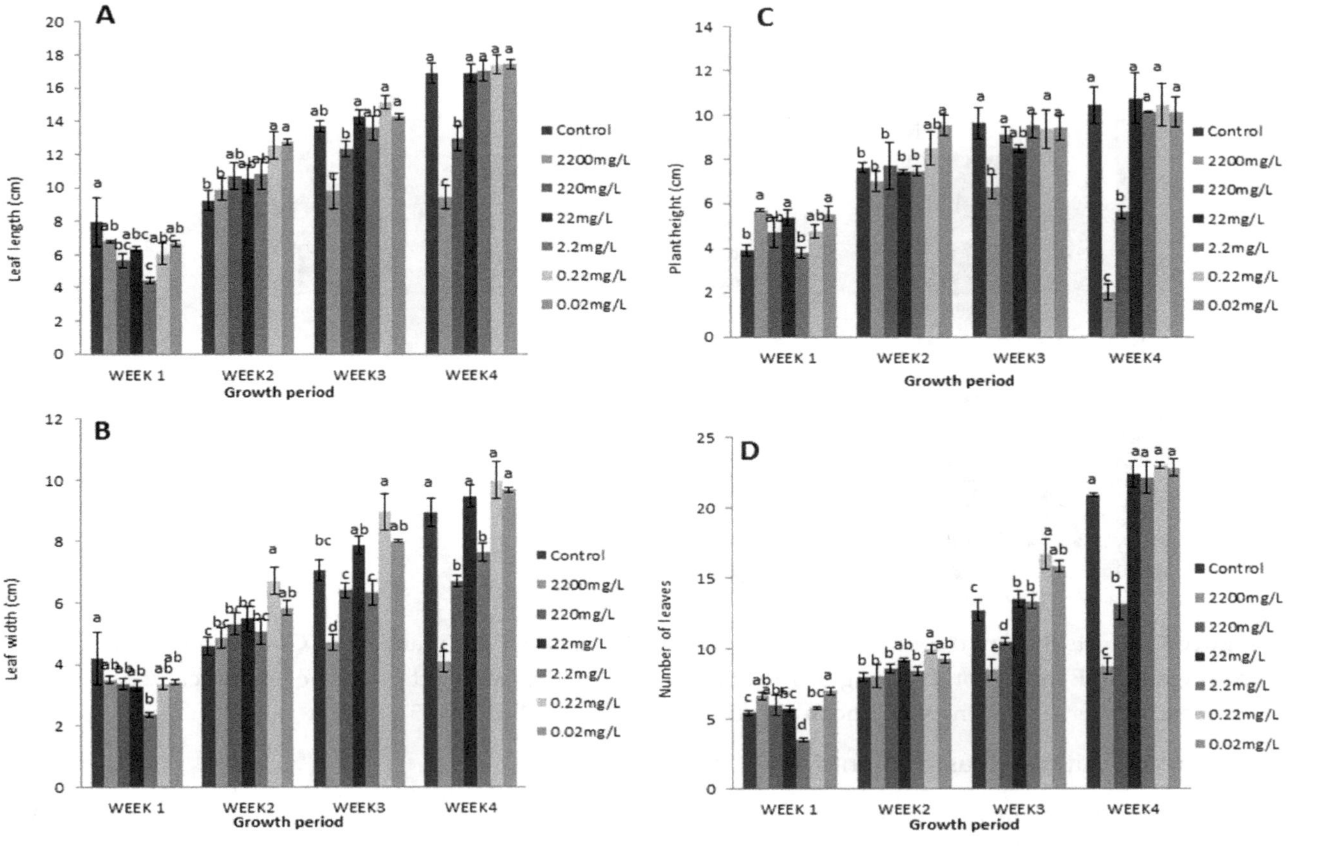

Figure 4. Effects of L-methionine concentrations on (**A**) leaf length, (**B**) plant height, (**C**) leaf width, and (**D**) the number of leaves per plant. Means followed by the same letters are not statistically different from each other, according to Duncan's multiple range test at ($p < 0.05$).

Similarly, relatively decreased leaf length (42.24% and 23.3%, respectively) and width (54.5% and 25.23%, respectively) were observed with increased treatment levels (Figure 4A,C). A strong increasing trend was also found for both traits in response to 0.2 mg/L of L-methionine. Overall, these results indicate that higher levels of L-methionine have an inhibiting effect on plant growth.

Leaf and plant area, root length, and fresh and dry weight of lettuce plants were improved by lower L-methionine concentrations (especially 0.2 mg/L) in advanced growth stages (Figure 5). However, mixed growth patterns were also present with different amino acid levels at each plant stage. A maximum reduction in leaf and plant area (75.5% and 74.653%, respectively) was found in 2200 mg/L treated plants, and a minimum (15.6% and 4.03%, respectively) in 0.2 mg/L plants (Figure 5A,B). Moreover, root length was found to be reduced at all levels except 0.2 mg/L, which caused a relative increase by 14.8% (Figure 5C). At the same time, the fresh and dry weight of roots were also lower with more concentrated treatment (Figure 5D), which suggests that amino acids are essential elements required as micronutrients for plant growth. An increased concentration leads to restricted plant biomass. Moreover, the decreased fresh and dry weight of lettuce plants indicates that nutrient stress and reduced photosynthetic activity were responsible for the lower accumulation of leaf organic material and growth rate.

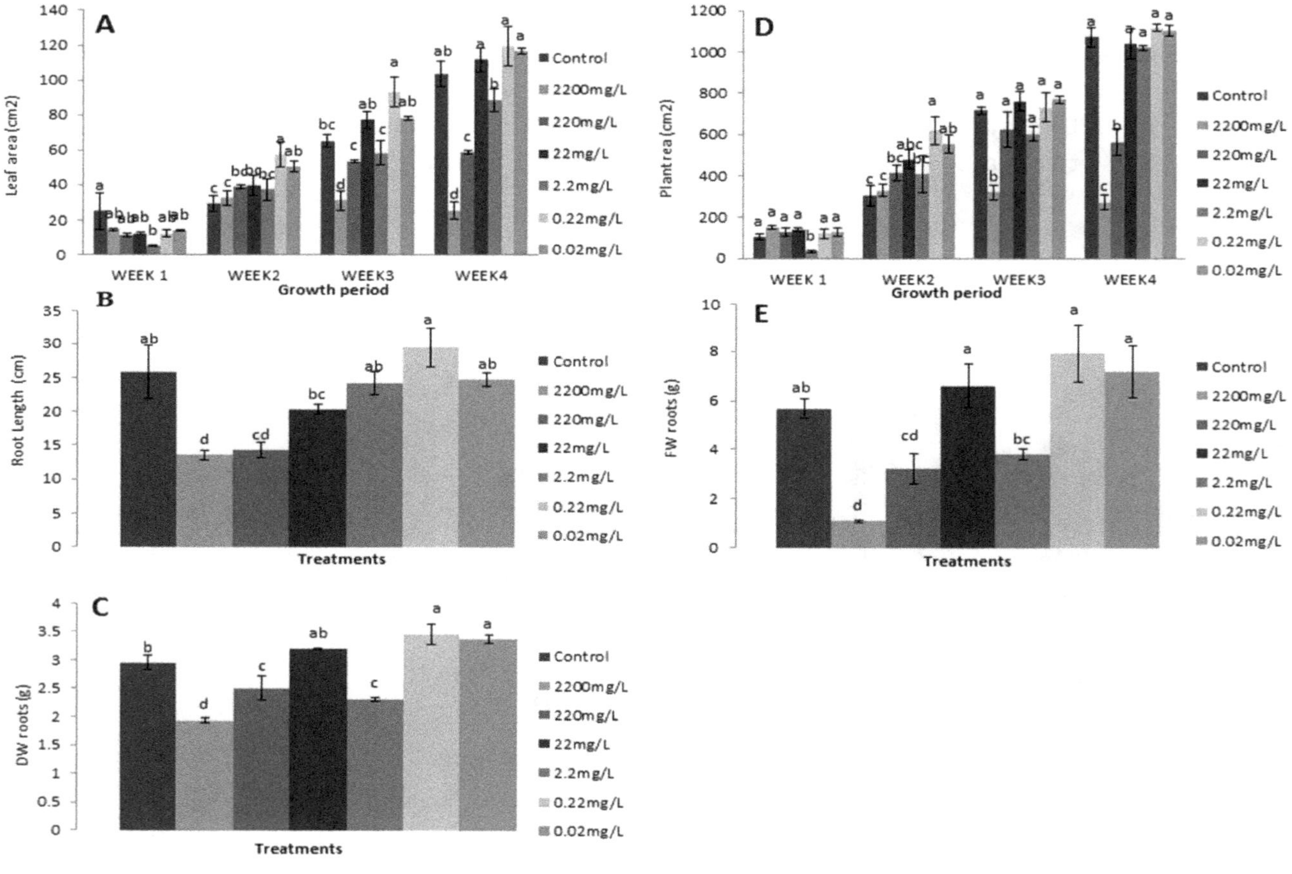

Figure 5. Effects of L-methionine concentrations on (**A**) leaf area, (**B**) plant area, (**C**) root length, (**D**) root fresh weight, (**E**) and root dry weight. Means followed by the same letters are not statistically different from each other, according to Duncan's multiple range test at ($p < 0.05$).

3.2.1. Relative Growth Measurements

Relative growth parameters measured in the second trial revealed that most of the traits were substantially increased by decreased L-methionine concentrations (Table 5). Surprisingly, root mass and leaf area ratios increased ($p < 0.05$) under higher concentration, which shows that plants can respond to a stress environment by maintaining their growth patterns. However, all other measured parameters had a decreasing tendency under a nutrient stress environment.

Table 5. Effects of concentrations of L-methionine on growth indices of lettuce plants.

Treatment (mg/L)	LAI ($cm^2\ cm^{-2}$)	RMR ($g\ g^{-1}$)	SLA ($cm^2\ g^{-1}$)	LAR ($cm^2\ g^{-1}$)	RGR ($g\ g^{-1}\ d^{-1}$)	RWC (%)
Control	0.24 a	0.04 b	1679.50 a	51.5 ab	0.33 c	91.2 a
2200	0.06 c	0.34 a	1259.19 ab	76.6 a	0.34 c	11.11 c
220	0.17 b	0.06 b	572.02 b	42.2 ab	0.23 c	72.22 b
22	0.25 a	0.02 b	1118.38 ab	36.9 b	0.27 c	93.72 a
2.2	0.24 a	0.02 b	1937.15 a	46.1 ab	0.52 c	92.96 a
0.22	0.26 a	0.01 b	1496.32 a	34.5 b	1.6 b	93.28 a
0.02	0.27 a	0.01 b	1641.48 a	45.5 ab	3.7 a	90.77 a

LAI, leaf area index; RMR, root mass ratio; SLA, specific leaf area; LAR, leaf area ratio; RGR, relative growth rate; RWC, relative water content. Means followed by the same letters are not statistically different from each other, according to Duncan's multiple range test ($p < 0.05$).

3.2.2. Photosynthetic Measurements

Enhanced photosynthetic activity, transpiration, and total chlorophyll content (Table 6) at lower L-methionine concentrations suggest that plants require this nutrient only in small amounts. More moderate transpiration activity occurred with all applied treatments, and reduced accumulation of chlorophyll in leaves causes restricted photosynthetic activity. There was no significant difference observed for stomatal conductance but it was relatively higher with 22 mg/L of L-methionine.

Table 6. Effects of L-methionine levels on net physiological and growth indicators of lettuce leaves.

Treatment (mg/L)	Net Photosynthesis Rate (Pn) ($\mu mol\ CO_2\ m^{-2}\ s^{-1}$)	Stomatal Conductance ($mol\ m^{-2}\ s^{-1}$)	Ci (μmol/mol)	Transpiration Rate (Tr) ($mol\ H_2O\ m^{-2}\ s^{-1}$)	Total Chlorophyll Content (SPAD)
Control	5.36 b	0.10 abc	363.22 bc	1.94 b	24.7 cd
2200	1.67 e	0.03 c	379.51 ab	0.64 d	23.7 d
220	2.37 de	0.05 bc	398.91 a	1.11 c	29.9 a
22	3.07 d	0.15 a	395.1 a	1.35 c	26.2 bc
2.2	4.47 c	0.06 bc	347.86 c	1.27 c	27.3 b
0.22	7.16 a	0.12 ab	369.23 bc	2.41 a	28.1 ab
0.02	6.61 a	0.09 abc	349.91 c	1.92 b	27.3b

Means followed by the same letters are not statistically different from each other, according to Duncan's multiple range test ($p < 0.05$).

3.2.3. Nutrient Contents

Data in Table 7 describe the effects of L-methionine on macro- and microelements. The increasing trend of essential element accumulation including N, P, and K at reduced L-methionine levels indicates that these elements affect plant metabolism and help to adapt to modified environmental cues, which directly or indirectly affects plant metabolism. For example, a significant increase in nitrogen (N) content in leaf tissues increases photosynthesis efficiency, which is key to increasing crop yield. Plant metabolism is maintained by these elements with lower fractions of amino acids to regulate plant growth and development. Mixed fractions of other elements at different concentrations signify their importance in plant health regulation. For example, in addition to essential elements, S, Mg, Fe, Cu, Mn, and Na accumulation was higher at all levels.

Any change in amino acid concentration leads to stress conditions, and plants respond differently at different levels by changing their growth patterns.

Moreover, a significant ($p < 0.05$) decrease in vitamin C content and leaf dry matter content and percentage also highlights the importance of micronutrients in plant metabolism.

Table 8 shows that vitamin C content decreased significantly ($p < 0.05$) with 2200 mg/L, but increased by 14.21% with 0.22 mg/L as compared to control. In contrast, a significant increase in leaf dry matter content was found with 2200 mg/L and in dry matter percentage ($p < 0.05$) with L-methionine application of 2200 mg/L and 220 mg/L. Moreover, a decrease in the fresh and dry weight of lettuce plants indicates that reduced photosynthetic activity was responsible for the lower accumulation of leaf organic material and reduced growth.

Table 7. Effects of L-methionine concentrations on essential macro- and microelements of lettuce leaves.

Treatment (mg/L)	N	P	K	S	Ca	Mg	Fe	Cu	Na	Zn	Al
	(% DW)	(mg/g)									
Control	5.3 c	17.5 d	128.3 c	9.7 d	67.1 c	11.3 d	2.5 c	0.03 c	4.3 d	0.5 c	7.6 b
2200	2.1 e	27 bc	164.1 c	23.8 a	106.1 ab	19.2 abc	5.5 a	0.08 a	15.4 a	0.8 b	9.9 ab
220	2.3 e	27.1 bc	233.1 b	19.6 b	127.9 a	24.5 a	4.4 ab	0.04 bc	11 b	0.4 c	10.4 a
22	4.2 d	25.2 c	254.3 b	12.4cd	83.2 bc	16.27 cd	3.2 bc	0.04 bc	6 c	0.4 c	8.7 ab
2.2	5.7 bc	39.0 a	351.8 a	14.1 c	108.8 bc	22.7 ab	3.0 c	0.05 b	11 b	1.1 a	9.3 ab
0.22	6.7 a	34.9 ab	336.4 a	13 cd	107.1 ab	19.8 abc	3.2 bc	0.04 bc	9.4 b	0.4 c	9 ab
0.02	6.1 b	29.5 bc	246.5 b	12.5 cd	99.2 abc	17.3 bc	3.4 bc	0.05 b	6.6 b	0.4 c	9.1 ab

Macronutrients: N, nitrogen; P, phosphorus; K, potassium; S, sulfur; Ca, Calcium; Mg, magnesium. Micronutrients: Fe, iron; Cu, copper; Mo, molybdenum; Na, sodium; Zn, zinc; Al, aluminum. Means followed by the same letters are not statistically different from each other, according to Duncan's multiple range test at ($p < 0.05$).

Table 8. Effects of L-methionine concentrations on vitamin C content, leaf dry matter content (LDMC), dry matter percentage (DM%), fresh weight (FW), and dry weight (DW) of lettuce leaves.

Treatment (mg/L)	Vitamin C (mg 100 g^{-1})	LDMC (g g^{-1})	DM (%)	Mean Fresh Weight (g)	Mean Dry Weight (g)
Control	0.21 a	0.13 b	4.7 b	37.97 bc	4.8 c
2200	0.11 b	0.49 a	48.8 a	0.86 d	0.8 d
220	0.19 a	0.12 b	11.9 ab	10.89 cd	2.3 d
22	0.19 a	0.06 b	5.69 b	58.89 ab	5.4 bc
2.2	0.21 a	0.06 b	5.62 b	44.29 ab	7.97 a
0.22	0.23a	0.05 b	5.29 b	70.94 a	8.7 a
0.02	0.21 a	0.04 b	4.23 b	60.89 ab	7.3 ab

Means followed by the same letters are not statistically different from each other, according to Duncan's multiple range test ($p < 0.05$).

As indicated by the outcomes shown in Tables 3 and 8, there was no significant change ($p < 0.05$) observed in plants with or without amino acid treatment. It was observed that all individual amino acid treatments, except for one L-methionine concentration, led to no significant ($p < 0.05$) impact on vitamin C content. The special case was L-methionine at 2200 mg L^{-1}, which prompted a decrease in vitamin C content, essentially contrasting with the control plants ($p < 0.05$).

4. Discussion

Amino acid application is a common practice for horticultural crops worldwide, with the majority of treatments making use of biostimulants with a mixture of amino acids [53]. In our study, we checked the activity of a single amino acid that regulates the nutrient contents involved in growth variables.

Previous investigations have demonstrated that plant developmental cues respond distinctively to the provision of amino acids [54–56]. It is likely that the effects of amino acids on plants rely on the kind of amino acids supplied [56] and the plant cultivars [57]

From our results, we can presume that amino acids (L-methionine, L-tryptophan, and L-glycine) not only make nutrients available to plants but also act as signal transducing molecules [31], as small doses are sufficient for plant development response, while these molecules can act as signals of several beneficial plant physiological processes. Studies demonstrate that amino acids in the form of a foliar spray on plants is a promising technique [38]. In this manner, L-methionine induces more prominent absorption of sulfur and nitrogen in plants, which also depends on the amount applied [26,58–60].

Plants utilize amino acids according to their nutritional needs and genetic background, as well as environmental and developmental cues [61,62]. This might be the reason why amino acid reactions were not consistent in both experiments. Therefore, we can assume that the decreasing effect of L-tryptophan on yield might be due to the inhibitory impact of auxin on vegetative growth. In this association, the reduction of lettuce yield per plant caused by the inhibition effect might be due to the detrimental effect of auxin accumulation stress on growth, the aggravation of mineral nutrient uptake, and the improvement of plant respiration [41]. The distinctive response was reported by Abbas et al. [54], in an investigation on L-tryptophan applied to chickpea at rates of 10^{-1} M, 10^{-2} M, and 10^{-3} M. They found random results with different parameters: root length was increased only in the control compared to the three treatments, the number of nodules increased only with 10^{-1} M, and nodule fresh and dry weight decreased with 10^{-3} M treatment and increased with the other two compared to control, while control remained nonsignificant. The most pods and highest plant weight were shown with 10^{-2} M, but pod weight per plant was significantly affected by all treatments due to the production of phytochromes suitable to chickpea. This experiment may provide evidence for the substantiating inconsistency of lettuce observed in our previous experiment.

However, our results contrast those described in [63], in which numbers of strawberry leaves per plant were significantly higher with the application of L-tryptophan than control.

A few reasons can clarify the positive effects of L-methionine. First, it has a role in maintaining the structure of proteins required for cell division, cell differentiation, and growth. Second, it provides sufficient sulfur and nitrogen according to plant needs. Third, the ability of L-methionine to be converted into polyamines and enlarge by entering the hormone structures [64] allows nitrogen movement between cells and organs [65]. It also functions as a buffer and behaves as a source of carbon and energy [66], and as a precursor of spermidine and gibberellin biosynthesis [43,67], growth regulators, and many secondary metabolites [43]. L-methionine also acts as a growth regulator of cytokinin, brassinosteroids, and auxin, increasing the initiation of roots; helps with the absorption of more nutrients by the plant [39,67], which may stimulate endogenous hormone homeostasis [68,69]; and is required for the development of hairy roots [67] at optimum levels.

Increased L-methionine levels influence phytohormones, which ultimately increases the chlorophyll content and chloroplast development or cytokinins [70,71]. An expected requirement for the prompting of L-methionine application might be the proximity of phytohormones (e.g., auxins and cytokinin). The phytohormones and signaling compounds may improve the photosynthetic activity, leading to better yield. Another possible mechanism involved with the amino acid effect could be related to the stimulation of root growth of treated plants, which may improve water and nutrient uptake capability, leading to yield productivity [68,69], as well as enhanced cell formation and increased fresh and dry matter [72], with increased growth behavior [69,70,73].

Our results demonstrate that high L-methionine concentration reduced plant growth due to damage to the photosynthetic apparatus [61] and blocking of nutrient uptake. Higher levels of this nutrient cause blockage of photosynthesis in stressed environments [65]. Padgett (1996) applied L-methionine to the root zones of chrysanthemum plants, producing a physiological disorder called methionosis, with the typical pattern of a metabolite–antimetabolite relationship [47]. It is thought that in this case, L-methionine, especially because of the large amounts applied, may function as an antimetabolite that interferes with normal amino acid metabolism. In other words, amino acids should be meticulously applied, as they could reduce the percentage of dry weight because they cause swollen, water-filled tissues due to depressed vegetative growth [65,74].

Thus, we speculate that application of L-methionine with 0.22 mg/L in the nutrient solution was sufficient, but other concentrations were too high and might have been a source of stress. Our results are consistent with previous studies [42,75] proposing that the improvement of novel "bio-sound items" ought to continue based on a foundational approach established in chemical synthesis, natural chemistry or biochemistry, and biotechnology connected to genuine plant physiological, agrarian, and environmental constraints. It was proposed that these items should work at low dosages, be biologically and ecologically friendly, and have reproducible advantages in horticultural plant development. The high amount of L-methionine also reduced vitamin C content. Therefore, we conclude that plant metabolism is affected by external N and thus can reflect the changes in N absorption, transport, and metabolism [76]. Similar findings have been reported in Chinese cabbage and lettuce [77]. The optimal required concentration is essential for optimal growth. It was confirmed from the previous study that when amino acids are added alone, care must be taken, as they can inhibit cell growth [63]. In general, the use of amino acids on plants can improve their capacity of transporting mineral components [66].

In view of the synthetic pathway of vitamin C and the synthesis of ascorbic acid requiring plant climatic changes and the conditions of plant sustenance, it may be hypothesized that whatever factor builds the sugar (or glucose) content in plant tissues can thus increase the vitamin C content [78]. It has been reported that amino acids [43] and nitrogen fertilizers do not impact the vitamin C content in broccoli. Conversely, in cauliflower, when nitrogen fertilizers are extended from 80 kg to 120 kg per ha, the ascorbic acid content is reduced by 7% [79].

Optimizing the amino acid content can bring about different morphogenetic responses; higher concentrations generally inhibit growth in *Cicer arietinum* [80]. The available information from various studies suggests that optimal levels of various amino acids may be species- or genotype-dependent, which needs to be determined before recommending their use [81]. Increasingly, plant ecologists working at all levels have become interested in the role of amino acid nutrition in the lives of plants and determining the proper amount of amino acids suitable for plant growth [82].

The depressive effect of L-tryptophan and high amounts of L-methionine on yield may be attributed to the inhibitory effect of auxin accumulation on vegetative growth, the disturbance in mineral uptake, and/or the enhancement of plant respiration [83].

5. Conclusions

Taking into consideration the discussion above, it can be inferred that L-methionine increases the chlorophyll content of plants and contributes to the saving of energy, thus boosting the plant yield. L-methionine led to significant increases in observed physiological factors in lettuce leaves at lower concentrations because at high concentrations it affects auxin uptake, which can kill plants. In brief, L-methionine at a concentration of 0.2 mg/L showed the best effect on the growth of lettuce plants. Therefore, we can say that L-methionine can contribute as a suitable substitute for fertilizers to increase crop yield. Future research should concentrate on assessing the mechanisms of how amino acids can influence the genetic transcription of various parameters, including supplement transporters, hormone production, and antioxidant metabolism. Along these lines, it will be possible to acquire the best understanding of the role of amino acids as biostimulants in lettuce plants.

Author Contributions: Conceptualization, W.J.; methodology, Q.L. and Y.G.; software, S.K.; validation, S.K.; formal analysis, S.K. and B.N.S.; investigation, S.K.; data curation, S.K.; writing—original draft preparation, S.K.; writing—review and editing, H.W., P.L., H.Y., and Q.L.; visualization, H.Y.; S.K.; supervision, W.J.; project administration, Y.G.; funding acquisition, W.J.

References

1. Samuolienė, G.; Urbonavičiūtė, A.; Duchovskis, P.; Bliznikas, Z.; Vitta, P.; Žukauskas, A. Decrease in nitrate concentration in leafy vegetables under a solid-state illuminator. *HortScience* **2009**, *44*, 1857–1860. [CrossRef]
2. Son, K.H.; Oh, M.M. Leaf shape, growth, and antioxidant phenolic compounds of two lettuce cultivars grown under various combinations of blue and red light-emitting diodes. *HortScience* **2013**, *48*, 988–995.
3. Christopoulou, M.; Wo, S.R.C.; Kozik, A.; McHale, L.K.; Truco, M.J.; Wroblewski, T.; Michelmore, R.W. Genome-wide architecture of disease resistance genes in lettuce. *G3 Genes Genomes Genet.* **2015**, *5*, 2655–2669. [CrossRef] [PubMed]
4. Amoozgar, A.; Mohammadi, A.; Sabzalian, M. Impact of light-emitting diode irradiation on photosynthesis, phytochemical composition and mineral element content of lettuce cv. Grizzly. *Photosynthetica* **2017**, *55*, 85–95. [CrossRef]

5. Nawaz, M.A.; Wang, L.; Jiao, Y.; Chen, C.; Zhao, L.; Mei, M.; Yu, Y.; Bie, Z.; Huang, Y. Pumpkin rootstock improves nitrogen use efficiency of watermelon scion by enhancing nutrient uptake, cytokinin content, and expression of nitrate reductase genes. *Plant Growth Regul.* **2017**, *82*, 233–246. [CrossRef]
6. Nawaz, M.A.; Chen, C.; Shireen, F.; Zheng, Z.; Sohail, H.; Afzal, M.; Ali, M.A.; Bie, Z.; Huang, Y. Genome-wide expression profiling of leaves and roots of watermelon in response to low nitrogen. *BMC Genom.* **2018**, *19*, 456. [CrossRef] [PubMed]
7. Scheible, W.-R.; Morcuende, R.; Czechowski, T.; Fritz, C.; Osuna, D.; Palacios-Rojas, N.; Schindelasch, D.; Thimm, O.; Udvardi, M.K.; Stitt, M. Genome-wide reprogramming of primary and secondary metabolism, protein synthesis, cellular growth processes, and the regulatory infrastructure of *Arabidopsis* in response to nitrogen. *Plant Physiol.* **2004**, *136*, 2483–2499. [CrossRef]
8. Yuan, L.; Yuan, Y.; Du, J.; Sun, J.; Guo, S. Effects of 24-epibrassinolide on nitrogen metabolism in cucumber seedlings under $Ca(NO_3)_2$ stress. *Plant Physiol. Biochem.* **2012**, *61*, 29–35. [CrossRef] [PubMed]
9. Amalfitano, C.; Gomez, L.D.; Frendo, P.; De Pascale, S.; Pepe, O.; Simister, R.; Ventorino, V.; Agrelli, D.; Borrelli, C.; McQueen-Mason, S.J.; et al. Plant–Rhizobiumsymbiosis, seed nutraceuticals, and waste quality for energy production of *Vicia faba* L. as affected by crop management. *Chem. Biol. Technol. Agric.* **2018**, *5*, 15. [CrossRef]
10. Jämtgård, S.; Näsholm, T.; Huss-Danell, K. Nitrogen compounds in soil solutions of agricultural land. *Soil Biol. Biochem.* **2010**, *42*, 2325–2330. [CrossRef]
11. Marschner, H. *Marschner's Mineral Nutrition of Higher Plants*; Academic Press: New York, NY, USA, 2011.
12. Caruso, G.; Conti, S.; La Rocca, G. Influence of crop cycle and nitrogen fertilizer form on yield and nitrate content in different species of vegetables. *Adv. Hort. Sci.* **2011**, *25*, 81–89.
13. Silva, L.F.; Hower, J.C.; Izquierdo, M.; Querol, X. Complex nanominerals and ultrafine particles assemblages in phosphogypsum of the fertilizer industry and implications on human exposure. *Sci. Total Environ.* **2010**, *408*, 5117–5122. [CrossRef]
14. Verger, P.J.; Boobis, A.R. Reevaluate pesticides for food security and safety. *Science* **2013**, *341*, 717–718. [CrossRef]
15. du Jardin, P. Plant biostimulants: Definition, concept, main categories and regulation. *Sci. Hortic.* **2015**, *196*, 3–14. [CrossRef]
16. Paschalidis, K.A.; Roubelakis-Angelakis, K.A. Spatial and temporal distribution of polyamine levels and polyamine anabolism in different organs/tissues of the tobacco plant. Correlations with age, cell division/expansion, and differentiation. *Plant Physiol.* **2005**, *138*, 142–152. [CrossRef] [PubMed]
17. Liu, H.; Su, X.; Li, X.; Zhao, X.; Zang, L.; Pan, W. Development of prolonged release microspheres of metformin hydrochloride using ion exchange resins. *J. Chin. Pharm. Sci.* **2006**, *15*, 155.
18. Pang, J.; Ross, J.; Zhou, M.; Mendham, N.; Shabala, S. Amelioration of detrimental effects of waterlogging by foliar nutrient sprays in barley. *Funct. Plant Biol.* **2007**, *34*, 221–227. [CrossRef]
19. Paungfoo-Lonhienne, C.; Thierry, G.; Lonhienne, A.; Doris, R.; Nicole, R.; Michael, C.; Richard, I.W.; Gamage, H.K.; Bernard, J.C.; Peer, M.S.; et al. Plants can use protein as a N source without assistance from other organisms. *Proc. Natl. Acad. Sci. USA* **2008**, *105*, 4524–4529. [CrossRef] [PubMed]
20. Rouphael, Y.; Colla, G. Synergistic Biostimulatory Action: Designing the Next Generation of Plant Biostimulants for Sustainable Agriculture. *Front. Plant Sci.* **2018**, *9*, 1655. [CrossRef]
21. Rouphael, Y.; Spíchal, L.; Panzarová, K.; Casa, R.; Colla, G. High-Throughput Plant Phenotyping for Developing Novel Biostimulants: From Lab to Field or From Field to Lab? *Front. Plant Sci.* **2018**, *9*, 1197. [CrossRef]
22. Kowalczyk, K.; Zielony, T.; Gajewski, M. Effect of Aminoplant and Asahi on yield and quality of lettuce grown on rockwool. *Biostimulators Mod. Agric. Veg. Crops* **2008**, 35–43.
23. Zhao, Y. Auxin biosynthesis and its role in plant development. *Annu. Rev. Plant Biol.* **2010**, *61*, 49–64. [CrossRef] [PubMed]
24. Maeda, H.; Dudareva, N. The shikimate pathway and aromatic amino acid biosynthesis in plants. *Annu. Rev. Plant Biol.* **2012**, *63*, 73–105. [CrossRef] [PubMed]
25. Calvo, P.; Nelson, L.; Kloepper, J. Agricultural uses of plant biostimulants. *Plant Soil* **2014**, *383*, 3–41. [CrossRef]
26. Forde, B.G.; Roberts, M.R. Glutamate receptor-like channels in plants: A role as amino acid sensors in plant defence? *F1000Prime Rep.* **2014**, *6*, 37. [CrossRef]

27. Miller, A.J.; Fan, X.; Shen, Q.; Smith, S.J. Amino acids and nitrate as signals for the regulation of nitrogen acquisition. *J. Exp. Bot.* **2007**, *59*, 111–119. [CrossRef]
28. Walch-Liu, P.; Forde, B.G. L-Glutamate as a novel modifier of root growth and branching: What's the sensor? *Plant Signal. Behav.* **2007**, *2*, 284–286. [CrossRef]
29. Weiland, M.; Mancuso, S.; Baluska, F. Signalling via glutamate and GLRs in *Arabidopsis thaliana*. *Funct. Plant Biol.* **2016**, *43*, 1–25. [CrossRef]
30. Halpern, M.; Bar-Tal, A.; Ofek, M.; Minz, D.; Muller, T.; Yermiyahu, U. The Use of Biostimulants for Enhancing Nutrient Uptake. In *Advances in Agronomy*; Sparks, D.L., Ed.; Academic Press: New York, NY, USA, 2015; Volume 130, pp. 141–174.
31. Teixeira, W.F.; Fagan, E.B.; Soares, L.H.; Umburanas, R.C.; Reichardt, K.; Neto, D.D. Foliar and seed application of amino acids affects the antioxidant metabolism of the soybean crop. *Front. Plant Sci.* **2017**, *8*, 327. [CrossRef]
32. Ertani, A.; Pizzeghelio, D.; Altissimo, A.; Nardi, S. Use of meat hydrolyzate derived from tanning residues as plant biostimulant for hydroponically grown maize. *J. Plant Nutr. Soil Sci.* **2013**, *176*, 287–296. [CrossRef]
33. Hildebrandt, T.M.; Nesi, A.N.; Araújo, W.L.; Braun, H.-P. Amino acid catabolism in plants. *Mol. Plant* **2015**, *8*, 1563–1579. [CrossRef] [PubMed]
34. Persson, J.; Högberg, P.; Ekblad, A.; Högberg, M.N.; Nordgren, A.; Näsholm, T. Nitrogen acquisition from inorganic and organic sources by boreal forest plants in the field. *Oecologia* **2003**, *137*, 252–257. [CrossRef]
35. Gioseffi, E.; de Neergaard, A.; Schjørring, J.K. Interactions between uptake of amino acids and inorganic nitrogen in wheat plants. *Biogeosciences* **2012**, *9*, 1509–1518. [CrossRef]
36. Koukounaras, A.; Tsouvaltzis, P.; Siomos, A.S. Effect of root and foliar application of amino acids on the growth and yield of greenhouse tomato in different fertilization levels. *J. Food Agric. Environ.* **2013**, *11*, 644–648.
37. El-Aal, M.A.; Eid, R.S. Effect of foliar spray with lithovit and amino acids on growth, bioconstituents, anatomical and yield features of soybean plant. *Plant Biotechnol.* **2018**, 187–201.
38. SH SADAK, M.; Abdelhamid, M.T.; Schmidhalter, U. Effect of foliar application of aminoacids on plant yield and some physiological parameters in bean plants irrigated with seawater. *Acta Biol. Colomb.* **2014**, *20*, 141–152.
39. El-Awadi, M.; El-Bassiony, A.; Fawzy, Z.; El-Nemr, M. Response of snap bean (*Phaseolus vulgaris* L) plants to nitrogen fertilizer and foliar application with methionine and tryptophan. *Nat. Sci.* **2011**, *9*, 87–94.
40. Yong, J.W.H.; Letham, S.D.; Wong, C.S.; Graham, D.F. Rhizobium-induced elevation in xylem cytokinin delivery in pigeon pea induces changes in shoot development and leaf physiology. *Funct. Plant Biol.* **2014**, *41*, 1323–1335. [CrossRef]
41. Paciorek, T.; Zazímalová, E.; Ruthardt, N.; Petrásek, J.; Stierhof, Y.D.; Kleine-Vehn, J.; Morris, D.A.; Emans, N.; Jürgens, G.; Geldner, N.; et al. Auxin inhibits endocytosis and promotes its own efflux from cells. *Nature* **2005**, *435*, 1251–1256. [CrossRef]
42. Yakhin, O.I.; Aleksandr, A.L.; Ildus, A.Y.; Patrick, H.B. Biostimulants in Plant Science: A Global Perspective. *Front. Plant Sci.* **2017**, *7*, 2049. [CrossRef]
43. Shekari, G.; Javanmardi, J. Effects of Foliar Application Pure Amino Acid and Amino Acid Containing Fertilizer on Broccoli (*Brassica oleracea* L. var. italica) Transplant. *Adv. Crop Sci. Tech.* **2017**, *5*, 280. [CrossRef]
44. Jiang, W.J. *New Techniques for Soilless Cultivation of Vegetables*, rev. ed.; Beijing Jindun Publishing House: Beijing, China, 2007; pp. 197–200.
45. Hunt, R. Plant growth curves. In *The Functional Approach to Plant Growth Analysis*; Edward Arnold Ltd.: London, UK, 1982.
46. Yoshida, S.; Kitano, M.; Eguchi, H. growth of lettuce plants (*Lactuca sativa* L.) under control of dissolved O_2 concentration in hydroponics. *Biotronics* **1997**, *26*, 39–45.
47. Woltz, S.; Jackson, C. Production of yellow strapleaf of chrysanthemum & similar disorders by amino acid treatment. *Plant Physiol.* **1961**, *36*, 197. [PubMed]
48. León, A.P.; Martín, J.P.; Chiesa, A. Vermicompost application and growth patterns of lettuce (*Lactuca sativa* L.). *Agric. Tropica Subtrop.* **2012**, *45*, 134–139. [CrossRef]
49. Medek, D.E.; Ball, M.C.; Schortemeyer, M. Relative contributions of leaf area ratio and net assimilation rate to change in growth rate depend on growth temperature: Comparative analysis of subantarctic and alpine grasses. *New Phytol.* **2007**, *175*, 290–300. [CrossRef] [PubMed]

50. Gao, Y.; Yu, H.; Liu, P.; Ma, C.; Li, Q.; Jiang, W. Ending composting during the thermophilic phase improves cultivation substrate properties and increasing winter cucumber yield. *Waste Manag.* **2018**, *79*, 260–272. [CrossRef]
51. Ahn, Y.S. Plant analysis for evaluating plant nutrition. In *International Training Workshop on Soil Test and Plant Analysis*; RDA& FFTC/ASPAC, 1987.
52. Snedecor, G.; Cochran, W. *Statistical Method*, 7th ed.; The Iowa State University Press: Ames, IA, USA, 1980; pp. 39–63.
53. Colla, G.; Nardi, S.; Cardarelli, M.; Ertani, A.; Lucini, L.; Canaguier, R.; Rouphael, Y. Protein hydrolysates as biostimulants in horticulture. *Sci. Hortic.* **2015**, *196*, 28–38. [CrossRef]
54. Liu, X.Q.; Chen, H.Y.; Ni, Q.X.; Kyu, S.L. Evaluation of the role of mixed amino acids in nitrate uptake and assimilation in leafy radish by using 15N-labeled nitrate. *Agric. Sci. China* **2008**, *7*, 1196–1202. [CrossRef]
55. Tsouvaltzis, P.; Koukounaras, A.; Siomos, A.S. Application of amino acids improves lettuce crop uniformity and inhibits nitrate accumulation induced by the supplemental inorganic nitrogen fertilization. *Int. J. Agric. Biol.* **2014**, *16*, 951–955.
56. Wang, H.J.; Wu, L.H.; Wang, M.Y.; Zhu, Y.H.; Tao, Q.N.; Zhang, F.S. Effects of amino acids replacing nitrate on growth, nitrate accumulation, and macroelement concentrations in pakchoi (*Brassica chinensis* L.). *Pedosphere* **2007**, *17*, 595–600. [CrossRef]
57. Mobini, M.; Khoshgoftarmanesh, A.H.; Ghasemi, S. The effect of partial replacement of nitrate with arginine, histidine, and a mixture of amino acids extracted from blood powder on yield and nitrate accumulation in onion bulb. *Sci. Hortic.* **2014**, *176*, 232–237. [CrossRef]
58. Vincill, E.D.; Bieck, A.M.; Spalding, E.P. Ca2+ conduction by an amino acid-gated ion channel related to glutamate receptors. *Plant Physiol.* **2012**, *159*, 40–46. [CrossRef] [PubMed]
59. Santi, C.; Zamboni, A.; Varanini, Z.; Pandolfini, T. Growth stimulatory effects and genome-wide transcriptional changes produced by protein hydrolysates in maize seedlings. *Front. Plant Sci.* **2017**, *8*, 433. [CrossRef] [PubMed]
60. Price, C.A. Leaf Gui: Analyzing the geometry of veins and areoles using image segmentation algorithms. In *High-Throughput Phenotyping in Plants*; Springer: New York, NY, USA, 2012; pp. 41–49.
61. Singh, R.; Parihar, P.; Prasad, S.M. Sulfur and calcium simultaneously regulate photosynthetic performance and nitrogen metabolism status in As-challenged *Brassica juncea* L. seedlings. *Front. Plant Sci.* **2018**, *9*, 772. [CrossRef] [PubMed]
62. Sebastian, A.; Prasad, M. Vertisol prevent cadmium accumulation in rice: Analysis by ecophysiological toxicity markers. *Chemosphere* **2014**, *108*, 85–92. [CrossRef]
63. El-Shiaty, O.H.; El-Sharabasy, S.F.; Abd El-Kareim, A.H. Effect of some amino acids and biotin on callus and proliferation of date palm (*Phoenix dactylifera* L.) Sewy cultivar. *Arab. J. Biotechnol.* **2004**, *7*, 265–272.
64. Kakkar, R.; Nagar, P.; Ahuja, P.; Rai, V. Polyamines and plant morphogenesis. *Biol. Plant.* **2000**, *43*, 1–11. [CrossRef]
65. Padgett, P.E.; Leonard, R.T. Free amino acid levels and the regulation of nitrate uptake in maize cell suspension cultures. *J. Exp. Bot.* **1996**, *47*, 871–883. [CrossRef]
66. Shafeek, M.; Helmy, Y.; Magda, A.; Shalaby, F.; Nadia, M.O. Response of onion plants to foliar application of sources and levels of some amino acid under sandy soil conditions. *J. Appl. Sci. Res.* **2012**, *8*, 5521–5527.
67. Davies, P.J. *Plant Hormones: Biosynthesis, Signal Transduction, Action!*; Springer: New York, NY, USA, 2004.
68. Colla, G.; Cardarelli, M.; Bonini, P.; Rouphael, Y. Foliar Applications of Protein Hydrolysate, Plant and Seaweed Extracts Increase Yield but Differentially Modulate Fruit Quality of Greenhouse Tomato. *HortScience* **2017**, *52*, 1214–1220. [CrossRef]
69. Rouphael, Y.; Colla, G.; Giordano, M.; El-Nakhel, C.; Kyriacou, M.C.; De Pascale, S. Foliar applications of a legume-derived protein hydrolysate elicit dosedependent increases of growth, leaf mineral composition, yield and fruit quality in two greenhouse tomato cultivars. *Sci. Hortic.* **2017**, *226*, 353–360. [CrossRef]
70. Bahari, A.; Pirdashti, H.; Yaghubi, M. The effects of amino acid fertilizers spraying on photosynthetic pigments and antioxidant enzymes of wheat (*Triticum aestivum* L.) under salinity stress. *Int. J. Agron. Plant Prod.* **2013**, *4*, 787–793.
71. Anne, C.; Thomas, S. Regulation of chloroplast development and function by cytokinin. *J. Exp. Bot.* **2015**, *66*, 4999–5013.

72. Fawzy, Z.; El-Shal, Z.; Yunsheng, L.; Zhu, O.; Sawan, O.M. Response of garlic (*Allium Sativum* L.) plants to foliar spraying of some bio-stimulants under sandy soil condition. *J. Appl. Sci. Res.* **2012**, *8*, 770–776.
73. Romero, I.; Téllez, J.; Yamanaka, L.E.; Steindel, M.; Romanha, A.J.; Grisard, E.C. Transsulfuration is an active pathway for cysteine biosynthesis in *Trypanosoma rangeli*. *Parasites Vectors* **2014**, *7*, 197. [CrossRef] [PubMed]
74. Abbas, S.H.; Sohail, M.; Saleem, M.; Mahmood, T.; Aziz, I.; Qamar, M.; Majeed, A.; Arif, M. Effect of L-tryptophan on plant weight and pod weight in chickpea under rainfed conditions. *SciTech Dev.* **2013**, *32*, 277–280.
75. Herve, J.J. Biostimulants, a new concept for the future; prospects offered by the chemistry of synthesis and biotechnology. *C. R. Acad. Agric. Fr.* **1994**, *80*, 91–102.
76. Yunna, Z.; Gang, L.; Houcheng, L.; Guangwen, S.; Riyuan, C.; Shiwei, S. Effects of partial replacement of nitrate with different nitrogen forms on the yield, quality and nitrate content of Chinese kale. *Commun. Soil Sci. Plant. Anal.* **2018**, 1384–1393.
77. Chen, G.L.; Gao, X.R.; Zhang, X.B. Effect of partial replacement of nitrate by amino acid and urea on nitrate content of non-heading Chinese cabbage and lettuce in hydroponic condition. *Agric. Sci. China* **2002**, 444–449.
78. Valpuesta, V.; Botella, M. Biosynthesis of L-ascorbic acid in plants: New pathways for an old antioxidant. *Trends Plant Sci.* **2004**, *9*, 573–577. [CrossRef]
79. Podsedek, A. Natural antioxidants and antioxidant capacity of Brassica vegetables: A review. *LWT-Food Sci. Technol.* **2007**, *40*, 1–11. [CrossRef]
80. John, S.J.; Guha-Mukherjee, S. *Plant Molecular Biology and Biotechnology*, 1st ed.; Narosa Pub. House: New Delhi, India, 1997; pp. 17–28.
81. El-sharabasy, S.; Fatma, I.; Gehan, H.; El-Dawayaty, M. Effect of different amino acids at different concentrations on multiplication and rooting stage of in vitro propagation of strawberries (Fragaria X Ananassa Duch cv. Chandler). *Egypt. J. Genet. Cytol.* **2015**, *44*, 31–34.
82. Rosati, A.; Day, K.; DeJong, T. Distribution of leaf mass per unit area and leaf nitrogen concentration determine partitioning of leaf nitrogen within tree canopies. *Tree Physiol.* **2000**, *20*, 271–276. [CrossRef] [PubMed]
83. Thimann, K.V. Auxins and the inhibition of plant growth. *Biol. Rev.* **1939**, *14*, 314–337. [CrossRef]

3

Foliar Applications of Biostimulants Promote Growth, Yield and Fruit Quality of Strawberry Plants Grown under Nutrient Limitation

Sebastian Soppelsa [1,2], Markus Kelderer [2], Claudio Casera [2], Michele Bassi [2], Peter Robatscher [2], Aldo Matteazzi [2] and Carlo Andreotti [1,*]

1 Faculty of Science and Technology, Free University of Bozen-Bolzano, Piazza Università, 5-39100 Bozen-Bolzano, Italy
2 Laimburg Research Centre, Laimburg 6, 39051 Pfatten-Vadena, Italy
* Correspondence: carlo.andreotti@unibz.it

Abstract: Biostimulants have been found effective in enhancing plant resistance toward stressful conditions. The aim of the present study was to evaluate the efficacy of selected biostimulants to overcome the negative effects of nutrient limitation on the growth performances and on the fruit quality of soilless cultivated strawberry plants. The condition of nutrient limitation was imposed by supplying the plants with only a single fertilization at transplantation and by excluding any further nutrient supply for the entire duration of the experiment (three months, from May to July). Strawberry plants were treated seven times during the period from preflowering up to berry maturation with different classes of biostimulants (humic acids, alfalfa hydrolysate, macroseaweed extract and microalga hydrolysate, amino acids alone or in combination with zinc, B-group vitamins, chitosan, and a commercial product containing silicon) at commercial dosages. The use of alfalfa hydrolysate, vitamins, chitosan, and silicon was able to promote biomass accumulation in roots (four to seven folds) and fruits (+20%) of treated plants, whereas the total leaf area increased by 15%–30%. Nutrient concentrations in leaves and roots showed variations for microelements (e.g., Fe, B, Zn, and Si) in response to biostimulant applications, whereas no significant differences were observed for macronutrient contents among treatments. Final berry yield was found around 20% higher in chitosan- and silicon-treated plants. Chitosan treatment significantly increased pulp firmness (by 20%), while a high nutritional value (e.g., phenolic compounds concentration) was observed in alfalfa- and seaweed-treated fruits (+18%–20% as compared to control). The overall outcomes of the present experiment show that selected biostimulants can be considered as a valid agronomic tool able to contrast the negative consequence of growing crops under insufficient nutritional conditions.

Keywords: soilless conditions; abiotic stress; alfalfa hydrolysate; chitosan; zinc; ascorbic acid; *Fragaria x ananassa*

---

1. Introduction

The nutritional status of strawberry plants is of major relevance for the achievement of the expected levels of productivity and overall fruit quality [1]. Growth, yield, and quality parameters were found to be positively correlated with the rate of mineral nutrients (macro and micro) used during the strawberry production cycle [1–3]. The current world strawberry production is often conducted under covered systems (greenhouses and plastic tunnels) and in combination with soilless technologies [4]. Soilless production systems require a particularly fine control of nutrient supply, but this control is difficult to be achieved especially when the use of mineral fertilizers is banned (i.e., in organic farming).

Under such conditions, nutrient deficiencies/imbalances often become a severe limiting factor for the overall economic sustainability of the cultivation [5,6].

With the aim to help both organic and integrated growers to overcome the problem of the insufficient nutritional status of soilless cultivated crops, different technical solutions are currently available on the market of fertilizer products, including the vast group of biostimulant compounds. These compounds were recently considered in the new EU regulation of fertilizer products and defined as products able to improve one or more of the following characteristics: (i) nutrient use efficiency; (ii) tolerance toward abiotic stresses; (iii) quality traits and (iv) the availability of confined nutrients in soil or the rhizosphere [7].

During the last few years, the international scientific literature has largely documented the beneficial effects of the use of biostimulants on several different crops. As for the use of biostimulant to contrast abiotic stresses, Goñi et al. [8] and Di Stasio et al. [9] recently reported the use of seaweed extracts from *Ascophyllum nodosum* to enhance the tolerance of tomato plants to drought and salinity stress, respectively. This positive effect was observed also with other biostimulants, such as protein hydrolysate on lettuce [10], chitosan on white clover [11], and silicon on tomato [12]. Regarding the stress caused by nutrient deficiency, biostimulants can contribute to soil nutrient availability, uptake, and assimilation of nutrients to the plants [13]. Mechanisms by which biostimulants affect nutrient uptake are generally related to an improvement of physicochemical properties of soil, nutrient solubility, root morphology, and root colonization by arbuscular mycorrhizal fungi [13,14]. Mattner et al. [15] found an increased root length and density in strawberry plants treated with seaweed extracts (*Duvillaea potatorum* and *A. nodosum*), which implies a more efficient use of nutrients by the crop. Extracts from other seaweed species (e.g., *Ecklonia maxima* and *Kappaphycus alvarezii*) also increased the uptake of several macro- and micronutrients, leading to a higher biomass accumulation and final yield in lettuce, soybean, and zucchini squash [16–18]. Spinelli et al. [19] observed the efficacy of seaweed extract in increasing root growth and enhancing tolerance to iron deficiency on strawberry plants, demonstrating that *A. nodosum* might be a valid substitute of Fe–chelate compounds (e.g., sequestrene). In addition to seaweed extracts, other biostimulant products based on amino acids (such as arginine, glycine, histidine, and phenylalanine) were also found to be able to reduce the negative effects of iron-induced chlorosis and the incidence of physiological disorders in horticultural products (tomato and apple) [20,21].

Biostimulants were also found to be effective in promoting yield in several vegetable and fruit species, while differently modulating the nutritional and functional properties of their edible products. Foliar applications of different biostimulants (protein hydrolysate, seaweed, and plant extracts) were all effective in increasing the yield of greenhouse tomato but showed different modulating effects on final fruit quality (e.g., protein hydrolysate-treated fruits showed a higher lycopene concentration, whereas treatments with plant extracts reduced the level of undesirable components such as nitrates) [22]. As concerns the fruit crops, the application of seaweed extracts increased the total soluble solids and organic acids content of mango fruits [23], whereas the use of chitosan and humic acids enhanced the strawberry fruit average size and marketability [24,25]. Also well-established is the induction effect of several biostimulants on the phenolic and flavonoid metabolism in horticultural crops. Treatments with *A. nodosum* induced a higher accumulation of total anthocyanins in strawberry fruits [26], whereas the application of plant and seaweed-based extracts increased total phenolics accumulation in spinach leaves [27,28]. Chitosan and alginate applications promoted the accumulation of anthocyanin and phenolic acids such as 3-O-glucosyl-resveratrol in *Vitis vinifera* [29]. Portu et al. [30] and El-Sayed [31] showed that the application of amino acids (e.g., phenylalanine) can enhance the content of anthocyanins, flavonols, and the total phenols in table grape. These biostimulants resulted in being able to be active on the phenylpropanoid pathway of the treated plants, as demonstrated by the enhanced expression of constitutive genes such as PAL, CHS, and CHI [32–37].

The purpose of this study was to understand if the use of selected biostimulant products could be considered to contrast the negative effects of nutrient limitation in soilless cultivated strawberry. With this aim, repeated preharvest applications of biostimulant compounds were performed and the effects on the growth and fruit yield measured. Moreover, changes in strawberry quality and nutritional value (total phenolic and anthocyanin content, antioxidant potential) were evaluated. Some of the selected substances were tested for the first time on strawberry plants and, to the best of our knowledge, this was the first study where the efficacy of several biostimulants was evaluated simultaneously during the same growing cycle.

2. Materials and Methods

2.1. Experimental Site and Biostimulant Applications

The trial was carried out in a greenhouse located at the Laimburg Research Centre, municipality of Vadena/Pfatten (46°22′ N; 11°17′ E; 237 m a.s.l.) in Alto-Adige/South Tyrol, Italy, during the period April–July 2016. Climatic conditions inside the greenhouse (temperature and humidity) were monitored and are reported in Figure 1.

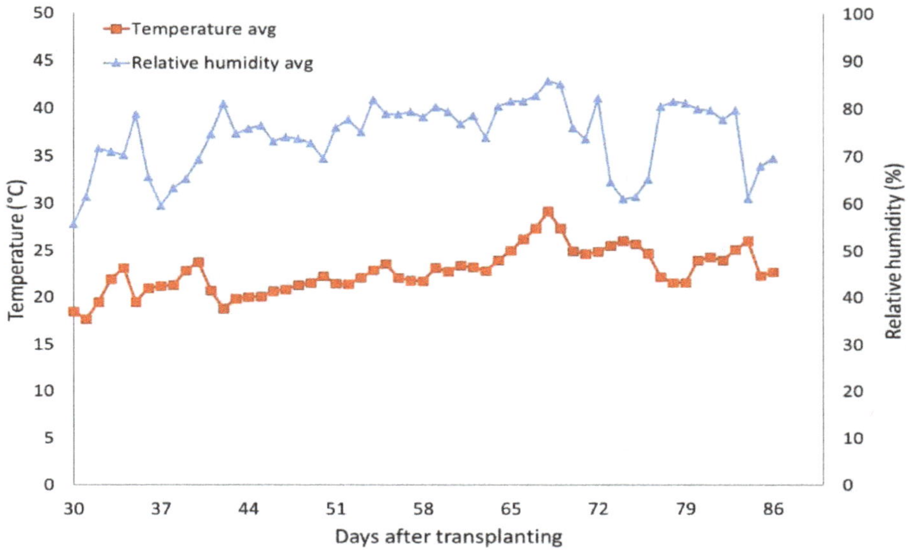

Figure 1. Climatic conditions (average daily temperature and relative humidity) inside the greenhouse during the evaluation period.

A total of 176 cold stored strawberry tray plants (*Fragaria* × *ananassa* Duch.) cv. Elsanta (the most commonly chosen cultivar for protected and open field cultivation in the area), very uniform in size and general conditions (Figure S1, Supplementary Material), were transplanted (4 plants per pot) on the 21st of April 2016 into rectangular plastic pots having the following size: 48 × 23 × 13 cm. A mixture of white peat and natural clay was used as growth medium. A week after the transplantation, plants were fertigated as used commercially for strawberry under greenhouse conditions with a starter fertilizer NPK (24-10-10 and microelements; Peters Excel, Scotts Company, Italy) at a rate of 1 g L^{-1} (EC 1.1 mS cm^{-1}; 0.15 L of solution $plant^{-1}$). Details on the substrate and on the nutrient solution composition are shown in Table 1. The plant material was characterized at the beginning of the experiment (at transplantation) by assessing the dry weight of the different organs (Table 2). For this purpose, four strawberry plants out of the initial group of plants were randomly sampled, washed, and divided into roots, crowns, and leaves. The plant organs were then put in an oven (ED 56, Binder GmbH, Tuttlingen, Germany) at 65 °C until they reached a stable weight and the dry mass recorded (g dry weight (DW) $plant^{-1}$). Moreover, mineral element concentration in leaves and roots was determined as described in Section 2.5 (Table 2). All plants were uniformly irrigated until runoff at two-day intervals using a watering can. During the growing cycle, no chemical pesticides were used, and no further fertilization with mineral nutrients was applied to plants for the whole experiment duration. Starting from the 20th of May (at preflowering stage, 30 DAT—days after transplanting), plants were foliar treated at weekly intervals with the biostimulant products. Seven applications were performed, in total using 10 different biostimulants. Details on the names, abbreviations and physicochemical characteristics of the utilized biostimulants are reported in Table 3. The applied concentrations were those reported in the label for commercial products, whereas those treatments still at prototype stage were applied according the information found in literature for similar compounds. The mechanism by which foliar applied biostimulants penetrate the leaf surface was not investigated in the present research, even though other studies conducted on similar products evidenced that the penetration occurs mainly through the cuticular layer and the stoma [38]. The penetration of these substances into leaves is a passive process and is influenced by several factors, including the concentration and the chemical nature of the applied molecules, the morphology of the leaf cuticle, and the degree of the stomatal opening [39]. Control plants were sprayed only with water. All the plants were sprayed until the runoff point using a hand sprayer (50 mL per plant). Pollination was performed with a brush by taking the pollen from the flower's own anthers. The experiment setup was organized as a completely randomized design with 4 replicates composed by 4 plants per replicate (i.e., 16 plants per treatment and 176 plants in total).

Table 1. Composition of the substrate and of the nutrient solution used for the experiment.

Physico-Chemical Composition of Substrate		Composition of the Nutrient Solution Used at Transplant	
pH	5.1	N (mg L^{-1})	240
Dry matter (%)	42.6	P_2O_5 (mg L^{-1})	100
Humidity (%)	57.4	K_2O (mg L^{-1})	100
Soluble salts (g L^{-1})	2.0	B (mg L^{-1})	0.2
NO_3 (mg L^{-1})	3.3	Cu (mg L^{-1})	0.1
NH_4 (mg L^{-1})	462.7	Fe (mg L^{-1})	1.2
P_2O_5 (mg L^{-1})	37.0	Mn (mg L^{-1})	0.5
K_2O (mg L^{-1})	245.0	Mo (mg L^{-1})	0.1
Mg (mg L^{-1})	121.0	Zn (mg L^{-1})	0.3
Na (mg L^{-1})	13.0		
B (mg L^{-1})	0.26		
Fe (mg L^{-1})	63.0		
Mn (mg L^{-1})	10.2		
Cu (mg L^{-1})	1.2		
Zn (mg L^{-1})	2.1		

Table 2. Characterization of the cold stored strawberry plants at the transplanting: dry weight of the organs and mineral element concentrations in leaves and roots.

Dry Weight of Plant Organs (g DW [a] plant^{-1})		Mineral Element Concentration	Leaves	Roots
Leaves	2.07 ± 0.31	N (%)	3.90 ± 0.28	2.65 ± 0.19
Crown	0.71 ± 0.23	P (%)	0.55 ± 0.10	0.43 ± 0.07
Roots	2.17 ± 0.58	K (%)	3.14 ± 0.20	0.76 ± 0.16
Total weight	4.95 ± 0.97	Ca (%)	1.39 ± 0.20	1.33 ± 0.34
		Mg (%)	0.42 ± 0.03	0.32 ± 0.06
		S (%)	0.52 ± 0.26	0.42 ± 0.13
		B (mg kg^{-1})	35.58 ± 13.20	22.93 ± 1.66
		Cu (mg kg^{-1})	19.98 ± 9.67	115.13 ± 65.82
		Fe (mg kg^{-1})	74.68 ± 32.96	149.83 ± 40.22
		Mn (mg kg^{-1})	168.35 ± 5.84	61.80 ± 23.66
		Zn (mg kg^{-1})	33.35 ± 5.32	85.68 ± 40.17
		Na (mg kg^{-1})	163.59 ± 91.87	299.73 ± 100.71
		Si (mg kg^{-1})	382.35 ± 61.93	181.83 ± 32.92

[a] DW: dry weight.

Table 3. Biostimulant characterization, properties, and concentration applied for the experiment.

Treatment	Active Ingredients	Moisture (%)	Ash (%)	Density (kg dm^{-3})	Organic Matter (%)	pH	Electrical Conductivity (dS m^{-1})	Total Organic C (% w w^{-1})	Total Organic N (% w w^{-1})	Free Amino Acids (% w w^{-1})	Total Amino Acids (% w w^{-1})	Other Characteristics	Concentration
CON	Water												
HAL	Humic acids	-	-	1.1	-	9.2	1.2	7.5	0.1	-	-	-	1.0 g L^{-1}
APH	Alfalfa protein hydrolysate	70.0	7.0	1.2	23.0	5.5	1.6	-	-	1.5	5.1	-	3.0 g L^{-1}
SEA	Macroseaweed extract	84.0	1.5	1.0	14.5	4.5	0.4	3	≤0.1	-	-	From *A. nodosum*	4.0 g L^{-1}
SPI	Microalga hydrolysate	-	-	1.2	-	5.5	1.5	16.8	3.9	6.5	-	From *Spirulina spp* (37% hydrol. degree)	4.0 g L^{-1}
MAA	Mix of amino acids	45.0	5.0	1.2	50.0	5.5	0.8	24.5	9	1.5	55.0	-	3.0 g L^{-1}
PHE	MAA combined with pure phenylalanine	55.0	5.0	1.2	40.0	5.5	-	19.6	7.2	1.8	45.0	Phenylalanine (1%)	3.0 g L^{-1}
ZIN	MAA combined with zinc	55.0	7.0	1.2	38.0	5.5	-	19.6	7.2	0.8	44.0	Zn (2%)	3.0 g L^{-1}
VIT	B-group vitamins (Sigma-Aldrich, USA)	-	-	-	-	-	-	-	-	-	-	B1-thiamine (33.3%), B2-riboflavin (33.3%), B6-pyridoxine (33.3%)	1.0 g L^{-1}
CHI	Chitosan—ChitoPlant Solution® (Agritalia, Italy)	98.3	0.01	-	1.7	5.2	-	-	-	-	-	-	10 mL L^{-1}
SIL	Siliforce® (ILSA S.p.A., Italy)	-	-	1.2	-	2.0	-	-	-	-	-	Si (8 g kg^{-1}), Zn (1.8%), Mo (0.2%)	0.3 mL L^{-1}

2.2. *Vegetative Growth and Leaf Gas Exchanges*

Plant growth dynamic as affected by biostimulant applications was determined every two weeks from preflowering (about 30 DAT) to the end of harvest (about 90 DAT) by measuring the area of two young trifoliate leaves per strawberry plant with a Li-Cor 3000 Leaf Area Meter (Li-Cor Inc., Lincoln, NE, USA). Biomass accumulation in different plant organs was calculated by subtracting the initial dry weight measured at transplanting stage (0 DAT, Table 2) from the final dry weight reached at the end of the growth cycle (90 DAT). The dry weight of the different plant organs at the end of production cycle was determined on four plants per each treatment as previously described in Section 2.1. Measurements included the determination of the final dry matter of organs that were already present at transplanting (roots, crown and leaves) as well as the weight of the newly formed organs (shoots, stolons, and fruits).

The leaf chlorophyll content was indirectly determined with a SPAD-502 Chlorophyll Meter (Konica Minolta, Tokyo, Japan) at biweekly intervals on two randomly selected trifoliate leaves per plant. Moreover, after the plants had received four applications of all the treatments, the net assimilation (A, μmol m^{-2} s^{-1}) and transpiration (E, mmol m^{-2} s^{-1}) rates of leaves were evaluated using a portable gas exchange analyzer (LCpro ADC, Hoddesdon Bioscientific, Ltd, UK). The gas exchange evaluations were conducted immediately before the fifth application of biostimulants (at 57 DAT), and 24 and 48 h after the application. Measurements were performed on a young, fully expanded single leaf of four randomly selected plants per treatment and were taken under saturating light conditions (1.800 μmol photons m^{-2} s^{-1}), around midday (11:00–13:00 h), using a broad leaf gas chamber with a window size of 6.25 cm^2 and a flow rate of 400 mL min^{-1}.

2.3. *Yield and Fruit Quality*

Ripe strawberry fruits (fully red color) were harvested every three days during the period from approximately 60 to 80 DAT. Harvested berries were counted and weighted with a digital scale in the laboratory. The cumulated yield per plant was determined by adding the weight of all the strawberries collected from the same plant during the harvesting period. Fruit quality was assessed on four strawberries per plant sampled at an intermediate pick during the harvesting period. Flesh firmness (kg cm^{-2}) was measured with a penetrometer equipped with a 6 mm diameter cap. The total soluble solids (°Brix) was determined with a portable refractometer (PAL-1, ATAGO, Tokyo, Japan), whereas the titratable acidity (g L^{-1} of citric acid) was measured with a titrator (TitroLine easy, SCHOTT, Mainz, Germany) by titrating strawberry pulp to pH 8.2 using 0.1 M NaOH. The external fruit color was assessed with a colorimeter (Minolta, model CR-400, Tokyo, Japan) by measuring the same four fruits at three different positions around the equatorial side of each fruit. The colorimetric coordinates (L*, a*, b*) were used to calculated the chroma and hue angle values with the formulas (1) and (2), respectively [40].

$$C^* = (a^{*2} + b^{*2})^{1/2} \tag{1}$$

$$h^\circ = arctan\ (b^*/a^*) \tag{2}$$

2.4. *Biochemical Analysis of Strawberry Fruits*

2.4.1. Sample Preparation and Extraction Procedure

Around 20 g of strawberry fruits per replicate was randomly collected at red mature stage and immediately frozen in liquid nitrogen and stored at −80 °C. The extraction was conducted using 25 mg of lyophilized sample, which was homogenized and extracted in 1.8 mL of extraction solution (80% methanol acidified with H_3PO_4, pH 1.0) and in 30 μL of 0.1 M NaF solution for 15 min at 5 °C. The extract was then filtered with PTFE filters (0.45 mm, Thermo Fisher Scientific), and the filtrate was stored at −80 °C until analysis. The extraction procedure specific for ascorbic acid analysis is described in Section 2.4.5.

2.4.2. Total Polyphenols Content (TPC)

Total phenolic content determination was performed on strawberry fruit extracts using the Folin–Ciocalteu assay following the methodology described in Meyers et al. [41] with some modifications. A total of 60 μL of the sample extract was diluted with 250 μL of deionized water. Then, 60 μL of Folin-Ciocalteu reagents was added, and the mixture was allowed to react in the dark for 6 min at 20 °C in agitation at 1500 rpm in the Thermomixer. A total of 630 μL of Na_2CO_3 (7.5% *w/v*) was added and incubated for 90 min at 20 °C always in the dark and under agitation at 1500 rpm. The total polyphenol content was determined at 740 nm using a spectrophotometer Cary 60 UV–Vis (Agilent Technologies, Santa Clara, CA, USA). Gallic acid standard solutions were used to calibrate the method (range 5–500 mg L^{-1}, $r^2 > 0.999$). The content of total polyphenols in each strawberry fruit extract was calculated and expressed as mg of gallic acid equivalents (GAE) per 100 g of dry weight (mg GAE 100 g^{-1} DW).

2.4.3. Determination of Total Anthocyanin Content (TAC)

Total anthocyanin content in strawberry extracts was determined according to Lee et al. [42] using the spectrophotometric pH differential method. Two dilutions of the same sample were prepared by adding 200 μL of extract to 800 μL of potassium chloride (0.25 M, pH 1) and to 800 μL of sodium acetate (0.4 M, pH 4.5), respectively. Absorbance was measured with a Cary 60 UV–Vis spectrophotometer at 520 and 700 nm at pH 1 and 4.5, where A = (A520−A700) pH 1−(A520−A700) pH 4.5. Total anthocyanin content was calculated using the Lambert–Beer law (ε = 26900 L mol^{-1} cm, MW = 449.2 g mol^{-1}) as mg cyanidin 3-glucoside equivalents (CGE) per 100 g of dry weight (mg CGE 100 g^{-1} DW).

2.4.4. Antioxidant Activity (ABTS)

Radical scavenging activity of the strawberry extracts was determined as described by Re et al. [43] with some modifications. ABTS was dissolved in water to a 7 mM concentration; the ABTS radical was obtained from reaction of ABTS stock solution with 2.45 mM potassium persulfate and allowing the mixture to stand in the dark at 4 °C for 16 h before use. The ABTS radical solution was diluted with deionized water to reach absorbance of 0.700 ± 0.02 at 734 nm. A total of 30 μL of each sample extract was added to 1.97 mL of diluted ABTS solution. The absorbance was measured at 734 nm on the Cary 60 UV–Vis spectrophotometer after 10 min in dark conditions. For each sample, the percentage of inhibition after 10 min of reaction was measured, and the concentration of sample (Trolox equivalent antioxidant capacity, TEAC) was calculated using the external calibration curve of the Trolox standard (Trolox, range 15.6–500 mg L^{-1}, $r^2 > 0.999$). The results were expressed as milligrams Trolox equivalents per 100 g of dry weight (mg Trolox 100 g^{-1} DW).

2.4.5. Ascorbic Acid Quantification

The analytical method for the ascorbic acid was based on Bassi et al. [44]. A total of 50 mg of freeze-dried sample was extracted using 1 mL of the extraction solution (700 μL deionized H_2O containing 8% (*v/v*) acetic acid and 3% (*w/v*) metaphosphoric acid added with 300 μL of methanol) [45], mixed at 3200 rpm for about 20 s at room temperature and filtered through a 0.20 μm PTFE filter. An HPLC Agilent 1260 Infinity system (Santa Clara, CA, USA) with a diode array (1260 DAD VL) detector, controlled through the software Agilent ChemStation™ (ver. C.01.03) (Agilent, Santa Clara, CA, USA), was used for the analysis of the ascorbic acid. The separation of the analyte was carried out at 25 °C using a Kinetex 5 μ C18 100 Å column (150 mm × 4.6 mm, 5 μm particle size; Phenomenex, Torrance, CA, USA) equipped with a precolumn (4.6 mm; Security Card, Phenomenex, Torrance, CA, USA). Detection wavelength was 260 nm. The mobile phases used were 5 mM KH_2PO_4, pH 4.8 (solvent A), and methanol (solvent B). The analytical gradient was as follows: 0 min, 100% A; 2.5 min, 100% A; 6 min, 80% A; 8 min, 100% A; and 13 min, 100% A. The flow rate was set at 1.0 mL min^{-1}. The temperature of the autosampler was 4 °C, and injection volume was 5 μL.

2.4.6. Chemicals

Ethanol (96%) was purchased from J.T. Baker (Center Valley, PA, USA). Acetic acid (96%) and potassium chloride were purchased from Merck (Kenilworth, NJ, USA), whereas hydrochloric acid (36%) and sodium acetate (anhydrous) were obtained from Fisher Chemical (Fisher Scientific, MA, USA). Phosphoric acid (≥99%), Folin–Ciocalteu's phenol reagent, sodium carbonate, 2,2′-azino-bis(3-ethylbenzothiazoline-6-sulphonic acid) diammonium salt (ABTS), sodium fluoride, ascorbic acid (99%), and Trolox (6-hydroxy-2,5,7,8-tetramethylchroman-2-carboxylic acid) were acquired from Sigma-Aldrich (St. Louis, MO, USA). Potassium persulfate ($K_2S_2O_8$) and gallic acid (≥99%) were purchased from Carl Roth (Karlsruhe, Germany). Methanol (HPLC-grade) was purchased from VWR Chemicals (Milan, Italy), and meta-phosphoric acid (≥99%) and monopotassium phosphate (≥99%) were acquired from Thermo Fisher Scientific Inc. (Waltham, MA, USA). The ultrapure water was prepared with a Millli-Q-water purification system (EMD Millipore Corporation, Billerica, MA, USA).

2.5. Mineral Element Analysis in Plant Tissues

The analysis of macro- and micronutrients was performed by collecting a 5 g DW sample of leaves and roots from 4 randomly selected plants at 0 DAT and from each of the 4 replicates per treatment at 90 DAT. Nitrogen content was determined with an elemental analyzer (LECO Truspec, LECO Corporation, St. Joseph, MI, USA); method according to Dumas [46]. The other macro- (P, K, Ca, Mg) and microelements (S, Fe, Cu, B, Zn, Mn, Na, Si) were analyzed with microwave-assisted acid digestion (EPA 3052 1996; [47]; Milestone Srl, Sorisole, BG, Italy Mod. UltraWave) using the inductively coupled plasma optical emission spectrometry (ICP-OES; [48]; Agilent, Santa Clara, CA, USA, Model 720).

2.6. Statistical Analysis

Significant differences between treatments were determined with a one-way analysis of variance (ANOVA) after that data were checked for normal distribution (Shapiro–Wilk normality test, $p > 0.05$) and equality of variances (Bartlett's test, $p > 0.05$). Mean separation was performed with the Dunnett test for $p < 0.05$. This procedure is recommended when working with several treatments [49], as in this case. Data expressed in percentage were arcsine-transformed prior to the application of the ANOVA. All analyses were carried out in R v. 3.3.1. (R Development Core Team, Vienna, Austria, 2016). Values were expressed as mean ± standard deviation (SD).

3. Results

3.1. Plant Growth Parameters

Biomass accumulation in strawberry leaves varied approximately between 1 and 5 g DW (Table 4 and Figure S2, supplementary material). Mix of amino acids (MAA)-treated plants showed a significantly lower leaf dry weight accumulation as compared to control. No significant differences were detected in the biomass accumulation in crown, shoots, and stolons of the differently treated plants (Table 4). Root biomass accumulation in untreated plants was generally low (0.6 g DW on average), whereas treatments with alfalfa protein hydrolysate (APH), macroseaweed extract (SEA), microalga hydrolysate (SPI), B-group vitamins (VIT), and chitosan (CHI) significantly enhanced biomass accumulation in the root system (e.g., approximately 7-fold in APH-treated plants). Fruit dry matter accumulation appeared to be significantly reduced in MAA- and amino acids combined with pure phenylalanine (PHE)-treated plants (values around 2 and 3 g DW plant^{-1} respectively, approximately −40% as compared to control) and increased in CHI- and Siliforce® (SIL)-treated plants (approximately by 20%). All biostimulant-treated plants (with the exception of SIL) resulted in being characterized by a higher relative share of dry matter allocation in the root system at the end of the growth cycle (+40%–80%) as compared to control (Figure 2).

Table 4. Accumulated biomass in different plant organs (g DW plant^{-1}) as affected by biostimulant applications.

Treatment	Leaves	Shoots and Stolons	Crowns	Roots	Fruits
CON	3.35 ± 1.32 [1]	0.87 ± 0.38	0.04 ± 0.25	0.59 ± 0.87	4.01 ± 0.19
HAL	3.17 ± 2.36	0.47 ± 0.20	0.07 ± 0.16	2.26 ± 1.72	3.96 ± 0.15
APH	3.47 ± 2.08	0.85 ± 0.30	0.13 ± 0.21	4.19 ± 0.34 ***	4.31 ± 0.15
SEA	2.05 ± 1.16	0.59 ± 0.22	0.08 ± 0.13	3.06 ± 1.49 *	4.26 ± 0.33
SPI	4.85 ± 1.81	0.83 ± 0.21	0.25 ± 0.25	2.91 ± 1.19 *	3.76 ± 0.20
MAA	0.87 ± 0.78 *	0.56 ± 0.39	0.10 ± 0.22	0.79 ± 0.97	1.98 ± 0.58 ***
PHE	1.70 ± 1.47	0.45 ± 0.37	0.00 ± 0.17	2.00 ± 1.70	3.01 ± 0.16 ***
ZIN	2.40 ± 1.00	0.65 ± 0.26	0.01 ± 0.22	1.56 ± 0.64	3.79 ± 0.08
VIT	5.30 ± 1.17	1.16 ± 0.21	0.33 ± 0.39	4.13 ± 2.29 **	4.32 ± 0.12
CHI	4.81 ± 0.85	0.86 ± 0.42	0.58 ± 0.37	3.75 ± 1.85 **	5.05 ± 0.39 ***
SIL	3.14 ± 1.42	0.69 ± 0.44	0.12 ± 0.31	1.49 ± 1.19	4.64 ± 0.57 *

[1] Means ± SD. Values followed by asterisk indicate significant differences between a single treatment group and control group, according to Dunnett's test ($n = 4$). *** $p < 0.001$; ** $p < 0.01$; * $p < 0.05$; values followed by no asterisk indicate absence of any significant differences. Treatments legend: CON, control; HAL, humic acids; APH, alfalfa protein hydrolysate; SEA, macroseaweed extract; SPI, microalga hydrolysate; MAA, mix of amino acids; PHE, MAA combined with pure phenylalanine; ZIN, MAA combined with zinc; VIT, B-group vitamins; CHI, chitosan; SIL, Siliforce®.

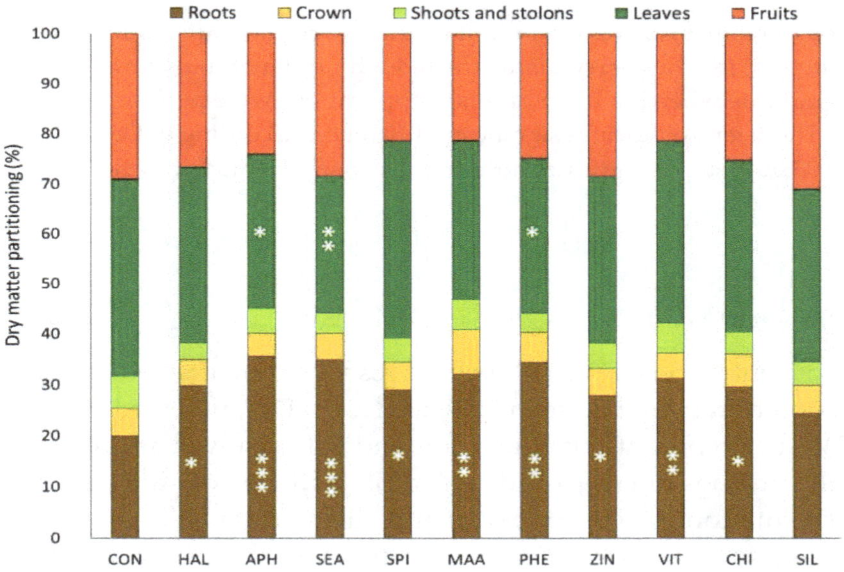

Figure 2. Dry matter distribution in strawberry plant organs treated with different biostimulant products and water (control). $n = 4$. The asterisk indicates significant differences between a single treatment group and control group within each plant organ, according to Dunnett's test. *** $p < 0.001$; ** $p < 0.01$; * $p < 0.05$; no asterisk indicates absence of any significant differences. Treatments legend: CON, control; HAL, humic acids; APH, alfalfa protein hydrolysate; SEA, macroseaweed extract; SPI, microalga hydrolysate; MAA, mix of amino acids; PHE, MAA combined with pure phenylalanine; ZIN, MAA combined with zinc; VIT, B-group vitamins; CHI, chitosan; SIL, Siliforce®.

Relative leaf dry matter allocation was significantly lower in SEA-, APH-, and PHE-treated plants, whereas in fruits, crown, shoots, and stolons, there were no significant differences as compared to the control. At different phases of the experimental cycle, the leaf area of plants treated with APH, VIT, CHI, and SIL was significantly larger than control (+15%–30%), whereas humic acids (HAL) and PHE applications significantly reduced the leaf surface (about 80 cm^2 compared to 100 cm^2 in control plants) at the end of the experiment (Figure 3).

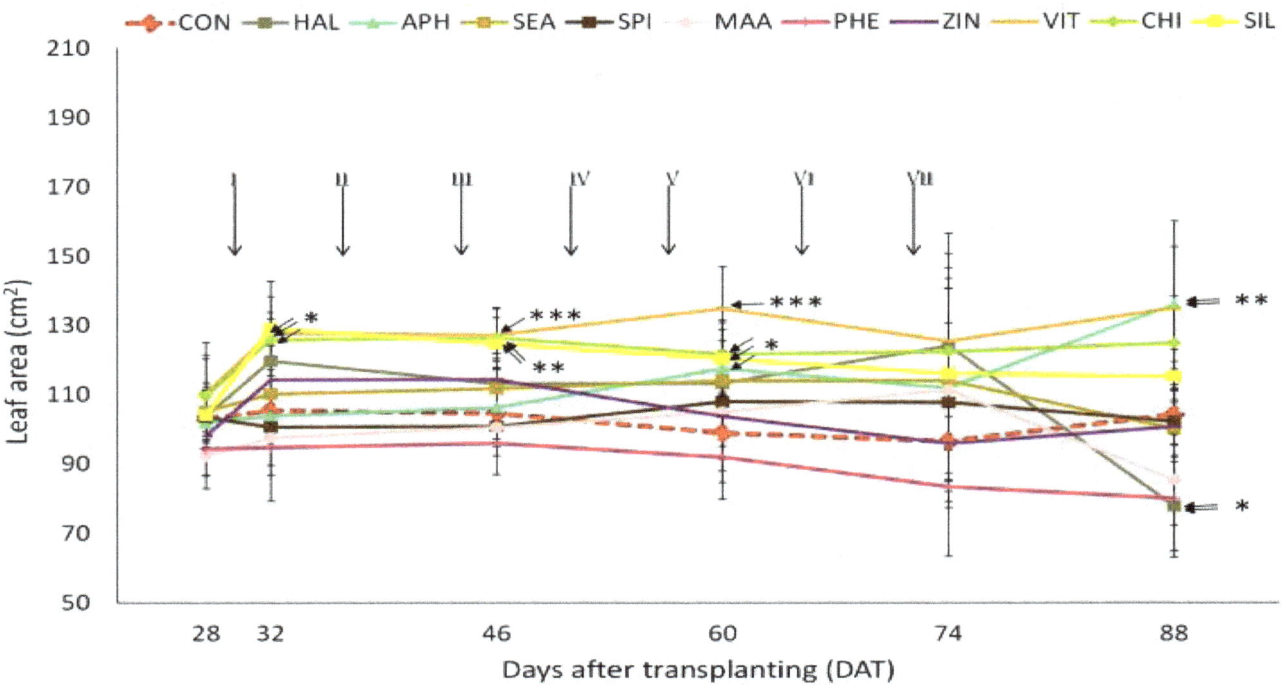

Figure 3. Leaf area dynamics during the period from 28 to 88 days after transplanting (DAT) (end of growth and production) in strawberry plants treated with different biostimulant products and water (control). Arrows indicate when the 7 biostimulant applications were performed. Vertical bars indicate mean ± SD, $n = 4$. The asterisk indicates significant differences according to Dunnett's test. *** $p < 0.001$; ** $p < 0.01$; * $p < 0.05$; no asterisk indicates absence of any significant differences.

3.2. Nutrient Concentration in Plant Tissues and SPAD Values

Macronutrient concentrations in strawberry leaves were not affected by the tested biostimulants (Table 5). Nitrogen content ranged between 2.4% and 2.7% DW, whereas phosphorus concentration was around 0.5% DW. Leaf content for potassium and calcium was approximately 2.6% and 2.2%, respectively. As for the magnesium (Mg) and sulfur (S), their concentrations were found around 0.4% (Mg) and 0.2% (S). The micronutrients zinc (Zn) and silicon (Si) presented higher concentrations in leaves of some of the tested biostimulants (Table 6). In detail, Zn was found significantly higher in leaves of MAA combined with zinc (ZIN)- and SIL-treated plants (+338% and +190% as compared to control, respectively), whereas Si content in strawberry leaves was significantly increased in plants treated with APH, ZIN and SIL, with values higher than 600 mg kg^{-1} DW compared to around 400 mg kg^{-1} DW of the control leaves. No significant differences were observed for the other micronutrient contents.

Macronutrient concentrations were not affected by biostimulant applications also at root level (Table 7). As far as the micronutrient concentrations are concerned, root boron content was found significantly higher (+20–30%) in control plants as compared to the majority of the biostimulant-treated plants (Table 8). Iron (Fe) content in root tissues was promoted by SPI and VIT applications (+273% and +202% as compared to control, respectively). Moreover, the Zn concentration was significantly increased in roots of ZIN-treated plants (140 mg kg^{-1} DW in comparison to 77 mg kg^{-1} DW in control roots). Finally, Si content was found to be higher in roots treated with HAL, SEA, SPI, ZIN, VIT, CHI, and SIL (around +75% as compared to control).

Table 5. Macronutrient concentrations in strawberry leaves at 90 DAT (days after transplanting) as affected by biostimulant applications.

Treatment	N (% DW)	P (% DW)	K (% DW)	Ca (% DW)	Mg (% DW)	S (% DW)
CON	2.59 ± 0.25 [1]	0.51 ± 0.04	2.52 ± 0.18	2.37 ± 0.15	0.38 ± 0.02	0.22 ± 0.03
HAL	2.46 ± 0.04	0.52 ± 0.05	2.72 ± 0.42	2.27 ± 0.27	0.39 ± 0.02	0.22 ± 0.03
APH	2.60 ± 0.27	0.58 ± 0.08	2.64 ± 0.12	2.26 ± 0.22	0.38 ± 0.01	0.22 ± 0.03
SEA	2.56 ± 0.17	0.55 ± 0.05	2.81 ± 0.38	2.25 ± 0.20	0.39 ± 0.03	0.24 ± 0.05
SPI	2.49 ± 0.35	0.56 ± 0.07	2.58 ± 0.23	2.42 ± 0.16	0.42 ± 0.04	0.24 ± 0.07
MAA	2.56 ± 0.09	0.62 ± 0.04	2.59 ± 0.12	2.14 ± 0.13	0.38 ± 0.03	0.20 ± 0.01
PHE	2.74 ± 0.39	0.52 ± 0.11	2.77 ± 0.08	2.42 ± 0.32	0.40 ± 0.03	0.24 ± 0.05
ZIN	2.47 ± 0.16	0.52 ± 0.09	2.75 ± 0.08	2.16 ± 0.24	0.37 ± 0.03	0.21 ± 0.05
VIT	2.38 ± 0.29	0.62 ± 0.02	2.79 ± 0.32	2.24 ± 0.18	0.42 ± 0.02	0.19 ± 0.03
CHI	2.36 ± 0.17	0.54 ± 0.05	2.69 ± 0.15	2.10 ± 0.15	0.39 ± 0.01	0.20 ± 0.01
SIL	2.38 ± 0.21	0.51 ± 0.06	2.74 ± 0.25	2.16 ± 0.15	0.37 ± 0.03	0.18 ± 0.03

[1] Means ± S.D. Values followed by asterisk indicate significant differences between a single treatment group and control group, according to Dunnett's test ($n = 4$). *** $p < 0.001$; ** $p < 0.01$; * $p < 0.05$; no asterisk indicates absence of any significant differences.

Table 6. Micronutrient concentrations in strawberry leaves at 90 DAT (days after transplanting) as affected by biostimulant applications.

Treatment	B (mg kg^{-1} DW)	Fe (mg kg^{-1} DW)	Mn (mg kg^{-1} DW)	Cu (mg kg^{-1} DW)	Zn (mg kg^{-1} DW)	Na (mg kg^{-1} DW)	Si (mg kg^{-1} DW)
CON	66.58 ± 10.93	81.65 ± 4.89	381.88 ± 107.48	7.00 ± 0.00	28.93 ± 3.01	240.12 ± 84.80	416.88 ± 61.75
HAL	58.75 ± 3.78	111.15 ± 28.87	331.58 ± 47.57	10.10 ± 3.59	34.53 ± 10.27	319.56 ± 45.58	538.53 ± 158.67
APH	60.55 ± 7.41	102.68 ± 32.96	355.08 ± 60.70	7.35 ± 0.64	25.98 ± 1.79	251.92 ± 126.22	782.65 ± 68.74 ***
SEA	55.93 ± 3.32	197.75 ± 143.89	300.90 ± 101.74	7.15 ± 0.30	28.73 ± 6.49	363.06 ± 74.44	424.30 ± 18.28
SPI	62.08 ± 16.99	114.60 ± 13.64	378.10 ± 104.30	7.00 ± 0.00	31.75 ± 2.32	327.93 ± 127.08	471.50 ± 33.98
MAA	61.18 ± 5.57	800.55 ± 1348.74	367.60 ± 57.87	9.53 ± 3.18	34.88 ± 7.92	243.16 ± 66.33	457.83 ± 113.69
PHE	61.75 ± 13.01	94.93 ± 16.11	375.75 ± 117.59	7.38 ± 0.75	31.38 ± 4.81	440.30± 262.85	480.68 ± 146.04
ZIN	59.43 ± 6.13	310.85 ± 332.71	285.73 ± 83.13	7.65 ± 1.23	127.43 ± 13.95 ***	270.27 ± 68.68	624.68 ± 98.52 **
VIT	62.40 ± 5.67	91.75 ± 13.21	347.75 ± 59.23	7.58 ± 0.67	36.15 ± 4.11	234.82 ± 79.06	398.65 ± 71.72
CHI	59.50 ± 4.46	115.40 ± 43.02	393.08 ± 38.55	7.13 ± 0.25	30.90 ± 0.85	218.92 ± 20.99	487.33 ± 110.15
SIL	60.40 ± 4.43	93.48 ± 23.08	366.53 ± 87.71	7.23 ± 0.45	83.80 ± 13.61 ***	254.18 ± 126.09	689.13 ± 41.81 ***

[1] Means ± S.D. Values followed by asterisk indicate significant differences between a single treatment group and control group, according to Dunnett's test ($n = 4$). *** $p < 0.001$; ** $p < 0.01$; * $p < 0.05$; no asterisk indicates absence of any significant differences.

Table 7. Macronutrient concentrations in strawberry roots at 90 DAT (days after transplanting) as affected by biostimulant application.

Treatment	N (% DW)	P (% DW)	K (% DW)	Ca (% DW)	Mg (% DW)	S (% DW)
CON	1.44 ± 0.12 [1]	0.15 ± 0.03	0.49 ± 0.18	1.60 ± 0.22	0.21 ± 0.02	0.35 ± 0.03
HAL	1.28 ± 0.17	0.13 ± 0.02	0.46 ± 0.16	1.74 ± 0.17	0.22 ± 0.01	0.37 ± 0.08
APH	1.45 ± 0.14	0.14 ± 0.05	0.44 ± 0.22	1.85 ± 0.29	0.23 ± 0.05	0.45 ± 0.09
SEA	1.41 ± 0.15	0.14 ± 0.03	0.40 ± 0.08	1.97 ± 0.16	0.23 ± 0.02	0.50 ± 0.08
SPI	1.35 ± 0.17	0.13 ± 0.02	0.43 ± 0.09	1.77 ± 0.16	0.22 ± 0.03	0.40 ± 0.07
MAA	1.65 ± 0.18	0.18 ± 0.03	0.49 ± 0.07	1.68 ± 0.21	0.22 ± 0.03	0.46 ± 0.13
PHE	1.39 ± 0.17	0.14 ± 0.02	0.48 ± 0.18	1.86 ± 0.23	0.20 ± 0.02	0.45 ± 0.13
ZIN	1.45 ± 0.08	0.15 ± 0.02	0.50 ± 0.14	1.82 ± 0.17	0.24 ± 0.04	0.43 ± 0.07
VIT	1.44 ± 0.08	0.14 ± 0.02	0.45 ± 0.09	1.91 ± 0.35	0.26 ± 0.01	0.57 ± 0.28
CHI	1.26 ± 0.19	0.13 ± 0.03	0.39 ± 0.03	1.93 ± 0.17	0.23 ± 0.01	0.46 ± 0.12
SIL	1.40 ± 0.07	0.17 ± 0.03	0.59 ± 0.18	1.69 ± 0.25	0.23 ± 0.02	0.41 ± 0.13

[1] Means ± S.D. Values followed by asterisk indicate significant differences between a single treatment group and control group, according to Dunnett's test ($n = 4$). *** $p < 0.001$; ** $p < 0.01$; * $p < 0.05$; no asterisk indicates absence of any significant differences.

Table 8. Micronutrient concentrations in strawberry roots at 90 DAT (days after transplanting) as affected by biostimulant application.

Treatment	B (mg kg^{-1} DW)	Fe (mg kg^{-1} DW)	Mn (mg kg^{-1} DW)	Cu (mg kg^{-1} DW)	Zn (mg kg^{-1} DW)	Na (mg kg^{-1} DW)	Si (mg kg^{-1} DW)
CON	17.28 ± 2.29	484.63 ± 94.84	56,25 ± 23,07	80.08 ± 53.24	77.08 ± 43.06	702.14 ± 81.07	350.25 ± 78.52
HAL	13.10 ± 3.45 *	1227.43 ± 538.00	51,43 ± 22,80	98.13 ± 73.45	67.48 ± 56.12	662.19 ± 118.41	528.63 ± 97.22 *
APH	11.43 ± 2.66 **	620.83 ± 215.52	81,40 ± 12,66	52.65 ± 19.62	44.98 ± 30.09	624.75 ± 77.51	444.93 ± 92.27
SEA	10.90 ± 0.98 ***	724.53 ± 214.65	55,35 ± 38,78	146.40 ± 84.29	42.95 ± 15.41	681.68 ± 81.97	538.88 ± 105.99 *
SPI	12.35 ± 2.64 **	1807.48 ± 1745.77 **	75,45 ± 31,44	68.90 ± 52.69	57.78 ± 38.74	624.80 ± 141.86	683.63 ± 182.23 ***
MAA	13.73 ± 3.36	648.95 ± 150.92	70,58 ± 19,96	207.95 ± 222.50	127.05 ± 94.10	579.94 ± 100.10	515.35 ± 127.19
PHE	12.98 ± 1.78 *	691.78 ± 309.30	97,50 ± 88,66	83.70 ± 113.54	40.78 ± 15.81	586.40 ± 61.30	461.80 ± 117.93
ZIN	12.33 ± 2.46 **	808.08 ± 157.22	70,25 ± 38,06	69.63 ± 52.43	139.78 ± 32.20 *	643.41 ± 191.04	689.03 ± 93.68 ***
VIT	8.53 ± 1.98 ***	1461.73 ± 956.78 *	100,13 ± 73,10	228.33 ± 63.72	95.98 ± 35.03	600.22 ± 90.50	631.18 ± 99.86 **
CHI	10.33 ± 2.76 ***	1079.63 ± 661.56	133,63 ± 39,71	80.23 ± 41.43	38.28 ± 8.27	655.94 ± 19.66	623.28 ± 187.63 **
SIL	12.68 ± 1.91 *	519.48 ± 183.10	55,30 ± 29,18	69.08 ± 50.39	67.98 ± 22.73	565.72 ± 98.53	590.98 ± 57.86 **

[1] Means ± S.D. Values followed by asterisk indicate significant differences between a single treatment group and control group, according to Dunnett's test ($n = 4$). *** $p < 0.001$; ** $p < 0.01$; * $p < 0.05$; no asterisk indicates absence of any significant differences.

SPAD values were not affected by biostimulant applications during most of the growing cycle. Only at the end of the experiment (88 DAT) was a significant higher SPAD value measured in plants treated with APH (+17% as compared to control) (Figure 4).

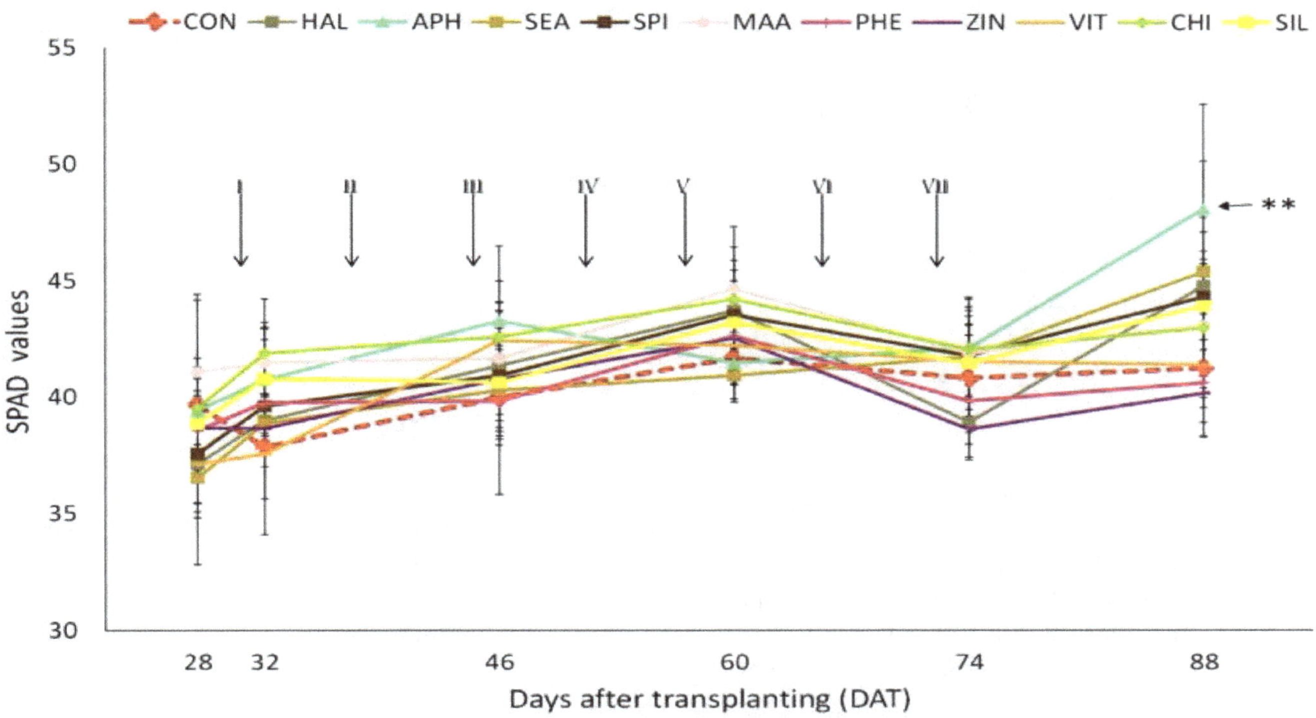

Figure 4. Chlorophyll content (as SPAD values) dynamics from 28 to 88 DAT (end of growth and production) in strawberry plants treated with different biostimulant products and water (control). Arrows indicate when the 7 biostimulant applications were performed. Vertical bars indicate mean ± SD, $n = 4$. The asterisk indicates significant differences according to Dunnett's test. *** $p < 0.001$; ** $p < 0.01$; * $p < 0.05$; no asterisk indicates absence of any significant differences.

3.3. Leaf Gas Exchanges

Before the fifth application (exactly at 57 DAT), a significantly higher leaf photosynthetic rate was measured for HAL-, ZIN-, and SIL-treated leaves (Figure 5A). After 24 h from the application, only SIL-treated leaves were still characterized by a significantly higher photosynthetic rate as compared to control (+32%), whereas at 48 h from the treatments, none of the biostimulants tested showed a significantly higher rate than control plants. Biostimulant-treated plants were characterized by a higher intensity of transpiration (almost double as compared to control) immediately before and after the fifth application (Figure 5B). Only HAL-treated plants did not show any significant difference with control 24 and 48 h after the application.

3.4. Yield, Harvest and Fruit Quality

CHI- and SIL-treated plants were characterized by a significantly higher average fruit weight as compared to control (Table 9). Their number of fruits per plant did not differ with control, leading to a significant 10–15 g yield increment per plant. Conversely, applications with amino acids (MAA and PHE) decreased the mean fruit weight, the number of fruits per plant, and the final yield (−49% and −22% as compared to control).

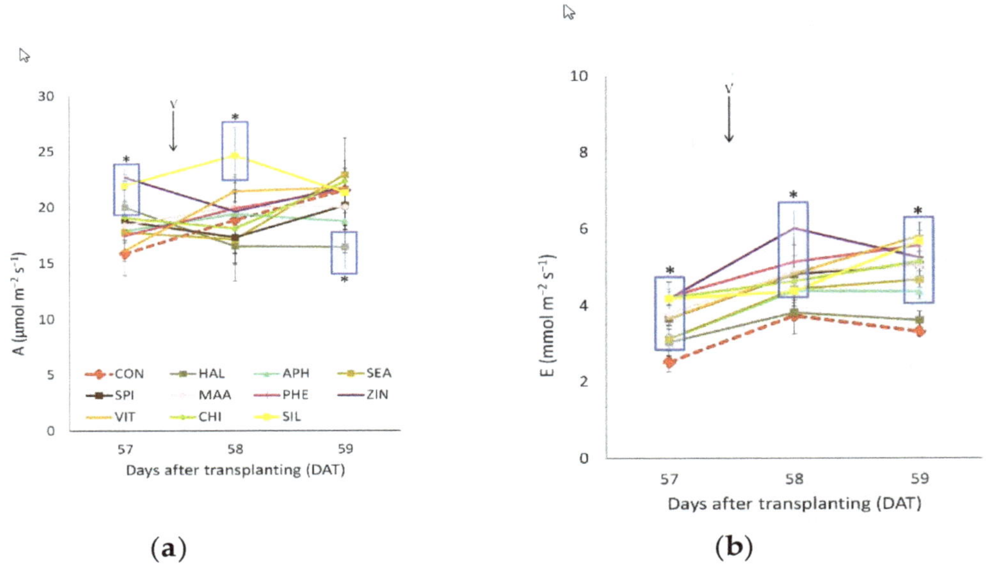

Figure 5. Photosynthetic (**a**) and transpiration (**b**) rates in strawberry leaves treated with different biostimulant products and water (control). 57 DAT: immediately before spray; 58 DAT: 24 h after spray; 59 DAT: 48 h after spray. Arrow indicates the fifth biostimulant application. Vertical bars indicate mean ± SD, $n = 4$. Asterisk (*), and the blue rectangle in the background indicates the group of treatments that significantly differed from control according to Dunnett's test ($p < 0.05$); no asterisk indicates absence of any significant differences.

Table 9. Yield parameters (final yield, number of fruits per plant and mean fruit weight) as affected by biostimulant products.

Treatment	Total Yield (g plant^{-1} FW [a])	Number Fruits Plant^{-1} (N°)	Mean Fruit Weight (g FW)
CON	50.84 ± 3.57 [1]	7.75 ± 0.25	6.55 ± 0.24
HAL	50.95 ± 1.95	7.42 ± 0.72	6.97 ± 0.42
APH	51.67 ± 2.90	7.00 ± 0.25	7.43 ± 0.46 *
SEA	53.57 ± 1.17	8.25 ± 0.87	6.52 ± 0.40
SPI	51.05 ± 7.71	7.17 ± 0.38	7.16 ± 0.07
MAA	25.95 ± 4.69 ***	4.50 ± 0.66 ***	5.74 ± 0.76 *
PHE	39.59 ± 2.73 **	6.58 ± 0.38 *	6.08 ± 0.43
ZIN	47.51 ± 1.89	7.17 ± 0.58	6.82 ± 0.84
VIT	54.58 ± 2.94	8.25 ± 0.25	6.60 ± 0.09
CHI	64.88 ± 3.02 ***	8.42 ± 0.80	7.72 ± 0.22 **
SIL	59.57 ± 6.29 *	7.92 ± 0.88	7.54 ± 0.11 *

[a] FW: fresh weight; [1] Means ± SD. Values followed by asterisk indicate significant differences between a single treatment group and control group, according to Dunnett's test ($n = 4$). *** $p < 0.001$; ** $p < 0.01$; * $p < 0.05$; no asterisk indicates absence of any significant differences.

The majority of strawberries (around 60% of the total yield per plant) were picked during the middle period of the harvesting time, corresponding to the second of a 3-week harvest period (Figure 6). VIT- and MAA-treated plants were characterized by a larger share of fruits picked at the beginning of the harvest, therefore showing an overall earlier ripening of the strawberries. Conversely, SEA-, SPI-, and CHI-treated plants showed a delayed maturation process as suggested by the higher share of fruits that were picked during the last period of the harvest.

Figure 6. Percentage of the total yield per plant harvested during the three harvesting windows (beginning, middle, and end). $n = 4$. Asterisk indicates significant differences between a single treatment group and control group within a harvesting window, according to Dunnett's test. *** $p < 0.001$; ** $p < 0.01$; * $p < 0.05$; no asterisk indicates absence of any significant differences.

Strawberry flesh firmness ranged between 0.5 and 0.8 kg cm^{-2} (Table 10). Fruits from CHI-treated plants were characterized by significantly firmer pulp as compared to control (+18%), whereas SPI-treated strawberries were significantly softer and sourer (10.1 g L^{-1} TA). Total soluble solids were generally increased by the biostimulant applications, with MAA-treated strawberries being those presenting the highest sugars accumulation (8.7 °Brix TSS, Table 8). Chroma values (C*) were significantly higher in fruits treated with HAL, APH, SEA, and ZIN (around +18% in comparison to control), indicating fruits with a brighter red coloration (Table 10), whereas no significant differences in fruit hue angle were detected.

Total phenolic content varied between 2100 and 2800 mg GAE 100 g^{-1} DW (Table 11). SEA and APH applications significantly enhanced the final total phenolic content of mature strawberry fruits (+20% as compared to control). Total anthocyanin content (TAC) in fruits was not significantly affected by biostimulant applications (Table 11), even though HAL, APH, and SEA were among those treatments characterized by the highest anthocyanin accumulation. The fruit antioxidant potential measured as ABTS was not altered significantly by the treatments, whereas the ascorbic acid concentration was significantly reduced by both biostimulant products containing the zinc mineral element such as ZIN and SIL (−24% and −19% as compared to control, respectively) (Table 11).

Table 10. Fruit quality traits (FF—firmness; TSS—total soluble solids; TA—titratable acidity and color parameters: C*—chroma; h°—hue angle) as affected by biostimulant products.

Treatment	FF (kg cm^{-2})	TSS (° Brix)	TA (g L^{-1})	C *	h°
CON	0.71 ± 0.02 [1]	7.07 ± 0.06	9.11 ± 0.09	27.41 ± 1.36	18.31 ± 3.46
HAL	0.69 ± 0.02	7.73 ± 0.61	9.32 ± 0.08	31.58 ± 2.71 *	17.72 ± 3.90
APH	0.73 ± 0.02	7.07 ± 0.51	9.48 ± 0.18	34.51 ± 2.54 ***	17.28 ± 3.70
SEA	0.72 ± 0.01	7.73 ± 0.50	9.74 ± 0.19	34.72 ± 3.74 ***	18.68 ± 2.01
SPI	0.58 ± 0.09 **	7.03 ± 0.15	10.19 ± 0.99 *	29.89 ± 1.07	16.13 ± 2.84
MAA	0.73 ± 0.04	8.70 ± 0.70 ***	8.66 ± 0.40	30.06 ± 0.90	16.98 ± 0.63
PHE	0.72 ± 0.03	7.23 ± 0.25	9.60 ± 0.29	28.69 ± 3.23	16.23 ± 0.80
ZIN	0.71 ± 0.03	7.97 ± 0.2 *	9.44 ± 0.11	32.86 ± 2.37 **	17.03 ± 4.61
VIT	0.74 ± 0.05	7.80 ± 0.44 *	8.79 ± 0.55	30.25 ± 1.99	14.38 ± 3.18
CHI	0.84 ± 0.01 **	6.97 ± 0.15	8.72 ± 1.00	29.05 ± 1.81	16.94 ± 2.84
SIL	0.73 ± 0.11	7.37 ± 0.15	9.34 ± 0.27	29.75 ± 1.04	17.08 ± 1.55

[1] Means ± SD. Values followed by asterisk indicate significant differences between a single treatment group and control group, according to Dunnett's test ($n = 4$). *** $p < 0.001$; ** $p < 0.01$; * $p < 0.05$; no asterisk indicates absence of any significant differences.

Table 11. Fruit nutraceutical values (TPC—total phenolic content; TAC—total anthocyanin content; ABTS—antioxidant potential; AA—ascorbic acid) as affected by biostimulant products.

Treatment	TPC (mg GAE 100 g^{-1} DW)	TAC (mg CGE 100 g^{-1} DW)	ABTS (mg Trolox 100 g^{-1} DW)	AA (mg AA 100 g^{-1} DW)
CON	2303.95 ± 116.11 [1]	354.89 ± 55.57	2886.04 ± 110.02	556.69 ± 37.18
HAL	2455.53 ± 406.32	435.01 ± 85.90	2865.18 ± 402.51	617.55 ± 102.80
APH	2803.17 ± 488.83 *	461.48 ± 56.26	3420.95 ± 252.11	572.33 ± 12.77
SEA	2734.73 ± 261.74 *	478.59 ± 126.19	3178.61 ± 296.83	548.29 ± 29.01
SPI	2211.79 ± 105.48	408.96 ± 13.62	2799.13 ± 44.49	582.32 ± 53.40
MAA	2327.76 ± 122.79	369.94 ± 139.69	3318.91 ± 463.59	587.54 ± 42.27
PHE	2506.83 ± 240.52	475.10 ± 83.04	2990.51 ± 316.85	539.14 ± 51.51
ZIN	2098.88 ± 152.98	404.44 ± 18.33	3011.75 ± 599.56	422.90 ± 56.38 *
VIT	2114.04 ± 270.33	396.17 ± 100.75	2831.57 ± 55.35	469.69 ± 10.50
CHI	2400.41 ± 40.91	417.22 ± 31.63	3047.21 ± 53.23	599.14 ± 84.73
SIL	2687.08 ± 197.94	307.97 ± 37.55	3220.11 ± 91.69	452.43 ± 94.26 *

[1] Means ± SD. Values followed by asterisk indicate significant differences between a single treatment group and control group, according to Dunnett's test ($n = 4$). *** $p < 0.001$; ** $p < 0.01$; * $p < 0.05$; no asterisk indicates absence of any significant differences.

4. Discussion

Abiotic stresses, including nutrient limitation, are generally the cause of a reduced plant growth and final yield [50]. Under the imposed experimental conditions, soilless cultivated strawberry plants were characterized by vegetative and reproductive performances that were below the standard for similar cultivation systems and cultivar [51,52]. This was also suggested by values of the leaf area to yield ratio that were found below 3 $cm^2\ g^{-1}$, under the threshold considered as indicative of a good vegetative–reproductive balance for strawberry [53]. Despite the limited supply of nutrients to the plants, which consisted of a single fertigation done one week after transplant (Table 1), macronutrient levels in leaves and roots (Tables 5 and 7) were found within the range of sufficiency for strawberry plants as reported in the literature [54,55]. Only nitrogen level was found to be lower (1.4% on average) than the concentration range found in strawberry roots of the cultivar 'Selva' grown under standard conditions (around 2.4%) [56]. Moreover, none of the applied biostimulants induced significant changes in the final macronutrient concentrations. Differently, some of the micronutrients (e.g., Fe, Zn, and Si) were found at a higher concentration in leaves and roots of plants treated with specific biostimulant compounds (Tables 6 and 8). More in detail, SPI-treated plants presented a higher Fe accumulation in the root system, probably as a consequence of the rhizosphere acidification promoted by selected metabolites included in the seaweed extract (e.g., kahydrin) that resulted in a more efficient mobilization and assimilation of the acid-soluble ions, including iron. A similar mechanism of Fe chelation following seaweed application was described by Spinelli et al. [19] on strawberry. ZIN and SIL applications resulted in higher concentrations of Zn and Si in leaves and roots (Tables 6 and 8). This result can be explained as the consequence of the direct supply of Zn and Si obtained by the application of both biostimulant treatments. Moreover, prior studies have highlighted the importance of amino acids (included in the ZIN formulation) as metal-chelating agents and carriers of micronutrients [57]. Regarding the lower root boron content observed in treated plants, Kaya et al. [58] reported that Si supply to tomato plants reduced boron concentration in plant tissues as a consequence of the B immobilization caused by the formation of boron–silicate complexes in the soil. Moreover, B availability is generally decreased under higher soil pH [59]. Although the substrate pH at the end of the experiment was not measured, biostimulant applications could have slightly enhanced soil pH, therefore limiting B availability for the plants [39,60].

Under the described growing and nutrient conditions, the use of selected biostimulants had a positive effect on strawberry plant growth and fruit yield. APH treatment enhanced biomass accumulation in roots, leaf area, and chlorophyll concentration (Table 4; Figures 3 and 4). These results reflect those of Ertani et al. [61] and Rouphael et al. [62] on pepper and tomato, respectively. The mode of action of the protein hydrolysates is likely to be linked to different plant physiological mechanisms, which include the stimulation of key enzymes involved in both the primary and secondary metabolisms, the hormone-like function of several components of the protein hydrolysate, and the indirect stimulation of the biological activities of the plant-associate microbes [63]. B-group vitamins were also effective in increasing leaf area and root growth (Figure 3 and Table 4). Vitamin B1 (thiamine) is an important cofactor involved in many primary metabolic processes (glycolysis, pentose phosphate pathway, and tricarboxylic acid cycle) [64,65], explaining therefore the enhanced growth performances shown by the treated plants under suboptimal growing conditions. CHI and SIL applications improved both above- and belowground growth and yield performances (Figure 3; Tables 4 and 9). These findings are consistent with those obtained by Mukta et al. [66] and by Hajiboland et al. [67] on chitosan- and silicon-treated strawberry plants and are probably linked with the enhanced stress tolerance shown by the treated plants [68,69]. As shown in Figure 5, biostimulants containing zinc (ZIN and SIL) and HAL were found to be able to enhance leaf photosynthesis. Zinc–metalloenzymes are important for the activity of the carboxylation process key enzyme (Rubisco) [70,71]. Considering that Zn-deficiency reduces the photosynthetic activity of plants [72], the enhanced photosynthetic performances characterizing ZIN- and SIL-treated plants could be therefore explained with their higher zinc concentration in leaves (Table 6).

Biostimulants have been also claimed to improve the quality attributes of horticultural products. The definition of quality of fruit and vegetable is considered extremely dynamic and involves a large number of intrinsic and extrinsic characteristics, including socioeconomic aspects linked to consumers' perceptions and acceptance of the products [73]. The potential role of plant biostimulants in enhancing the quality of greenhouse vegetables has been recently reviewed by Rouphael et al. [74]. In the present experiment, primary fruit quality attributes were slightly affected by the biostimulant applications (Table 10). Chitosan was found to be able to enhance flesh firmness with potential consequences on the shelf life period extension of treated fruits. Similar results were also obtained by Bhaskara Reddy et al. [75] and Hernández-Muñoz et al. [76] on strawberry (different cultivars). Treatments with amino acids and B-group vitamins led to an earlier ripening of the fruits (Figure 6) and to strawberries with higher sugar content (Table 10), a finding that was also described by Khan et al. [77] on grape fruits. Alfalfa protein hydrolysate and seaweed extracts improved the final coloration and the phenolic concentration in strawberry fruit at harvest (Tables 10 and 11). These results confirm the capacity of both biostimulants to interfere with the phenylpropanoid pathway, leading to more colored and stress-resistant crops [21,27,78,79]. Finally, a reduction in ascorbic acid level in fruits treated with biostimulants containing zinc (ZIN and SIL) was recorded (Table 11). This effect might be due to the inhibitory action of selected metals (e.g., zinc and copper) on the enzyme activity of ascorbic acid metabolism as also shown by Olkhovych et al. [80] on other species.

5. Conclusions

The aim of the present research was to examine if the application of different biostimulants could help the growth and yield performances of soilless cultivated strawberry plants under limited nutrient availability. The main findings suggest that selected biostimulants (i.e., alfalfa protein hydrolysate, B-group vitamins, chitosan, and silicon) could be effective in stimulate vegetative growth and final fruit yield. Moreover, biostimulants based on seaweed extracts, protein hydrolysate, and chitosan were found able to improve strawberry commercial (firmness and external color) and nutritional (phenolic compounds) quality. The outcomes of this research allow therefore a positive overall evaluation of biostimulants as agronomic tools able to contrast the negative consequence of growing crops under insufficient nutritional conditions. Further work needs to be done to deepen our understanding of the uptake mechanisms and use efficacy of the biostimulants under the current strawberry industrial production systems, according to their way of application and interaction with the different farming techniques (open field or protected cultivations, fresh or cold stored plants, genotypes, etc.).

Author Contributions: S.S. contributed in the set-up of the experimental protocol, performed the experiment, processed the fruit samples, analyzed the data, and participated in the writing of the paper; M.K. contributed to the set-up of the experiment and to the interpretation of the results; C.C. gave technical assistance for the implementation of the study; M.B. contributed to the analysis of phenols, anthocyanins, antioxidant potential, and ascorbic acid; P.R. contributed to the set-up of the analytical work related to the quality indices of the fruits; A.M. contributed to the analysis of mineral elements; C.A. coordinated the research, worked on the statistical analysis, and contributed to the writing of several sections of the manuscript.

References

1. Tagliavini, M.; Baldi, E.; Lucchi, P.; Antonelli, M.; Sorrenti, G.; Baruzzi, G.; Faedi, W. Dynamics of nutrients uptake by strawberry plants (*Fragaria × ananassa* Dutch.) grown in soil and soilless culture. *Eur. J. Agron.* **2005**, *23*, 15–25. [CrossRef]
2. Albregts, E.E.; Howard, C.M.; Chandler, C.K. Strawberry responses to K rate on a fine sand soil. *Hort. Sci.* **1991**, *26*, 135–138. [CrossRef]
3. Shirko, R.; Nazarideljou, M.J.; Akbar, M.A.; Naser, G. Photosynthetic reaction, mineral uptake, and fruit quality of strawberry affected by different levels of macronutrients. *J. Plant Nutr.* **2018**, *41*, 1807–1820. [CrossRef]

4. Hukkanen, A.T.; Kokko, H.I.; Buchala, A.J.; McDougall, G.J.; Stewart, D.; Kärenlampi, S.O.; Karjalainen, R.O. Benzothiadiazole induces the accumulation of phenolics and improves resistance to powdery mildew in strawberries. *J. Agric. Food Chem.* **2007**, *55*, 1862–1870. [CrossRef] [PubMed]
5. de Ponti, T.; Rijk, B.; van Ittersum, M.K. The crop yield gap between organic and conventional agriculture. *Agric. Syst.* **2012**, *108*, 1–9. [CrossRef]
6. van Eysinga, J.R.; van Caem, H.E. Nutrition of glasshouse strawberries with nitrogen, phosphorus and potassium. *Sci. Hortic.* **1977**, *7*, 359–368. [CrossRef]
7. European Council. Available online: https://www.consilium.europa.eu/ (accessed on 11 May 2019).
8. Goñi, O.; Quille, P.; O'Connell, S. *Ascophyllum nodosum* extract biostimulants and their role in enhancing tolerance to drought stress in tomato plants. *Plant Physiol. Biochem.* **2018**, *126*, 63–73. [CrossRef]
9. Di Stasio, E.; Van Oosten, M.J.; Silletti, S.; Raimondi, G.; dell'Aversana, E.; Carillo, P.; Maggio, A. *Ascophyllum nodosum*-based algal extracts act as enhancers of growth, fruit quality, and adaptation to stress in salinized tomato plants. *J. Appl. Phycol.* **2018**, *30*, 2675–2686. [CrossRef]
10. Lucini, L.; Rouphael, Y.; Cardarelli, M.; Canaguier, R.; Kumar, P.; Colla, G. The effect of a plant-derived biostimulant on metabolic profiling and crop performance of lettuce grown under saline conditions. *Sci. Hortic.* **2015**, *182*, 124–133. [CrossRef]
11. Li, Z.; Zhang, Y.; Zhang, X.; Merewitz, E.; Peng, Y.; Ma, X.; Huang, L.; Yan, Y. Metabolic pathways regulated by chitosan contributing to drought resistance in white clover. *J. Proteome Res.* **2017**, *16*, 3039–3052. [CrossRef]
12. Shi, Y.; Zhang, Y.; Han, W.; Feng, R.; Hu, Y.; Guo, J.; Gong, H. Silicon enhances water stress tolerance by improving root hydraulic conductance in *Solanum lycopersicum* L. *Front. Plant Sci.* **2016**, *7*, 196. [CrossRef]
13. De Pascale, S.; Rouphael, Y.; Colla, G. Plant biostimulants: Innovative tool for enhancing plant nutrition in organic farming. *Eur. J. Hortic. Sci.* **2018**, *82*, 277–285. [CrossRef]
14. Halpern, M.; Bar-Tal, A.; Ofek, M.; Minz, D.; Muller, T.; Yermiyahu, U. The use of biostimulants for enhancing nutrient uptake. In *Advances in Agronomy*; Sparks, D.L., Ed.; Elsevier Science Publishing Co Inc.: San Diego, CA, USA, 2015; Volume 130, pp. 141–174. ISBN 978-0-12-802137-8.
15. Mattner, S.W.; Milinkovic, M.; Arioli, T. Increased growth response of strawberry roots to a commercial extract from *Durvillaea potatorum* and *Ascophyllum nodosum*. *J. Appl. Phycol.* **2018**, *30*, 2943–2951. [CrossRef]
16. Crouch, I.J.; Beckett, R.P.; Van Staden, J. Effect of seaweed concentrate on the growth and mineral nutrition of nutrient-stressed lettuce. *J. Appl. Phycol.* **1990**, *2*, 269–272. [CrossRef]
17. Rathore, S.S.; Chaudhary, D.R.; Boricha, G.N.; Ghosh, A.; Bhatt, B.P.; Zodape, S.T.; Patolia, J.S. Effect of seaweed extract on the growth, yield and nutrient uptake of soybean (*Glycine max*) under rainfed conditions. *South Afr. J. Bot.* **2009**, *75*, 351–355. [CrossRef]
18. Rouphael, Y.; De Micco, V.; Arena, C.; Raimondi, G.; Colla, G.; De Pascale, S. Effect of *Ecklonia maxima* seaweed extract on yield, mineral composition, gas exchange, and leaf anatomy of zucchini squash grown under saline conditions. *J. Appl. Phycol.* **2017**, *29*, 459–470. [CrossRef]
19. Spinelli, F.; Fiori, G.; Noferini, M.; Sprocatti, M.; Costa, G. A novel type of seaweed extract as a natural alternative to the use of iron chelates in strawberry production. *Sci. Hortic.* **2010**, *125*, 263–269. [CrossRef]
20. Ghasemi, S.; Khoshgoftarmanesh, A.H.; Afyuni, M.; Hadadzadeh, H. Iron(II)–amino acid chelates alleviate salt-stress induced oxidative damages on tomato grown in nutrient solution culture. *Sci. Hortic.* **2014**, *165*, 91–98. [CrossRef]
21. Soppelsa, S.; Kelderer, M.; Casera, C.; Bassi, M.; Robatscher, P.; Andreotti, C. Use of biostimulants for organic apple production: Effects on tree growth, yield, and fruit quality at harvest and during storage. *Front. Plant Sci.* **2018**, *9*, 1342. [CrossRef]
22. Colla, G.; Cardarelli, M.; Bonini, P.; Rouphael, Y. Foliar applications of protein hydrolysate, plant and seaweed extracts increase yield but differentially modulate fruit quality of greenhouse tomato. *Hortscience* **2017**, *52*, 1214–1220. [CrossRef]
23. Mohamed, A.Y.; El-Sehrawy, O.A. Effect of seaweed extract on fruiting of Hindy Bisinnara mango trees. *J. Am. Sci.* **2013**, *9*, 537–544.
24. El-Miniawy, S.; Ragab, M.; Youssef, S.; Metwally, A. Response of strawberry plants to foliar spraying of chitosan. *Res. J. Agric. Biol. Sci.* **2013**, *9*, 366–372.
25. Neri, D.; Lodolini, E.M.; Savini, G.; Sabbatini, P.; Bonanomi, G.; Zucconi, F. Foliar application of humic acids on strawberry (*cv Onda*). In *Proceedings of the Acta Horticulturae*; International Society for Horticultural Science (ISHS): Leuven, Belgium, November 2002; pp. 297–302.

26. Roussos, P.A.; Denaxa, N.-K.; Damvakaris, T. Strawberry fruit quality attributes after application of plant growth stimulating compounds. *Sci. Hortic.* **2009**, *119*, 138–146. [CrossRef]
27. Fan, D.; Hodges, D.M.; Zhang, J.; Kirby, C.W.; Ji, X.; Locke, S.J.; Critchley, A.T.; Prithiviraj, B. Commercial extract of the brown seaweed *Ascophyllum nodosum* enhances phenolic antioxidant content of spinach (*Spinacia oleracea* L.) which protects *Caenorhabditis elegans* against oxidative and thermal stress. *Food Chem.* **2011**, *124*, 195–202. [CrossRef]
28. Rouphael, Y.; Giordano, M.; Cardarelli, M.; Cozzolino, E.; Mori, M.; Kyriacou, M.; Bonini, P.; Colla, G. Plant-and seaweed-based extracts increase yield but differentially modulate nutritional quality of greenhouse spinach through biostimulant action. *Agronomy* **2018**, *8*, 126. [CrossRef]
29. Cai, Z.; Kastell, A.; Mewis, I.; Knorr, D.; Smetanska, I. Polysaccharide elicitors enhance anthocyanin and phenolic acid accumulation in cell suspension cultures of Vitis vinifera. *Plant Cell Tissue Organ Cult. Pctoc* **2012**, *108*, 401–409. [CrossRef]
30. Portu, J.; López-Alfaro, I.; Gómez-Alonso, S.; López, R.; Garde-Cerdán, T. Changes on grape phenolic composition induced by grapevine foliar applications of phenylalanine and urea. *Food Chem.* **2015**, *180*, 171–180. [CrossRef]
31. El-Sayed, M.E.A. Improving fruit quality and marketing of "Crimson Seedless" grape using some preharvest treatments. *J. Hortic. Sci. Ornam. Plants* **2013**, *5*, 218–226.
32. Ertani, A.; Schiavon, M.; Muscolo, A.; Nardi, S. Alfalfa plant-derived biostimulant stimulate short-term growth of salt stressed *Zea mays* L. plants. *Plant Soil* **2012**, *364*, 145–158. [CrossRef]
33. Landi, L.; Feliziani, E.; Romanazzi, G. Expression of defense genes in strawberry fruits treated with different resistance inducers. *J. Agric. Food Chem.* **2014**, *62*, 3047–3056. [CrossRef]
34. Schiavon, M.; Pizzeghello, D.; Muscolo, A.; Vaccaro, S.; Francioso, O.; Nardi, S. High Molecular Size Humic Substances Enhance Phenylpropanoid Metabolism in Maize (*Zea mays* L.). *J. Chem. Ecol.* **2010**, *36*, 662–669. [CrossRef]
35. Vera, J.; Castro, J.; Gonzalez, A.; Moenne, A. Seaweed polysaccharides and derived oligosaccharides stimulate defense responses and protection against pathogens in plants. *Mar. Drugs* **2011**, *9*, 2514–2525. [CrossRef]
36. Xie, L.; Wang, Z.H.; Cheng, X.H.; Gao, J.J.; Zhang, Z.P.; Wang, L.J. 5-Aminolevulinic acid promotes anthocyanin accumulation in Fuji apples. *Plant Growth Regul.* **2013**, *69*, 295–303. [CrossRef]
37. Youssef, K.; Sanzani, S.M.; Ligorio, A.; Ippolito, A.; Terry, L.A. Sodium carbonate and bicarbonate treatments induce resistance to postharvest green mould on citrus fruit. *Postharvest Biol. Technol.* **2014**, *87*, 61–69. [CrossRef]
38. Saa, S.; Olivos-Del Rio, A.; Castro, S.; Brown, P.H. Foliar application of microbial and plant based biostimulants increases growth and potassium uptake in almond (*Prunus dulcis* [Mill.] D. A. Webb). *Front. Plant Sci* **2015**, *6*. [CrossRef]
39. Rouphael, Y.; Colla, G. Synergistic biostimulatory action: Designing the next generation of plant biostimulants for sustainable agriculture. *Front. Plant Sci.* **2018**, *9*, 1655. [CrossRef]
40. Sacks, E.J.; Shaw, D.V. Color change in fresh strawberry fruit of seven genotypes stored at 0C. *HortScience* **1993**, *28*, 209–210. [CrossRef]
41. Meyers, K.J.; Watkins, C.B.; Pritts, M.P.; Liu, R.H. Antioxidant and antiproliferative activities of strawberries. *J. Agric. Food Chem.* **2003**, *51*, 6887–6892. [CrossRef]
42. Lee, J.; Durst, R.W.; Wrolstad, R.E. Determination of total monomeric anthocyanin pigment content of fruit juices, beverages, natural colorants, and wines by the pH differential method: Collaborative study. *J. Aoac Int.* **2005**, *88*, 1269–1278.
43. Re, R.; Pellegrini, N.; Proteggente, A.; Pannala, A.; Yang, M.; Rice-Evans, C. Antioxidant activity applying an improved ABTS radical cation decolorization assay. *Free Radic. Biol. Med.* **1999**, *26*, 1231–1237. [CrossRef]
44. Bassi, M.; Lubes, G.; Bianchi, F.; Agnolet, S.; Ciesa, F.; Brunner, K.; Guerra, W.; Robatscher, P.; Oberhuber, M. Ascorbic acid content in apple pulp, peel, and monovarietal cloudy juices of 64 different cultivars. *Int. J. Food Prop.* **2018**, 1–9. [CrossRef]
45. *AOAC Vitamin C in juices and vitamin preparations. Official Method 967.21*; AOAC Official Methods of Analysis, 18th; Association of Official Analytical Chemists: Gaithersburg, MD, USA, 2005; p. 45.1.14.
46. ISO 16634-1:2008. *Food Products—Determination of the Total Nitrogen Content by Combustion According to the Dumas Principle and Calculation of the Crude Protein Content—Part 1: Oilseeds and Animal Feeding Stuffs*; International Organization for Standardization: Geneva, Switzerland, 2008.

47. EPA 3052. *Microwave assisted acid digestion of siliceous and organically based matrics*; U.S. Environmental Protection Agency: Washington, DC, USA, 1996.
48. EPA 6010D. *Inductively coupled plasma–Optical emission spectrometry*; U.S. Environmental Protection Agency: Washington, DC, USA, 2018.
49. Dunnett, C.W. Multiple comparisons between several treatments and a specified treatment. In *Proceedings of the Linear Statistical Inference*; Caliński, T., Klonecki, W., Eds.; Springer: New York, NY, USA, 1985; pp. 39–47.
50. Cramer, G.R.; Urano, K.; Delrot, S.; Pezzotti, M.; Shinozaki, K. Effects of abiotic stress on plants: A systems biology perspective. *Bmc Plant Biol.* **2011**, *11*, 163. [CrossRef]
51. Jarosz, Z.; Konopinska, J. Effect of substrate type and nitrogen fertilization upon yielding and chemical composition of 'Elsanta' strawberry cultivar grown in unheated foil tunnel. *Acta Sci. Pol. Hortorum Cultus* **2010**, *9*, 87–96.
52. Saied, A.S.; Keutgen, A.J.; Noga, G. The influence of NaCl salinity on growth, yield and fruit quality of strawberry cvs. 'Elsanta' and 'Korona'. *Sci. Hortic.* **2005**, *103*, 289–303. [CrossRef]
53. Casierra-Posada, F.; Torres, I.D.; Blanke, M.M. Fruit quality and yield in partially defoliated strawberry plants in the tropical highlands. *Gesunde Pflanz.* **2013**, *65*, 107–112. [CrossRef]
54. Almaliotis, D.; Velemis, D.; Bladenopoulou, S.; Karapetsas, N. Leaf nutrient levels of strawberries (cv Tudla) in relation to crop yield. In *Proceedings of the Acta Horticulturae*; International Society for Horticultural Science (ISHS): Leuven, Belgium, 2002; pp. 447–450.
55. Campbell, C.R.; Miner, G.S. Strawberry, annual hill culture. In *Reference sufficiency ranges for plant analysis in the southern region of the United States*; Southern Cooperative Series Bulletin 394; North Carolina Department of Agriculture & Consumer Services: Raleigh, NC, USA, 2000; ISBN 1-58161-394-6.
56. Jamali, B.; Eshghi, S.; Tafazoli, E. Mineral composition of 'Selva' strawberry as affected by time of application of nitric oxide under saline conditions. *Hortic. Env. Biotechnol.* **2015**, *56*, 273–279. [CrossRef]
57. Popko, M.; Michalak, I.; Wilk, R.; Gramza, M.; Chojnacka, K.; Górecki, H. Effect of the new plant growth biostimulants based on amino acids on yield and grain quality of winter wheat. *Molecules* **2018**, *23*, 470. [CrossRef]
58. Kaya, C.; Tuna, A.L.; Guneri, M.; Ashraf, M. Mitigation effects of silicon on tomato plants bearing fruit grown at high boron levels. *J. Plant Nutr.* **2011**, *34*, 1985–1994. [CrossRef]
59. Tsadilas, C.D.; Kassioti, T.; Mitsios, I.K. Influence of liming and nitrogen forms on boron uptake by tobacco. *Commun. Soil Sci. Plant Anal.* **2005**, *36*, 701–708. [CrossRef]
60. Hellequin, E.; Monard, C.; Quaiser, A.; Henriot, M.; Klarzynski, O.; Binet, F. Specific recruitment of soil bacteria and fungi decomposers following a biostimulant application increased crop residues mineralization. *PLoS ONE* **2019**, *13*, e0209089. [CrossRef]
61. Ertani, A.; Sambo, P.; Nicoletto, C.; Santagata, S.; Schiavon, M.; Nardi, S. The use of organic biostimulants in hot pepper plants to help low input sustainable agriculture. *Chem. Biol. Technol. Agric.* **2015**, *2*, 11. [CrossRef]
62. Rouphael, Y.; Colla, G.; Giordano, M.; El-Nakhel, C.; Kyriacou, M.C.; De Pascale, S. Foliar applications of a legume-derived protein hydrolysate elicit dose-dependent increases of growth, leaf mineral composition, yield and fruit quality in two greenhouse tomato cultivars. *Sci. Hortic.* **2017**, *226*, 353–360. [CrossRef]
63. Colla, G.; Hoagland, L.; Ruzzi, M.; Cardarelli, M.; Bonini, P.; Canaguier, R. Biostimulant action of protein hydrolysates: Unraveling their effects on plant physiology and microbiome. *Front. Plant Sci.* **2017**, *8*, 2202. [CrossRef]
64. Goyer, A. Thiamine in plants: Aspects of its metabolism and functions. *Phytochemistry* **2010**, *71*, 1615–1624. [CrossRef]
65. Rapala-Kozik, M. Vitamin B1 (thiamine): A cofactor for enzymes involved in the main metabolic pathways and an environmental stress protectant. In *Advances in Botanical Research*; Rébeillé, F., Douce, R., Eds.; Academic Press: Cambridge, MA, USA, 2011; Volume 58, pp. 37–91. ISBN 0065-2296.
66. Mukta, J.; Rahman, M.; As Sabir, A.; Gupta, D.R.; Surovy, M.Z.; Rahman, M.; Islam, M.T. Chitosan and plant probiotics application enhance growth and yield of strawberry. *Biocatal. Agric. Biotechnol.* **2017**, *11*, 9–18. [CrossRef]
67. Hajiboland, R.; Moradtalab, N.; Eshaghi, Z.; Feizy, J. Effect of silicon supplementation on growth and metabolism of strawberry plants at three developmental stages. *N. Z. J. Crop Hortic. Sci.* **2018**, *46*, 144–161. [CrossRef]

68. Agostinho, F.; Tubana, B.; Martins, M.; Datnoff, L. Effect of different silicon sources on yield and silicon uptake of rice grown under varying phosphorus rates. *Plants* **2017**, *6*, 35. [CrossRef]
69. Artyszak, A. Effect of silicon fertilization on Crop yield quantity and quality: A literature review in Europe. *Plants* **2018**, *7*, 54. [CrossRef]
70. DiMario, R.J.; Clayton, H.; Mukherjee, A.; Ludwig, M.; Moroney, J.V. Plant carbonic anhydrases: Structures, locations, evolution, and physiological roles. *Mol. Plant* **2017**, *10*, 30–46. [CrossRef]
71. Sturikova, H.; Krystofova, O.; Huska, D.; Adam, V. Zinc, zinc nanoparticles and plants. *J. Hazard. Mater.* **2018**, *349*, 101–110. [CrossRef]
72. Mattiello, E.M.; Ruiz, H.A.; Neves, J.C.L.; Ventrella, M.C.; Araújo, W.L. Zinc deficiency affects physiological and anatomical characteristics in maize leaves. *J. Plant Physiol.* **2015**, *183*, 138–143. [CrossRef]
73. Kyriacou, M.C.; Rouphael, Y. Towards a new definition of quality for fresh fruits and vegetables. *Sci. Hortic.* **2018**, *234*, 463–469. [CrossRef]
74. Rouphael, Y.; Kyriacou, M.C.; Petropoulos, S.A.; De Pascale, S.; Colla, G. Improving vegetable quality in controlled environments. *Sci. Hortic.* **2018**, *234*, 275–289. [CrossRef]
75. Bhaskara Reddy, M.; Belkacemi, K.; Corcuff, R.; Castaigne, F.; Arul, J. Effect of pre-harvest chitosan sprays on post-harvest infection by Botrytis cinerea and quality of strawberry fruit. *Postharvest Biol. Technol.* **2000**, *20*, 39–51. [CrossRef]
76. Hernández-Muñoz, P.; Almenar, E.; Valle, V.D.; Velez, D.; Gavara, R. Effect of chitosan coating combined with postharvest calcium treatment on strawberry (*Fragaria* × *ananassa*) quality during refrigerated storage. *Food Chem.* **2008**, *110*, 428–435. [CrossRef]
77. Khan, A.S.; Ahmad, B.; Jaskani, M.J.; Ahmad, R.; Malik, A.U. Foliar application of mixture of amino acids and seaweed (*Ascophyllum nodosum*) extract improve growth and physicochemical properties of grapes. *Int. J. Agric. Biol.* **2012**, *14*, 383–388.
78. Cluzet, S.; Torregrosa, C.; Jacquet, C.; Lafitte, C.; Fournier, J.; Mercier, L.; Salamagne, S.; Briand, X.; Esquerre'-Tugaye', M.-T.; Dumas, B. Gene expression profiling and protection of *Medicago truncatula* against a fungal infection in response to an elicitor from green algae *Ulva* spp. *Plant Cell Env.* **2004**, *27*, 917–928. [CrossRef]
79. Ertani, A.; Pizzeghello, D.; Francioso, O.; Sambo, P.; Sanchez-Cortes, S.; Nardi, S. *Capsicum chinensis* L. growth and nutraceutical properties are enhanced by biostimulants in a long-term period: Chemical and metabolomic approaches. *Front. Plant. Sci.* **2014**, *5*. [CrossRef]
80. Olkhovych, O.; Volkogon, M.; Taran, N.; Batsmanova, L.; Kravchenko, I. The Effect of copper and zinc nanoparticles on the growth parameters, contents of ascorbic acid, and qualitative composition of amino acids and acylcarnitines in *Pistia stratiotes* L. (Araceae). *Nanoscale Res. Lett.* **2016**, *11*. [CrossRef]

4

Effects of Water Stress and Modern Biostimulants on Growth and Quality Characteristics of Mint

Hosam O. Elansary [1,2,*], Eman A. Mahmoud [3], Diaa O. El-Ansary [4] and Mohamed A. Mattar [5,6,*]

[1] Plant Production Department, College of Food and Agriculture Sciences, King Saud University, Riyadh 11451, Saudi Arabia
[2] Floriculture, Ornamental Horticulture, and Garden Design Department, Faculty of Agriculture (El-Shatby), Alexandria University, Alexandria 21545, Egypt
[3] Department of Food Industries, Damietta University, Damietta 3417, Egypt; emanmail2005@du.edu.eg
[4] Precision Agriculture Laboratory, Department of Pomology, Faculty of Agriculture (El-Shatby), Alexandria University, Alexandria 21545, Egypt; diaa.elansary@alexu.edu.eg
[5] Department of Agricultural Engineering, College of Food and Agriculture Sciences, King Saud University, Riyadh 11451, Saudi Arabia
[6] Agricultural Engineering Research Institute (AEnRI), Agricultural Research Centre, Giza 12618, Egypt
* Correspondence: helansary@ksu.edu.sa (H.O.E.); mmattar@ksu.edu.sa (M.A.M.)

Abstract: Natural biostimulants combine different elicitors that may influence economic properties of herbal crops, such as mint. Mint (*Mentha longifolia* L.) plants were subjected to three water levels based on container substrate capacity (CSC; 100% CSC, 70% CSC, and 50% CSC) and/or applications of four biostimulants (CRADLE™, Mobilizer™, Nanozim De'Lite™ [ND], and Nanozim NXT™ [NN]). ND and NN exhibited higher vegetative growth and root dry weight than the control (without biostimulants) and other treatments. NN produced the highest fresh and dry mint yields under all water levels. Irrigation water-use efficiency (IWUE) of NN was highest (2.78 kg m^{-3}) with 70% CSC, whereas the control produced the lowest IWUE (1.85 kg m^{-3}) with 100% CSC. Biostimulants boosted physiological and metabolic responses, including gas exchange, leaf water potential, relative water content, and proline accumulation of stressed plants. NN treatment with 70% CSC had the highest essential oil (EO) ratio (3.35%). Under 70% and 50% CSC with NN treatment, the proportion of 1,8-cineol increased and that of pulegone decreased in EOs. Increased antioxidant activities, reduced H_2O_2 levels, and increased catalase and superoxide dismutase activities were observed. Applications of ND and NN during water stress conditions increased economic and medicinal properties of mint EOs with applications in the agricultural and pharmaceutical industries.

Keywords: *Mentha longifolia*; biostimulants; *Ascophyllum nodosum*; humic acid; antioxidants

---

1. Introduction

Mint plants have a long history as traditional medicinal plants [1]. *Mentha longifolia* L. belongs to the family Lamiaceae and naturally occurs in Egypt, Saudi Arabia, and most Arabian countries. The fresh/dried plants are mainly used as an herbal medicine for the treatment of indigestion, menstrual pain, coughs, asthma, fever, and headaches [2,3]. The fresh leaves are used in soft drinks and as garnishes for salads in some countries. The essential oil (EO) is used in the pharmaceutical, cosmetic, and food industries [4]. The EOs exhibit strong antimicrobial activity against several microorganisms [3].

Water stress is one of the major limiting factors for agriculture and food safety worldwide [5]. This stress causes reduced vegetative growth and great losses to farmers. Different studies have focused on the effects of water stress on the growth parameters and EO yield. Zade et al. [6] reported that water stress decreased peppermint plant fresh and dry weight, leaf number, plant height, and root dry weight but nonetheless increased EOs compared to that of normal irrigation in greenhouse and field experiments. Figueroa-Pérez et al. [7] showed that water stress decreased fresh and dry weights of peppermint but increased composition of plant secondary metabolites and antioxidant capacity. Ekren et al. [8] reported that plant height and yield of purple basil were negatively affected by water stress, whereas the EO content increased and irrigation water-use efficiencies were not significantly affected. Shormin et al. [9] showed that the harmful effects of water stress on Japanese mint yield could not be compensated by high nitrogen quantities. Farahani et al. [10] also reported the highest content of EO in balm occurred at 60% field capacity (FC). However, in other studies, Khorasaninejad et al. [11] showed that water stress had negative effects on some growth parameters and EO content of peppermint plants. Razmjoo et al. [12] found that this stress reduced some growth parameters and EO content of chamomile.

Several approaches have been applied to control water stress, such as the use of biostimulants. Modern biostimulants have been produced to increase the productivity and the quality of horticultural crops and help the plants tolerate stress conditions. Some of these biostimulants are mixtures of seaweed extracts, humic acid, and macro and micro elements, whereas other products contain mixtures of mycorrhiza and seaweed extracts, as well as other micro elements. Seaweed extracts work as elicitors for plant secondary metabolites, including EOs and may increase the pharmaceutical properties against microorganisms [13,14]. However, the effects of the mixtures of seaweed extracts and other elicitors, such as humic acid and specific minerals have not been investigated for mint plants. Further, water stress may cause significant changes in the EO composition and these changes might cause parallel changes in the antimicrobial properties of respective EOs.

In this investigation, our goal was to determine the effects of water stress and commercial biostimulants on the growth, physiology, secondary metabolites, and antioxidant activities of mint (*Mentha longifolia* L.). We propose that these natural biostimulants modulate growth, EO ratio, and EO constitutes, leading to enhanced bioactivity of mint plants. These effects indeed have the potential to have future agricultural industry applications.

2. Materials and Methods

2.1. Plant Material

Uniformly rooted cuttings of mint (*Mentha longifolia* L.) were brought from nurseries of the Alexandria University farm in February 2018 and 2019 (as two successive growing seasons). The species was identified and vouchered by Hosam Elansary in the Faculty of Agriculture, Alexandria University. The sandy soil (75.5%, 13.2%, and 11.3% of sand, silt and clay, respectively) samples were air dried and sieved with a 2 mm mesh. The soil had an FC of 20.5%, wilting point of 9.6%, electrical conductivity of 0.36 mS cm^{-1}, organic matter of 1.4%, pH of 6.2, total nitrogen of 0.085%, total phosphorus of 0.05%, and total sulfur of 0.03%. After proper soil preparation, the plants were grown in 2.1 L plastic pots containing the natural sandy soil supplemented with Crystalon® (65 kg N ha^{-1} as urea, 40 kg P_2O_5 ha^{-1} as triple superphosphate, 34 kg K_2O ha^{-1} as potassium sulfate, and 2 g L^{-1} media) in the greenhouse. The temperature inside the greenhouse ranged between 15.0 °C (night) and 27.3 °C (day) and the relative humidity ranged between 67% and 72% during the growing period. The photosynthetic active radiation was approximately 1000 μmol m^{-2} s^{-1} at noon. Daily watering by drip irrigation was applied to reach the full pot substrate FC. Pots were irrigated equally for 30 days after transplantation (DAT). Mint plants were harvested at 90 DAT. Container substrate capacity (CSC) is the maximum amount of water that can be retained by the substrate after the discharge because of gravity [15]. Before planting, the gravimetric method was used to determine CSC or FC by watering the plants to saturate the soil then the pots were left to drain for 60 min and the volume of drained water was quantified and the difference between the supplied and drained water volumes were considered the volumetric water retained by the soil (i.e., *CSC*). The amount of water applied (*AWA*) to compensate for the soil water deficit to reach the FC is calculated as follows:

$$AWA = (CSC - \theta_v)\ D\ A \tag{1}$$

where θ_v is soil water content at the irrigation event, *D* is the soil depth, and *A* is the surface area of the pot.

2.2. Treatments

The plants were subjected to three watering levels of CSC (100%, 70%, and 50%) after 30 DAT and/or single biostimulant of four commercial biostimulants, namely, CRADLE™, Mobilizer™, Nanozim NXT™, and Nanozim De'Lite™ (Biostadt, Mumbai, India). CRADLE (CR) powder is a mycorrhizal biofertilizer developed by InGene Organics, India and was used at g L^{-1} growing soil. Mobilizer (Mob) is a granular mycorrhizal biofertilizer mixed with kelp seaweed extract (*Macrocystis pyrifera*), humic acid, and amino acids and was applied at g L^{-1} growing soil. Nanozim De'Lite (ND) is a granular formulation of 25% ($w\ w^{-1}$) seaweed (*Ascophyllum nodosum*), 25% (w/w) carbohydrates, 2% ($w\ w^{-1}$) amino acid, and 1% (w/w) potassium (K_2O), and was used at 1 g L^{-1} with irrigation water. Nanozim NXT (NN) is a liquid mixture of 15% ($w\ w^{-1}$) seaweed (*Ascophyllum nodosum*), 5% (w/w) humic acid, 1% ($w\ w^{-1}$) potassium (K_2O), 0.01% ($w\ w^{-1}$) phosphorus (P_2O_5), 0.05% (w/w) alginic acid, 0.05% ($w\ w^{-1}$) hydrolyzed protein, and several micronutrients and was applied at 1.5 mL L^{-1} of irrigation water. The doses of the biostimulants and method of applications matched the manufacturer recommendations and untreated plants with biostimulants were considered the controls. Plants were grouped into three blocks containing 10 replicates per treatment [3 water levels (100%, 70%, and 50% CSC) × (4 biostimulants +1 control "without biostimulant") = 15 treatments] and totaling 450 plants (150 plants/block × 3 blocks) in a completely randomized design.

2.3. Measurements

2.3.1. Morphological and Physiological

Following 9 weeks of treatments, several morphological measurements were determined including leaf number (plant^{-1}), leaf area (cm^2 plant^{-1}), plant heights (cm), plant fresh weight (g), plant dry weight (g), and root dry weight (g). Irrigation water-use efficiency (IWUE, kg m^{-3}) was calculated by dividing the fresh weight of the plant (kg) by the total *AWA* (m^3) to each treatment during the growing period [16]. A digital area meter was used to determine the leaf area. The dry weights were determined following drying at 35 °C in an oven until reaching a constant weight.

Gas exchange measurements were performed on fully expanded leaves, under clear, sunny conditions using a portable photosynthesis system analyzer (ADC BioScientific, LCi, Bioscientific, Ltd., Hoddesdon, UK) and included photosynthetic rate (A), transpiration rate (E), and stomatal conductance (gs). Leaf midday water potential and midday relative water content were calculated at the end of the experiments at noon following the methods of Elansary et al. [17]. Leaf proline composition was also determined following the methods of Elansary et al. [18].

2.3.2. Essential Oil and Gas Chromatography/Mass Spectrometry (GC/MS)

The EOs were obtained by hydro-distillation of dried leaves for 1 h in Clevenger type glass equipment in the Department of Plant Production, King Saud University. The EO ratio was determined per treatment and the EOs were maintained dry by subjecting samples to anhydrous sodium sulfate, then stored at 4 °C. A Thermo Scientific, Trace GC Ultra was used coupled with a mass spectrometer (ISQ). A TG-1MS column (narrow bore, length 30 m × 0.32 mm ID, 0.25 μm film thickness) was used and the carrier gas was helium. The machine was programed with a starting temperature of 45 °C, then a gradual increase was made to 165 °C (4 °C min^{-1}), followed by an increase to 280 °C (15 °C min^{-1}), and ending with holding time of 15 min. A 2 μL sample of each EO was injected at 250 °C on a splitless mode flow (1 mL min^{-1}) for splitless time (3 min) followed by another split flow (10 mL min^{-1}). The FID was also accomplished in the same column and program. A homologous series of *n*-alkanes (C_{10}–C_{36}) was used to identify the compounds by the retention time and indices were coupled with a mass spectral search program (NIST Ver. 2.0) and WILEY libraries. Selected references from the literature were also used for comparison purposes [1,13].

2.3.3. Antioxidant Potential

Lipid peroxidation levels expressed as thiobarbituric acid reactive substances (TBARS), catalase (CAT) activity, H_2O_2, and superoxide dismutase (SOD) activities were quantified in frozen petal tissues [19].

2.4. Statistical Analyses

Data obtained from both years (2018 and 2019) were subjected to analyses of variance (ANOVA) with a completely randomized design to determine the significance of differences among treatments in SPSS Version 22 software. Standard errors (SE) were calculated from the data presenting for the mean of 20 replicates from both years. The least significant differences (LSD) method at $p \leq 0.05$ was used to compare all means at each watering level [20].

3. Results

3.1. Morphological Responses

Plants subjected to different water levels of *CSC* and/or biostimulants showed different morphological responses as shown in Table 1. The plants were subjected to three watering levels of *CSC* (100%, 70%, and 50%) and/or single biostimulants (CR, Mob, ND, and NN). Under 100% *CSC*, NN-treated plants showed the highest leaf number and leaf area and was followed by ND, Mob, and CR. ND and NN treatments at 70% *CSC* and 50% CSC showed the highest increase in leaf number and leaf area as compared to those of other treatments, as well as those of the control. The tallest plants and highest root dry weight were found treatments with NN. CR and Mob showed no significant differences under 50% *CSC*. On the other hand, when water stress increased (i.e., 70% *CSC* and 50% *CSC*), the values for leaf number (20.19% and 45.77%), leaf area (24% and 52.40%), plant height (17.30% and 41.35%), and root dry weight (15.70% and 49.59%) decreased significantly. Plant fresh and dry weights increased significantly in biostimulant-treated plants compared to those of the control, and the highest increase was for plants subjected to NN followed by ND under different water levels (Table 2). NN-treated plants with 70% *CSC* water level had the highest IWUE (average of 2.78 kg m^{-3}) in both growth seasons, which was statistically different from that of all other treatments.

3.2. Physiological and Metabolic Performance

Figure 1a–c shows the effects of water stress and biostimulants on gas exchange, namely, A, E, and gs of mint plants during the growing seasons in 2018 and 2019 (shown as averages). Under 100% *CSC*, the A of mint plants showed a ($p < 0.05$) significant increase in plants subjected to NN (8.35 μmol CO_2 m^{-2} s^{-1}) compared to that of other treatments (Figure 1a). The reduction in irrigation water to 70% *CSC* and 50% *CSC* caused significantly reduced A values in control plants. However, biostimulant treatment increased the values of A, wherein NN-treated plants under 70% *CSC* and 50% *CSC* had the highest values (7.4 and 4.9 μmol CO_2 m^{-2} s^{-1}, respectively), followed by ND-treated plants (7.3 and 4.6 μmol CO_2 m^{-2} s^{-1}, respectively).

Table 1. Average morphological responses for mint plants grown under greenhouse conditions during the growth season in 2018 and 2019, subjected to three water levels, 100%, 70%, and 50% *CSC*, and four biostimulants, CR, Mob, ND, and NN, as well as a control (without biostimulants).

Variable	Leaf Number (plant^{-1})						Leaf Area (cm^2 plant^{-1})						Plant Height (cm)						Root Dry Weight (g)					
Water Levels/Biosti.	Control	CR	Mob	ND	NN	Means	Control	CR	Mob	ND	NN	Means	Control	CR	Mob	ND	NN	Means	Control	CR	Mob	ND	NN	Means
100% CSC	57.00^{e}	62.10^{d}	67.20^{c}	72.30^{b}	77.20^{a}	67.16^{A}	228.00^{e}	247.10^{d}	268.80^{c}	288.20^{b}	307.80^{a}	267.98^{A}	27.60^{e}	29.10^{d}	30.20^{c}	32.10^{b}	35.30^{a}	30.86^{A}	1.05^{e}	1.13^{d}	1.19^{c}	1.29^{b}	1.38^{a}	1.21^{A}
70% CSC	47.00^{e}	51.20^{d}	53.10^{c}	56.80^{b}	59.90^{a}	53.60^{B}	178.60^{e}	193.56^{d}	200.78^{c}	216.80^{b}	228.62^{a}	203.67^{B}	23.10^{e}	24.40^{d}	25.70^{c}	26.90^{b}	27.50^{a}	25.52^{B}	0.82^{d}	1.01^{c}	1.04^{c}	1.12^{b}	1.21^{a}	1.02^{B}
50% CSC	29.00^{d}	34.80^{c}	35.90^{c}	39.30^{b}	43.10^{a}	36.42^{C}	101.50^{d}	121.20^{c}	125.65^{c}	137.55^{b}	151.85^{a}	127.55^{C}	16.20^{d}	17.30^{c}	17.80^{c}	18.90^{b}	20.30^{a}	18.10^{C}	0.56^{d}	0.59^{c}	0.60^{c}	0.64^{b}	0.67^{a}	0.61^{C}
Means	44.33^{E}	49.37^{D}	52.07^{C}	56.13^{B}	60.07^{A}		169.37^{E}	187.29^{D}	198.41^{C}	214.18^{B}	229.42^{A}		22.30^{E}	23.60^{D}	24.57^{C}	25.97^{B}	27.70^{A}		0.81^{E}	0.91^{D}	0.94^{C}	1.02^{B}	1.06^{A}	
Water levels Biosti. Water levels × Biosti.	*** ($LSD_{0.05}$ = 0.877) *** ($LSD_{0.05}$ = 1.132) *** ($LSD_{0.05}$ = 1.961)						*** ($LSD_{0.05}$ = 2.502) *** ($LSD_{0.05}$ = 3.230) *** ($LSD_{0.05}$ = 5.594)						*** ($LSD_{0.05}$ = 0.282) *** ($LSD_{0.05}$ = 0.365) *** ($LSD_{0.05}$ = 0.631)						*** ($LSD_{0.05}$ = 0.010) *** ($LSD_{0.05}$ = 0.013) *** ($LSD_{0.05}$ = 0.022)					

Data represent the mean calculated from n = 100, 60, or 20 for each water level, biostimulants or their interaction, respectively, in two growth seasons. *** ($p \leq 0.001$). Means differing in lowercase letters (a–e) within each row indicate significant differences at $p \leq 0.05$. Means differing in uppercase letters (A–E) within each mean row and each mean column indicate significant differences according to the least significant difference (LSD) test at $p \leq 0.05$. CSC (container substrate capacity), Control (without biostimulants), CR (CRADLE™), Mob (Mobilizer™), ND (Nanozim De'Lite™), NN (Nanozim NXT™), and Biosti. (biostimulant).

Table 2. Average plant fresh and dry weights and irrigation water-use efficiency (IWUE) for mint plants grown under greenhouse conditions during the growth season in 2018 and 2019, subjected to three water levels, 100%, 70%, and 50% *CSC*, and four biostimulants, CR, Mob, ND, and NN, as well as a control (without biostimulants).

Variable	Water Applied (mm)	Plant Fresh Weight (g)						Plant Dry Weight (g)						IWUE (kg m^{-3})					
Water Levels/Biosti.		Control	CR	Mob	ND	NN	Means	Control	CR	Mob	ND	NN	Means	Control	CR	Mob	ND	NN	Means
100% CSC	35.91	7.38^{d}	8.01^{c}	8.48^{b}	8.47^{b}	9.24^{a}	8.32^{A}	6.64^{e}	7.05^{d}	7.38^{c}	7.79^{b}	8.41^{a}	7.45^{A}	1.85^{d}	2.01^{c}	2.13^{b}	2.12^{b}	2.32^{a}	2.08^{B}
70% CSC	27.00	6.02^{d}	7.53^{c}	7.39^{c}	7.81^{b}	8.34^{a}	7.42^{B}	5.12^{d}	6.48^{c}	6.50^{c}	6.87^{b}	7.42^{a}	6.48^{B}	2.01^{d}	2.51^{c}	2.46^{c}	2.60^{b}	2.78^{a}	2.47^{A}
50% CSC	22.14	4.69^{b}	4.71^{b}	4.83^{b}	5.18^{a}	5.31^{a}	4.94^{C}	3.75^{d}	3.96^{c}	4.01^{c}	4.25^{b}	4.46^{a}	4.09^{C}	1.91^{b}	1.92^{b}	1.96^{b}	2.11^{a}	2.16^{a}	2.01^{C}
Means		6.03^{E}	6.75^{D}	6.90^{C}	7.15^{B}	7.63^{A}		5.17^{E}	5.83^{D}	5.96^{C}	6.31^{B}	6.76^{A}		1.92^{E}	2.15^{D}	2.18^{C}	2.28^{B}	2.42^{A}	
Water levels Biosti. Water levels ×Biosti.		*** ($LSD_{0.05}$ = 0.086) *** ($LSD_{0.05}$ = 0.111) *** ($LSD_{0.05}$ = 0.193)						*** ($LSD_{0.05}$ = 0.075) *** ($LSD_{0.05}$ = 0.097) *** ($LSD_{0.05}$ = 0.168)						*** ($LSD_{0.05}$ = 0.028) *** ($LSD_{0.05}$ = 0.036) *** ($LSD_{0.05}$ = 0.062)					

Data represent the mean calculated from n = 100, 60, or 20 for each water level, biostimulants or their interaction, respectively, in two growth seasons. *** ($p \leq 0.001$). Means differing in lowercase letters (a–e) within each row indicate significant differences at $p \leq 0.05$. Means differing in uppercase letters (A–E) within each mean row and each mean column indicate significant differences according to the least significant difference (LSD) test at $p \leq 0.05$. CSC (container substrate capacity), Control (without biostimulants), CR (CRADLE™), Mob (Mobilizer™), ND (Nanozim De'Lite™), NN (Nanozim NXT™), and Biosti. (biostimulant).

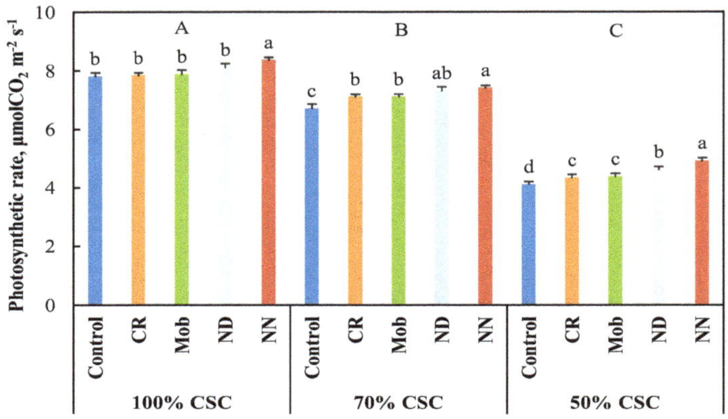

(a)

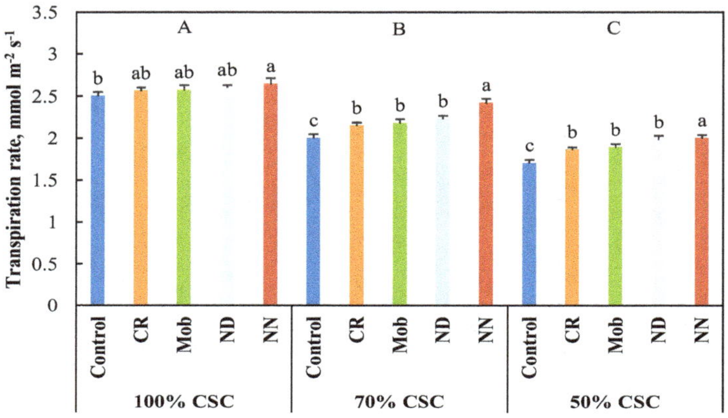

(b)

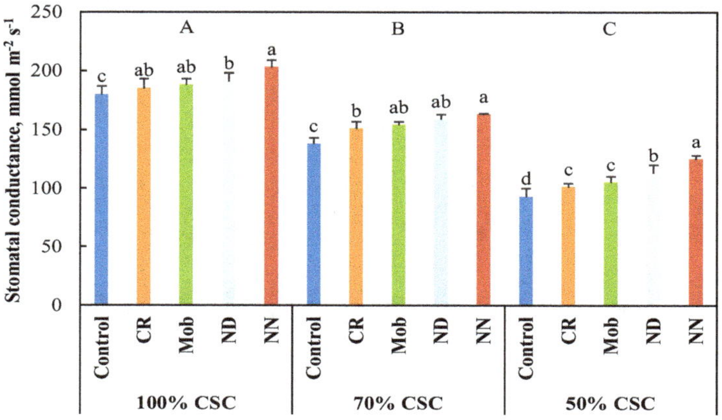

(c)

Figure 1. *Cont.*

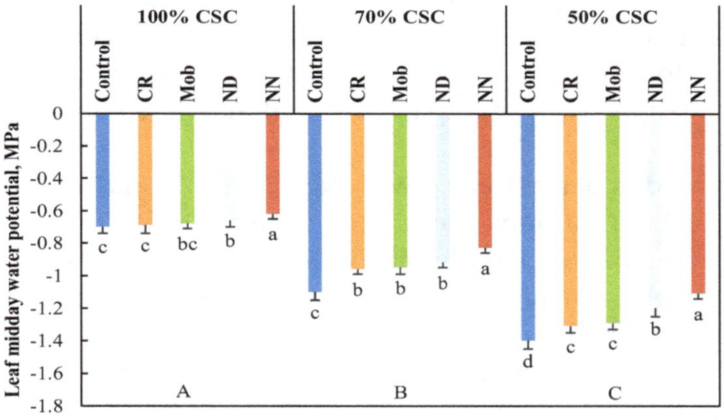

(d)

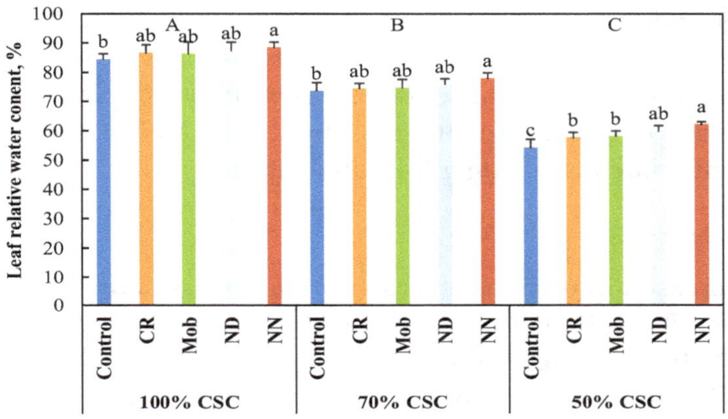

(e)

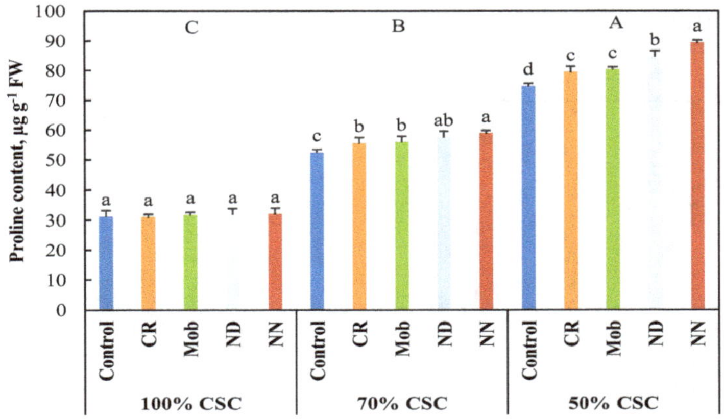

(f)

Figure 1. Photosynthetic rate (**a**), transpiration rate (**b**), stomatal conductance (**c**), leaf midday water potential (**d**), leaf relative water content (**e**), and proline content (**f**) responses of mint plants, average of two growing seasons, as affected by three water levels, 100%, 70%, and 50% *CSC* (container substrate capacity), and four biostimulants, CR (CRADLE™), Mob (Mobilizer™), ND (Nanozim De'Lite™), and NN (Nanozim NXT™), and a Control (without biostimulants). FW (fresh weight). Different capital letters on top indicate significant differences between water levels at $p \leq 0.05$. Different letters on top of columns indicate significant differences between biostimulants across water levels at $p \leq 0.05$. Bars indicate the means ± SE of the mean ($n = 20$).

The values of E were higher in plants subjected to biostimulants compared to that of the control (Figure 1b). Treating plants with CR, Mob, and ND showed no significant differences in E under different *CSC* treatments. However, the E in NN-treated plants was significantly ($p < 0.05$) increased by 5.6%, 21%, and 17.6%, respectively, compared with that of the control plants under 100%, 70%, and 50% *CSC*.

The values of gs increased in plants treated with NN (203, 163, and 125 mmol m^{-2} s^{-1}, respectively) compared to that of other biostimulant treatments under 100%, 70%, and 50% *CSC* (Figure 1c). Statistically significant differences in gs under 50% *CSC* were found only between the ND treatment and both CR and Mob treatments, but significant effects on gs to the same treatments were not observed for 100% and 70% *CSC*. Irrespective of the biostimulant treatments, there were significant ($p < 0.05$) differences between the water levels treatments, where the 100% *CSC* was superior.

Leaf midday water potential increased in plants subjected to different biostimulants under 70% *CSC* (average of −0.95 MPa) and 50% *CSC* (average of −1.26 MPa) compared to that of 100% *CSC* (average of −0.67 MPa), as shown in Figure 1d. The NN treatment had the lowest water potential under different water level treatments. The leaf relative water content increased significantly ($p < 0.05$) in plants subjected to NN under different water levels (Figure 1e), where NN-treated plants with 100% *CSC* had the highest value of 88.3%. The proline content (Figure 1f) increased in plants subjected stress conditions only (70% and 50% *CSC*) and the increases were higher in plants subjected to biostimulants than that of the control (without biostimulants) treatment. The NN-treated plants with 50% CSC had the highest proline content value of 89.1 μg g^{-1} fresh weight.

3.3. EO Ratio and Constitutes

The EO ratio was increased in response to biostimulant treatments as shown in Figure 2. Control treatments had the lowest EO ratio of 2.7%, 2.8%, and 2.1% fresh weight, respectively, under 100%, 70%, and 50% *CSC*. In NN-treated plants, the average EO ratio was increased by 21.57%, followed by that of ND (15.81%), in relation to that of the control plants under water level treatments. There was a significant difference in the EO ratio ($p < 0.05$) of mint plants between different biostimulant treatments within each water level, except between CR and Mob ($p > 0.05$) under water stress conditions of 70% and 50% *CSC*. Irrespective of the biostimulant treatments, the EO ratio under 70% *CSC* was not significantly higher than that of 100% *CSC*.

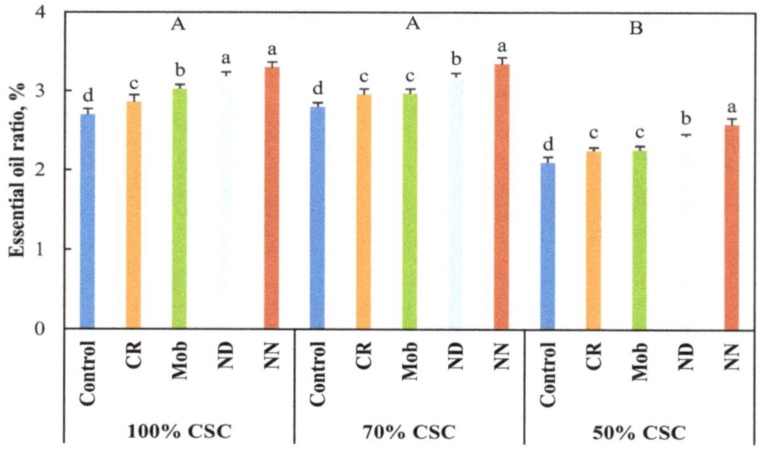

Figure 2. Essential oil ratio of mint plants, average of two growing seasons, as affected by three water levels: 100, 70, and 50% *CSC* (container substrate capacity), and four biostimulants: CR (CRADLE™), Mob (Mobilizer™), ND (Nanozim De'Lite™), and NN (Nanozim NXT™), in addition to Control (without biostimulants). Different capital letters on top are significant differences between water levels at $p \le 0.05$. Different letter on top columns indicate significant differences between biostimulants across water levels at $p \le 0.05$. Bars give the means ± SE of the mean ($n = 20$).

Major EO constitutes in all treatments were 1-menthone, isopulegone, pulegone, α-pinene, 1,8-cineol, and α-terpineol ratios as shown in Tables 3 and 4. The 1-menthone, isopulegone and pulegone were significantly ($p < 0.001$) reduced in plants subjected to water stress and biostimulant treatments (Table 3). Irrespective of the biostimulant treatments, 70% and 50% *CSC* plants exhibited decreased 1-menthone by 5.3% and 9.6%, respectively, compared to that of 100% *CSC* plants. Likewise, isopulegone for these plants was decreased by 5.3% and 8.6% and pulegone by 6.9% and 8.5%, respectively.

Irrespective of the water level treatments, ND and NN treatments significantly reduced 1-menthone, isopulegone, and pulegone, by 23.8% and 33.3%; 8.1% and 18.6%; 11.1% and 18%on average, respectively, compared to that of the control treatment. However, the application of biostimulants significantly ($p < 0.001$) increased the α-pinene, 1,8-cineol, and α-terpineol ratios under different water stress conditions (Table 4). These ratios were significantly ($p < 0.001$) different between 100%, 70%, and 50% *CSC* plants, where 100% CSC showed the lowest values. The NN-treated plants yielded the highest ratios for α-pinene (4.3%), 1,8-cineol (36.1%), and α-terpineol (3.4%) at 50% *CSC*.

3.4. Antioxidant Activities

There was a significantly ($p < 0.05$) reduced accumulation of lipid peroxidation and H_2O_2 in plants subjected to different biostimulants, as shown in Figure 3. The NN-treated plants yielded the lowest values of lipid peroxidation (57, 46, and 27 μmol TBARS g^{-1} fresh weight, respectively) and H_2O_2 (2.6, 4.6, and 5.9 μmol g^{-1} fresh weight, respectively) under 100%, 70%, and 50% *CSC*. However, control plants had the highest values of lipid peroxidation (64, 54, and 37 μmol TBARS g^{-1} fresh weight, respectively) and H_2O_2 (2.9, 5.2, and 7.2 μmol g^{-1} fresh weight, respectively). It was observed that there were no significant differences between the control and CR treatments in lipid peroxidation and H_2O_2 under different water levels, except for the 70% *CSC* for lipid peroxidation.

In 50% *CSC*, there were only significant differences between the Mob and ND treatments. On the contrary, there were significant ($p < 0.05$) increases in the activities of CAT and SOD of leaf extracts of biostimulant-treated plants under normal and water stress conditions (Figure 3). The highest CAT and SOD activity values were found in NN treatments (increasing 25.5% and 40.3%, respectively), followed by that of ND-treated plants (increasing 17.9% and 26.8%, respectively) comparing with those of the control treatment, which had the lowest values. The CAT activity was significantly ($p < 0.05$) increased by 30% and 58.2% on average, respectively, in plants subjected to water stress (70% and 50% *CSC*) compared to that of the normal (100% *CSC*) condition, whereas SOD activity was increased by 79.3%and 123.7% on average, respectively.

Table 3. Average 1-menthone, isopulegone, and pulegone for mint plants grown under greenhouse conditions during the growth season in 2018 and 2019, subjected to three water levels, 100%, 70%, and 50% *CSC*, and four biostimulants, CR, Mob, ND, and NN, as well as a control (without biostimulants).

Variable	1-Menthone (%)						Isopulegone (%)						Pulegone (%)					
Water Levels/Biosti.	Control	CR	Mob	ND	NN	Means	Control	CR	Mob	ND	NN	Means	Control	CR	Mob	ND	NN	Means
100% CSC	2.20 ^{a}	2.10 ^{c}	1.90 ^{c}	1.70 ^{d}	1.50 ^{e}	1.88 ^{A}	5.60 ^{a}	5.50 ^{a,b}	5.40 ^{b}	5.20 ^{c}	4.90 ^{d}	5.32 ^{A}	56.46 ^{a}	54.27 ^{b}	53.51 ^{b}	50.32 ^{c}	46.93 ^{d}	52.30 ^{A}
70% CSC	2.10 ^{a}	2.00 ^{b}	1.80 ^{c}	1.60 ^{d}	1.40 ^{e}	1.78 ^{B}	5.40 ^{a}	5.30 ^{b}	5.20 ^{c}	5.00 ^{d}	4.30 ^{e}	5.04 ^{B}	53.34 ^{a}	50.12 ^{b}	49.68 ^{b}	47.08 ^{c}	43.22 ^{d}	48.69 ^{B}
50% CSC	2.00 ^{a}	2.00 ^{a}	1.70 ^{b}	1.50 ^{c}	1.30 ^{d}	1.70 ^{C}	5.30 ^{a}	5.10 ^{b}	5.00 ^{b}	4.80 ^{c}	4.10 ^{d}	4.86 ^{C}	52.13 ^{a}	49.52 ^{b}	48.31 ^{c}	46.52 ^{d}	42.71 ^{e}	47.84 ^{C}
Means	2.10 ^{A}	2.04 ^{B}	1.80 ^{C}	1.60 ^{D}	1.40 ^{E}		5.44 ^{A}	5.30 ^{B}	5.20 ^{C}	5.00 ^{D}	4.43 ^{E}		53.98 ^{A}	51.30 ^{B}	50.50 ^{C}	47.98 ^{D}	44.29 ^{E}	
Water levels	*** ($LSD_{0.05}$ = 0.016)						*** ($LSD_{0.05}$ = 0.046)						*** ($LSD_{0.05}$ = 0.545)					
Biosti.	*** ($LSD_{0.05}$ = 0.021)						*** ($LSD_{0.05}$ = 0.059)						*** ($LSD_{0.05}$ = 0.703)					
Water levels × Biosti.	*** ($LSD_{0.05}$ = 0.036)						*** ($LSD_{0.05}$ = 0.102)						ns					

Data represent the mean calculated from n = 100, 60, or 20 for each water level, biostimulants or their interaction, respectively, in two growth seasons. ^{ns} (not significant) and *** ($p \leq 0.001$). Means differing in lowercase letters (a–e) within each row indicate significant differences at $p \leq 0.05$. Means differing in uppercase letters (A–E) within each mean row and each mean column indicate significant differences according to the least significant difference (LSD) test at $p \leq 0.05$. CSC (container substrate capacity), Control (without biostimulants), CR (CRADLE™), Mob (Mobilizer™), ND (Nanozim De'Lite™), NN (Nanozim NXT™), and Biosti. (biostimulant).

Table 4. Average α-pinene, 1,8-cineol, and α- terpineol ratios of mint plants grown under greenhouse conditions during the growth season in 2018 and 2019, subjected to three water levels, 100%, 70%, and 50% *CSC*, and four biostimulants, CR, Mob, ND, and NN, as well as a control (without biostimulants).

Variable	α-Pinene (%)						1,8-Cineol (%)						α- Terpineol (%)					
Water Levels/Biosti.	Control	CR	Mob	ND	NN	Means	Control	CR	Mob	ND	NN	Means	Control	CR	Mob	ND	NN	Means
100% CSC	0.90 ^{e}	1.80 ^{d}	3.40 ^{c}	3.60 ^{b}	4.00 ^{a}	2.74 ^{C}	24.28 ^{c}	25.09 ^{c}	25.20 ^{c}	27.31 ^{b}	29.11 ^{a}	26.20 ^{A}	1.70 ^{e}	1.90 ^{d}	2.30 ^{c}	2.50 ^{b}	2.70 ^{a}	2.22 ^{C}
70% CSC	1.50 ^{e}	2.50 ^{d}	3.80 ^{b}	3.70 ^{c}	4.20 ^{a}	3.14 ^{B}	25.62 ^{c}	26.62 ^{c}	26.70 ^{c}	28.61 ^{b}	35.71 ^{a}	28.65 ^{B}	1.90 ^{e}	2.10 ^{d}	2.40 ^{c}	2.70 ^{b}	3.10 ^{a}	2.44 ^{B}
50% CSC	2.10 ^{e}	3.10 ^{d}	3.70 ^{c}	4.10 ^{b}	4.30 ^{a}	3.46 ^{A}	26.82 ^{c}	27.79 ^{c}	27.92 ^{c}	29.82 ^{b}	36.11 ^{a}	29.70 ^{C}	2.10 ^{e}	2.30 ^{d}	2.50 ^{c}	3.00 ^{b}	3.40 ^{a}	2.66 ^{A}
Means	1.50 ^{E}	2.47 ^{D}	3.63 ^{C}	3.80 ^{B}	4.17 ^{A}		25.57 ^{D}	26.50 ^{C}	26.61 ^{C}	28.58 ^{B}	33.64 ^{A}		1.90 ^{E}	2.10 ^{D}	2.40 ^{C}	2.73 ^{B}	3.07 ^{A}	
Water levels	*** ($LSD_{0.05}$ = 0.047)						*** ($LSD_{0.05}$ = 0.544)						*** ($LSD_{0.05}$ = 0.018)					
Biosti.	*** ($LSD_{0.05}$ = 0.061)						*** ($LSD_{0.05}$ = 0.702)						*** ($LSD_{0.05}$ = 0.024)					
Water levels × Biosti.	*** ($LSD_{0.05}$ = 0.105)						*** ($LSD_{0.05}$ = 1.216)						*** ($LSD_{0.05}$ = 0.041)					

Data represent the mean calculated from n = 100, 60, or 20 for each water level, biostimulants or their interaction, respectively, in two growth seasons. *** ($p \leq 0.001$). Means differing in lowercase letters (a–e) within each row indicate significant differences at $p \leq 0.05$. Means differing in uppercase letters (A–E) within each mean row and each mean column indicate significant differences according to the least significant difference (LSD) test at $p \leq 0.05$. CSC (container substrate capacity), Control (without biostimulants), CR (CRADLE™), Mob (Mobilizer™), ND (Nanozim De'Lite™), NN (Nanozim NXT™), and Biosti. (biostimulant).

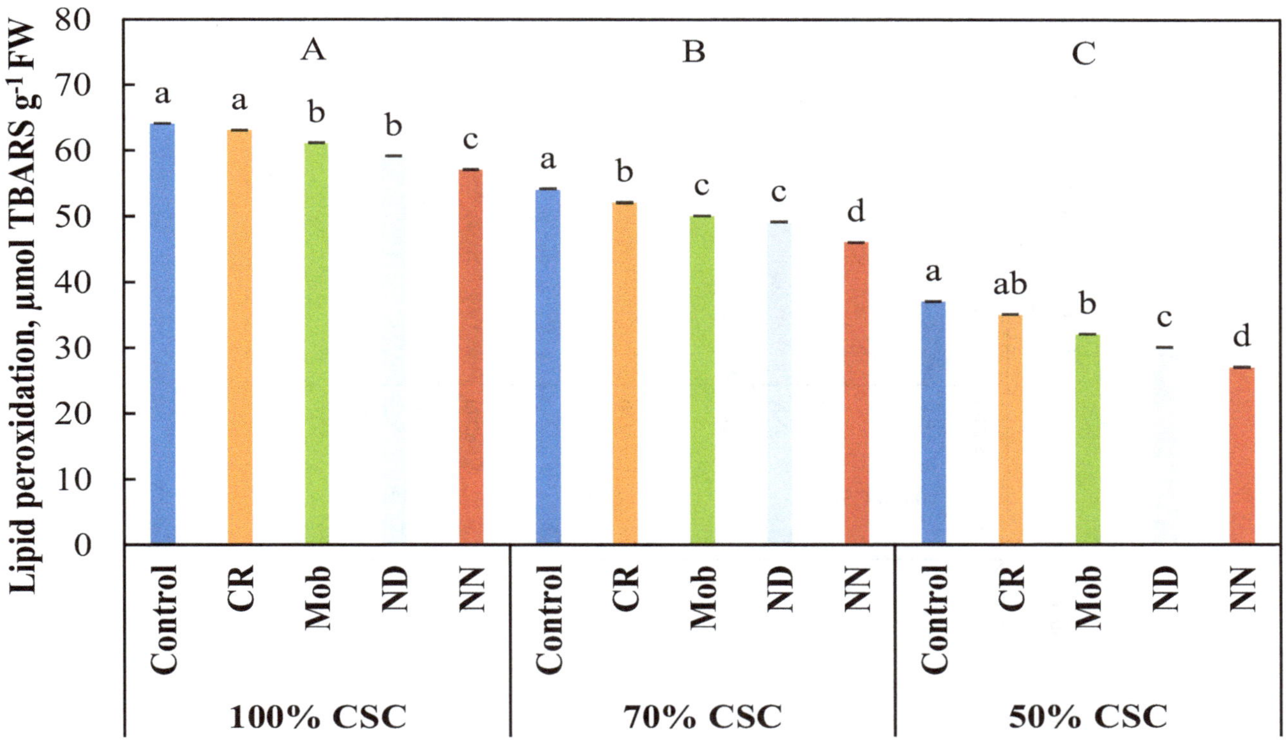

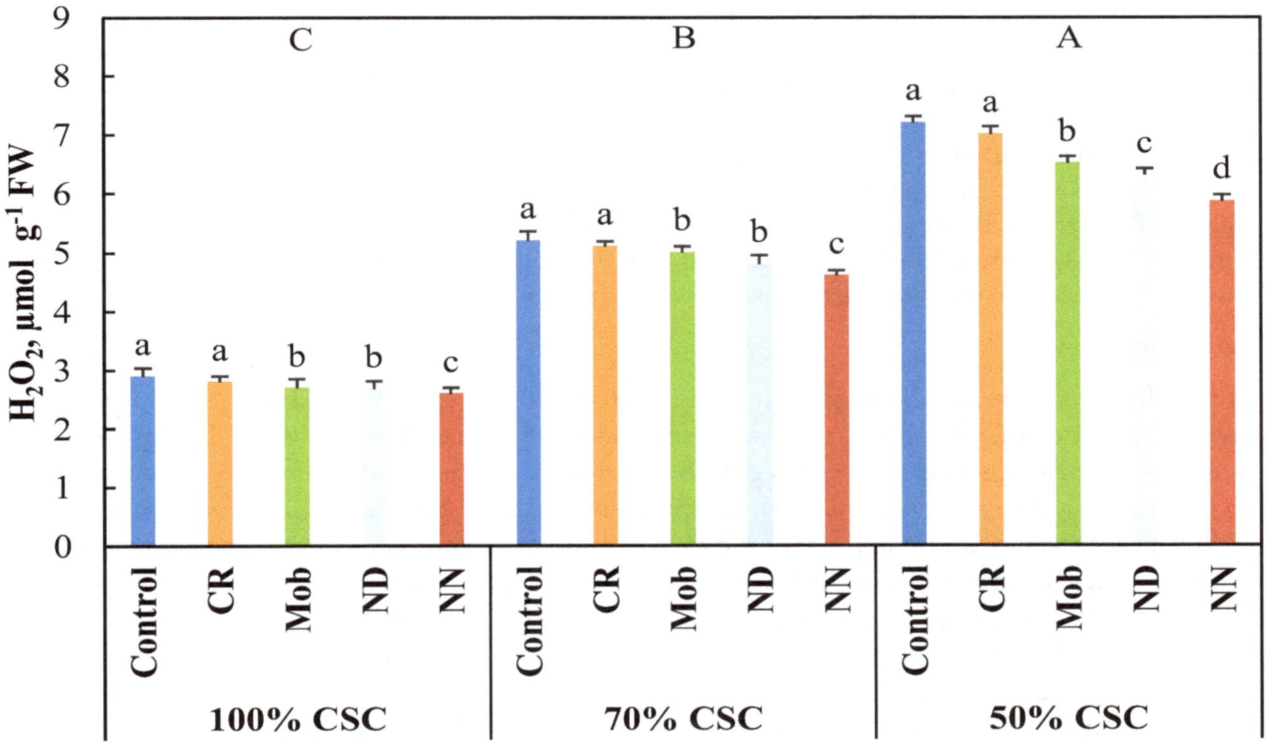

Figure 3. *Cont.*

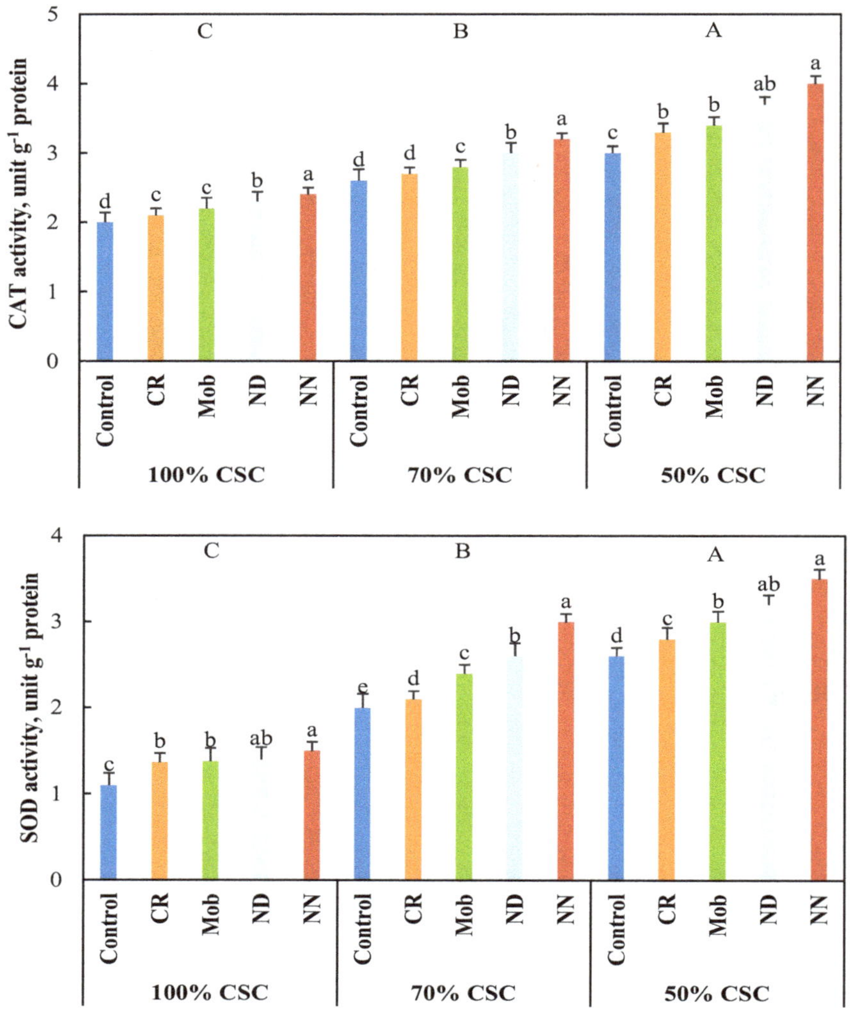

Figure 3. Lipid peroxidation, H_2O_2, catalase (CAT), and superoxide dismutase (SOD) activities of mint plants, average of two growing seasons, as affected by three water levels, 100%, 70%, and 50% *CSC* (container substrate capacity), and four biostimulants, CR (CRADLE™), Mob (Mobilizer™), ND (Nanozim De'Lite™), and NN (Nanozim NXT™), and a control (without biostimulants). FW (fresh weight). Different capital letters on top indicate significant differences between water levels at $p \leq 0.05$. Different letters on top columns indicate significant differences between biostimulants across water levels at $p \leq 0.05$. Bars provide the means ± SE of the mean ($n = 20$).

4. Discussion

Water stress is one of the major limiting factors of the growth and productivity of plants worldwide [21]. The amount of irrigation water applied influenced the biomass and EO yields of mint. The fresh and dry weights of mint were decreased with the irrigation water stress because of vegetative growth (i.e., leaf number and plant height), which decreased under water deficit conditions. Reduction in growth parameters as a consequence of drought has also been described in peppermint [6,7,11], Japanese mint [9], purple basil [8], balm [10], and chamomile [12]. The irrigation water level of 50% *CSC* had a negative effect on EO yield of mint. This is in agreement with earlier findings in peppermint [11] and chamomile [12], and in contrary to the results found in the previous studies from Ekren et al. [8] in purple basil and Farahani et al. [10] in balm.

The application of biostimulants under water stress conditions (70% and 50% *CSC*) showed enhanced growth by means of increased leaf number, plant height, root dry weight, fresh and dry weights, and IWUE. These morphological improvements are mainly attributed to the composition of these biostimulants. The most active biostimulant in this study was NN, which is composed of a unique mixture of important compounds: seaweed extract (15%), humic acid (5%), macro (potassium 1% and phosphorus) and micro elements, alginic acid, and hydrolyzed protein, as described in the materials and methods. The major constitutes of the NN biostimulant were seaweed extracts (*Ascophyllum nodosum*) and humic acid. The application of *Ascophyllum nodosum* extracts as plant biostimulants has been reported in several studies [19,22]. Humic acid may increase the leaf area, stem diameter, plant dry weight in different plants [23,24] and may ameliorate stress conditions in tomatoes [25]. Potassium, phosphorus, and microelements play critical roles in the growth and morphology of most plants [26,27]. However, the mixture was superior in the ameliorating effects against water stress in mint plants compared to other commercial biostimulants. The second biostimulant showing relatively high morphological performance was ND, which is mainly composed of *Ascophyllum nodosum* extracts (25%), carbohydrates (25%), (*w*/*w*) amino acid (2%) and potassium (1%). ND showed slightly lower morphological promoting effects than that of NN. CR and Mob are mainly composed of mycorrhizal biofertilizer. However, Mob contains additional components, including seaweed extract (*Macrocystis pyrifera*), humic acid, and amino acids, which may explain the slight increased vegetative performance of Mob compared to that of CR. Furthermore, *Macrocystis pyrifera* has been reported to have stimulatory effects on plant growth [28].

Gas exchange parameters (*gs*, *E*, and *A*) are important indicators of the physiological performance of plants under stress conditions. The increase in *gs* under stress conditions in response to external factors is strongly related to enhanced gas exchange through the stomata [29]. The increased gas exchange is normally reflected as enhanced transpiration and photosynthesis rates in the leaves [30]. There were increases in the gas exchange in plants treated with different biostimulants under water stress conditions, which indicated that these biostimulants acted as effective stress ameliorants. Leaf water potential and relative water content reductions might be associated with stress conditions [31,32]. They increased in this study in plants subjected to different biostimulants, indicating enhanced metabolic performance of treated plants. Furthermore, the increased proline composition in biostimulant-treated plants reflected enhanced stress tolerance as found in previous studies using other external elicitors [22,33].

The main constitutes of the EO were pulegone and 1,8-cineol. A previous study on the same species from Egypt reported comparable composition of both compounds [1]. There were fluctuations in the main constitutes of EO, as well as specific compounds, including pulegone, isopulegone, 1-menthone, 1,8-cineol, α-pinene, and α-terpineol. Secondary metabolites, such as cineole are usually associated with terpenes [34] and this explains the parallel increase in 1,8-cineol, α-terpineol, and pinene. 1-menthone and isopulegone are metabolites of pulegone. The pulegone is not favored in the EO composition of mint plants because of its carcinogenic effects at high doses [1], whereas, cineol is a favored compound in EOs because of its medicinal applications and pharmaceutical potential [35,36]. The application of NN showed the highest increase in 1,8-cineol and related terpenes ratios and lowest compositions of pulegone compared to that of the control and other biostimulant treatments. This result suggests that NN application may have future applications in medicinal plants, such as mints. The use of the NN biostimulant is a novel approach for enhancing the chemical composition of the EOs of mint plants by reducing hazardous compounds and increasing useful ones as found in this study.

In this study, the application of seaweed extract-based biostimulants mixed with humic acid and/or macro elements represented a novel tool for the enhancement of the medicinal properties of major medicinal plants, such as mints. The achievement of enhanced antioxidant activities of the EOs of mint might be of great importance for agricultural and related pharmaceutical industries. The oil of mints is routinely used in perfume and cosmetic preparations, as well as in the food industries, such as in chocolate and soft drinks. The development of new EO compositions with increased antioxidant properties will increase the additive value of the medicinal crop and will assist in reducing dependence on synthetic antioxidants to control ROS accumulation.

5. Conclusions

This study revealed an association between the application of specific biostimulants and the increase/decrease of the main EO composition (cineol and pulegone) of mint plants. The application of this finding is related to the agricultural, medicinal, and pharmaceutical industries. There were increases in the morphological characteristics, physiological performance, and EO ratio of biostimulant-treated plants. The morphological and physiological enhancements indicated increased tolerance to water stress. Further, biostimulant-treated plants showed higher antioxidant activities, reduced accumulation of H_2O_2, and increased CAT and SOD activities, which indicated an antioxidant stress tolerance activation mechanism in treated plants. The application of biostimulants to mint plants increased the quantity and quality of produced EOs and enhanced the medicinal properties, as well as that of the traditional medicinal crop. ND and NN are recommended under water stress conditions in mint.

Author Contributions: H.O.E., conceptualization, funding acquisition, investigation, supervision, project administration, methodology, writing—original draft; M.A.M., formal analysis, investigation, data curation, methodology, project administration, writing—review and editing; E.A.M., supervision, visualization; D.O.E.-A., supervision, visualization. All authors have read and agreed to the published version of the manuscript.

Acknowledgments: The authors extend their appreciation to the Deanship of Scientific Research at King Saud University for their financial support of the present study.

References

1. Elansary, H.O.; Ashmawy, N.A. Essential oils of mint between benefits and hazards. *J. Essent. Oil-Bear. Plants* **2013**, *16*, 429–438. [CrossRef]
2. Stringaro, A.; Vavala, E.; Colone, M.; Pepi, F.; Mignogna, G.; Garzoli, S.; Cecchetti, S.; Ragno, R.; Angiolella, L. Effects of *Mentha suaveolens* essential oil alone or in combination with other drugs in candida albicans. *Evid. Based Complement. Alternat. Med.* **2014**, *2014*, 9. [CrossRef] [PubMed]
3. Khoury, M.; Stien, D.; Eparvier, V.; Ouaini, N.; El Beyrouthy, M. Report on the medicinal use of eleven Lamiaceae species in Lebanon and rationalization of their antimicrobial potential by examination of the chemical composition and antimicrobial activity of their essential oils. *Evid. Based Complement. Altern. Med.* **2016**, *2016*, 17. [CrossRef] [PubMed]
4. Sarkic, A.; Stappen, I. Essential oils and their single compounds in cosmetics—A critical review. *Cosmetics* **2018**, *5*, 11. [CrossRef]
5. Mancosu, N.; Snyder, R.L.; Kyriakakis, G.; Spano, D. Water scarcity and future challenges for food production. *Water* **2015**, *7*, 975–992. [CrossRef]
6. Zade, N.S.E.; Sadeghi, A.; Moradi, P. Streptomyces strains alleviate water stress and increase peppermint (*Mentha piperita*) yield and essential oils. *Plant Soil* **2019**, *434*, 441–452. [CrossRef]
7. Figueroa-Pérez, M.G.; Rocha-Guzmán, N.E.; Pérez-Ramírez, I.F.; Mercado-Silva, E.; Reynoso-Camacho, R. Metabolite profile, antioxidant capacity, and inhibition of digestive enzymes in infusions of peppermint (*Mentha piperita*) grown under drought stress. *J. Agric. Food Chem.* **2014**, *62*, 12027–12033. [CrossRef]
8. Ekren, S.; Sönmez, Ç.; Özçakal, E.; Kurttaş, Y.S.K.; Bayram, E.; Gürgülü, H. The effect of different irrigation water levels on yield and quality characteristics of purple basil (*Ocimum basilicum* L.). *Agric. Water Manag.* **2012**, *109*, 155–161. [CrossRef]
9. Shormin, T.; Khan, M.A.H.; Alamgir, M. Response of different levels of nitrogen fertilizer and water stress on the growth and yield of Japanese mint (*Mentha arvensis* L.). *Bangladesh J. Sci. Ind. Res.* **2009**, *44*, 137–145. [CrossRef]
10. Farahani, H.A.; Valadabadi, S.A.; Daneshian, J.; Khalvati, M.A. Evaluation changing of essential oil of balm (*Melissa officinalis* L.) under water deficit stress conditions. *J. Med. Plants Res.* **2009**, *3*, 329–333.
11. Khorasaninejad, S.; Mousavi, A.; Soltanloo, H.; Hemmati, K.; Khalighi, A. The effect of drought stress on growth parameters, essential oil yield and constituent of Peppermint (*Mentha piperita* L.). *J. Med. Plant Res.* **2011**, *5*, 5360–5365.

12. Razmjoo, K.; Heydarizadeh, P.; Sabzalian, M.R. Effect of salinity and drought stresses on growth parameters and essential oil content of *Matricaria chamomila*. *Int. J. Agric. Biol.* **2008**, *10*, 451–454.
13. Elansary, H.O.; Yessoufou, K.; Shokralla, S.; Mahmoud, E.A.; Skalicka-Woźniak, K. Enhancing mint and basil oil composition and antibacterial activity using seaweed extracts. *Ind. Crops Prod.* **2016**, *92*, 50–56. [CrossRef]
14. El-Esawi, M.A.; Elansary, H.O.; El-Shanhorey, N.A.; Abdel-Hamid, A.M.E.; Ali, H.M.; Elshikh, M.S. Salicylic acid-regulated antioxidant mechanisms and gene expression enhance rosemary performance under saline conditions. *Front. Physiol.* **2017**, *8*, 716. [CrossRef]
15. Fonteno, W.; Hardin, C.; Brewster, J. *Procedures for Determining Physical Properties of Horticultural Substrates Using the NCSU Porometer*; Horticultural Substrates Laboratory, North Carolina State University: Raleigh, NC, USA, 1995.
16. Howell, T.A. Challenges in increasing water use efficiency in irrigated agriculture. In Proceedings of the International Symposium on Water and Land Management for Sustainable Irrigated Agriculture, University of C¸ Ukurova, Adana, Turkey, 4–8 April 2006; pp. 53–63.
17. Elansary, H.O.; Salem, M.Z.M. Morphological and physiological responses and drought resistance enhancement of ornamental shrubs by trinexapac-ethyl application. *Sci. Hortic.* **2015**, *189*, 1–11. [CrossRef]
18. Elansary, H.O.; Yessoufou, K. Growth regulators and mowing heights enhance the morphological and physiological performance of seaspray turfgrass during drought conditions. *Acta Physiol. Plant.* **2015**, *37*, 232. [CrossRef]
19. Elansary, H.O.; Yessoufou, K.; Abdel-Hamid, A.M.E.; El-Esawi, M.A.; Ali, H.M.; Elshikh, M.S. Seaweed extracts enhance salam turfgrass performance during prolonged irrigation intervals and saline shock. *Front. Plant Sci.* **2017**, *8*, 830. [CrossRef]
20. Snedecor, G.W.; Cochran, W.G. *Statistical Methods, New York*, 8th ed.; The Iowa State University Press Publishing: Ames, IA, USA, 1989; p. 503.
21. Osakabe, Y.; Osakabe, K.; Shinozaki, K.; Tran, L.P. Response of plants to water stress. *Front. Plant Sci.* **2014**, *5*, 86. [CrossRef]
22. Al-Ghamdi, A.A.; Elansary, H.O. Synergetic effects of 5-aminolevulinic acid and Ascophyllum nodosum seaweed extracts on Asparagus phenolics and stress related genes under saline irrigation. *Plant Physiol. Biochem.* **2018**, *129*, 273–284. [CrossRef]
23. Melo, R.O.d.; Oliveira, H.P.d.; Silveira, K.C.; Baldotto, L.E.B.; Baldotto, M.A. Initial performance of maize in response to humic acids and plant growth-promoting bacteria. *Revista Ceres* **2018**, *65*, 271–277. [CrossRef]
24. Schoebitz, M.; López, M.D.; Serrí, H.; Martínez, O.; Zagal, E. Combined application of microbial consortium and humic substances to improve the growth performance of blueberry seedlings. *J. Soil Sci. Plant Nutr.* **2016**, *16*, 1010–1023. [CrossRef]
25. Abdellatif, I.M.Y.; Abdel-Ati, Y.Y.; Abdel-Mageed, Y.T.; Hassan, M.A.M. Effect of humic acid on growth and productivity of tomato plants under heat stress. *J. Hortic. Res.* **2017**, *25*, 59–66. [CrossRef]
26. Razaq, M.; Zhang, P.; Shen, H.; Salahuddin. Influence of nitrogen and phosphorous on the growth and root morphology of Acer mono. *PLoS ONE* **2017**, *12*. [CrossRef] [PubMed]
27. Wang, M.; Zheng, Q.; Shen, Q.; Guo, S. The critical role of potassium in plant stress response. *Int. J. Mol. Sci.* **2013**, *14*, 7370–7390. [CrossRef]
28. Godlewska, K.; Michalak, I.; Tuhy, Ł; Chojnacka, K. Plant growth biostimulants based on different methods of seaweed extraction with water. *Biomed Res. Int* **2016**, *2016*, 5973760. [CrossRef]
29. Ekinci, M.; Yildirim, E.; Sahin, U. Changes in gas exchange capacity and selected physiological properties of squash seedlings (*Cucurbita pepo* L.) under well-watered and drought stress conditions AU—Ors, Selda. *Arch. Agron. Soil Sci.* **2016**, *62*, 1700–1710.
30. Kusumi, K.; Hirotsuka, S.; Kumamaru, T.; Iba, K. Increased leaf photosynthesis caused by elevated stomatal conductance in a rice mutant deficient in SLAC1, a guard cell anion channel protein. *J. Exp. Bot.* **2012**, *63*, 5635–5644. [CrossRef]
31. Argyrokastritis, I.G.; Papastylianou, P.T.; Alexandris, S. Leaf water potential and crop water stress index variation for full and deficit irrigated cotton in mediterranean conditions. *Agric. Agric. Sci. Procedia* **2015**, *4*, 463–470. [CrossRef]
32. Elansary, H.O.; El-Ansary, D.O. Trinexapac-ethyl application enhanced physiological performance of *Prunus persica* (L.) Batsch under drought conditions. *Bangladesh J. Bot.* **2017**, *46*, 1367–1373.

33. Elansary, H.O. Basil morphological and physiological performance under trinexapac-ethyl foliar sprays and prolonged irrigation intervals. *Acta Physiol. Plant.* **2015**, *37*, 92. [CrossRef]
34. Myburg, A.A.; Grattapaglia, D.; Tuskan, G.A.; Hellsten, U.; Hayes, R.D.; Grimwood, J.; Jenkins, J.; Lindquist, E.; Tice, H.; Bauer, D.; et al. The genome of Eucalyptus grandis. *Nature* **2014**, *510*, 356. [CrossRef] [PubMed]
35. Kim, K.Y.; Seo, H.J.; Min, S.S.; Park, M.; Seol, G.H. The effect of 1,8-cineole inhalation on preoperative anxiety: A randomized clinical trial. *J. Evid. Based Complement. Altern. Med.* **2014**, *2014*, 7. [CrossRef] [PubMed]
36. Abu-Darwish, M.S.; Cabral, C.; Ferreira, I.V.; Gonçalves, M.J.; Cavaleiro, C.; Cruz, M.T.; Al-bdour, T.H.; Salgueiro, L. Essential oil of common sage (*Salvia officinalis* L.) from Jordan: Assessment of safety in mammalian cells and its antifungal and anti-inflammatory potential. *J. Biomed Res. Int.* **2013**, *2013*, 9.

5

Changes in Biochemistry and Yield in Response to Biostimulants applied in Bean (*Phaseolus vulgaris* L.)

Anna Kocira [1,*], Joanna Lamorska [1], Rafał Kornas [1], Natalia Nowosad [1], Marzena Tomaszewska [1], Danuta Leszczyńska [2], Katarzyna Kozłowicz [3] and Sylwester Tabor [4]

1 Institute of Agricultural Sciences, State School of Higher Education in Chełm, Pocztowa 54, 22-100 Chełm, Poland; j.lamorska@gmail.com (J.L.); rkornas@pwsz.chelm.pl (R.K.); nnowosad@pwsz.chelm.pl (N.N.); mtomaszewska@pwsz.chelm.pl (M.T.)

2 Institute of Soil Science and Plant Cultivation, State Research Institute, Czartoryskich 8, 24-100 Puławy, Poland; leszcz@iung.pulawy.pl

3 Department of Biological Bases of Food and Feed Technologies, University of Life Sciences in Lublin, Głęboka 28, 20-612 Lublin, Poland; katarzyna.kozlowicz@up.lublin.pl

4 Department of Production Engineering, Logistics and Applied Computer Science, Agricultural University Kraków, Balicka 116B, 30-149 Kraków, Poland; sylwester.tabor@urk.edu.pl

* Correspondence: akocira@pwsz.chelm.pl

Abstract: Biostimulants are preparations that favorably impact the growth, development, and yield of plants. The research objective was to examine the effect of the frequency of use of Kelpak, Terra Sorb Complex and Fylloton biostimulants on improving the yield and nutritional properties of beans. Been seeds (variety Oczko) were sown in the first week of May in 2015, 2016, and 2017. During the growing season, Fylloton (1%), Terra Sorb Complex (0.5%), and Kelpak (1%) biostimulants were applied by single (BBCH 12-13) and double spraying of plants (BBCH 12-13, BBCH 61). All variants of treatment with biostimulants were compared with the control. Single application of Kelpak had a positive effect on increasing the number of pods. The double application of Kelpak increased the number and yield of seeds and protein contents. Double application of Fylloton increased the number of seeds, and application of Terra Sorb Complex increased the protein content in the beans. Application of all biostimulants increased the flavonoid content. Biostimulants containing seaweed (Kelpak–*Ecklonia maxima* extract) or amino-acid extracts (Fylloton–*Ascophyllum nodosum* extract and amino acids or Terra Sorb Complex–amino acids) increased the seed yield, while improving its quality by increasing the content of protein, polyphenols, and flavonoids. It was found that the double application of Kelpak biostimulant stimulated the yield and quality of beans to a greater extent.

Keywords: bean; biostimulants; amino acids; seaweed extract; yield; protein; phenols; flavonoids

1. Introduction

Agricultural production seeks technological solutions to improve the quality of yields. Therefore, biostimulants are increasingly popular as preparations that favorably impact the growth, development, and yield of plants [1,2], and they are safe for humans and environmentally friendly at the same time. Du Jardin [3] defines biostimulants as "any substance or microorganism applied to plants with the aim of enhancing nutrition efficiency, abiotic stress tolerance, and/or crop quality traits, regardless of its nutrient content". In addition, in order to develop legal provisions regarding the registration of biostimulants based on their specificity of operation, the European Biostimulants Industry Council (EBIC) was created. Currently, however, their registration is based on legal provisions on fertilizers and pesticides, and, for some of them, there is a marketing gap in many European Union (EU) member states [3–6].

Depending on the origin, there are natural or synthetic biostimulants. The former are obtained from biological material, and the latter are structurally similar and functionally identical to biological material [7]. The group of natural biostimulants includes preparations based on free amino acids, humic compounds, seaweed or fruit extracts, chitin and its derivative, chitosan, or microbial inoculants (free-living bacteria, fungi, and arbuscular mycorrhizal fungi) [8–10]. Of this group, biostimulants containing seaweed extract and protein hydrolysates are the most important category of substances that stimulate plant growth and development [11,12].

Biostimulants affect the metabolic processes occurring in the plant, stimulating the synthesis or activity of phytohormones, facilitating the growth of the root system, and improving the uptake, translocation, and utilization of nutrients, which determines the quality of the obtained yield [3,8–12]. Moreover, biostimulants increase plant resistance to abiotic stress factors such as drought, frost, salinity, and environmental contamination with heavy metals, which is probably caused by changes in the enzymatic activity of antioxidant compounds and their increased synthesis [8,13].

In bean cultivation, the most commonly used are extracts of brown algae, e.g., of the species *Ascophyllum nodosum*, *Laminaria* spp., *Ecklonia maxima*, *Sargassum* spp., and *Fucus* spp. [14–17]. The positive effect of seaweed extracts on plant growth, development, and yield is attributed to the presence of phytohormones and low-molecular-weight compounds [18,19]. Some authors suggest that the polysaccharides and polyphenols present in the extract are also important, since they are allelochemicals which increase plant resistance to stress conditions [20–22]. Generally, organic and mineral compounds occur in seaweed extracts, the content of which depends on the algae species, their harvest date, and the applied extraction process [8]. The most important bioactive ingredients include proteins, enzymes, and amino acids (glycoproteins, metalloproteins, exogenous amino acids such as aspartic acid, glutamic acid, alanine), phytohormones (auxins, cytokinins, gibberellins, abscisic acid, polyamines, betaine, ethylene, brassinosteroids, brasinolide, castasterone), polyphenols (florotanins, ecol, floroglucin), phytalexins, vitamins (C, B_2, B_{12}, D_3, E, K, niacin, panthotenic and folic acid), oligosaccharides, polysaccharides (agar, hyaluronic acid, alginic acid and its salts, carrageenans, fucans, mannitol, sorbitol, laminarin), macro- and microelements (Mg, Cu, Fe, Br, Zn, I, Mn), and essential unsaturated fatty acids (arachidonic, eicosapentenoic, γ-linolenic). Although vitamin A is not present in seaweed extracts, the presence of its precursors, i.e., carotene and possibly fucoxanthin was detected [14,23–31].

The positive effect of seaweed extracts on plants is visible in the stimulation of phytohormone synthesis, the uptake and translocation of nutrients, and soil conditioning, which is done by improving water–air conditions and the activity of beneficial soil microorganisms [32,33]. It was demonstrated that, even in low concentrations, seaweed extracts induce a number of physiological processes in the plant, contributing to their better growth, flowering, yield size, and quality, and improving the nutritional and storage quality of crop plants. In addition, the use of seaweed extracts increases the tolerance of plants to unfavorable growing conditions, for example, salinity, drought, or extreme temperatures [12,34,35]. The use of seaweed extract has a positive effect on plant growth and the size and quality of the obtained yield of tomato, eggplant, pepper, lettuce, beans, soybean, and wheat [13,25,36–46]. Undoubtedly, the beneficial effects of biostimulants on plants, under both optimal and stressful growing conditions, can be associated with stimulation of enzymatic activity related to carbon and nitrogen metabolism, the Krebs cycle, and glycolysis. Treatment of plants with these preparations may induce activity similar to that of phytohormones (auxin, gibberellin), which in effect improves their nutrition by modifying the structure of the root system [8,9,47].

The biostimulant Kelpak (Kelp Products Ltd.) is based on an extract from *Ecklonia maxima* (Osbeck) Papenfuss and contains auxins (11 $mg \cdot dm^{-3}$), cytokinins (0.031 $mg \cdot dm^{-3}$), alginates (1.5 $g \cdot L^{-1}$), amino acids (total 441.3 $mg \cdot 100\ g^{-1}$), mannitol (2261 $mg \cdot L^{-1}$), neutral sugars (1.08 $g \cdot L^{-1}$), and small amounts of macroelements (N 0.09%, P 90.7 $mg \cdot kg^{-1}$, K 7163.3 $mg \cdot kg^{-1}$, Ca 190.4 $mg \cdot kg^{-1}$, Mg 337.2 $mg \cdot kg^{-1}$, Na 1623.7 $mg \cdot kg^{-1}$) and microelements (mean composition: Mn 17.3 $mg \cdot kg^{-1}$, Fe 40.7 $mg \cdot kg^{-1}$, Cu 13.5 $mg \cdot kg^{-1}$, Zn 17.0 $mg \cdot kg^{-1}$, B 33.0 $mg \cdot kg^{-1}$) [14,48]. The very high auxin-to-cytokinin ratio is responsible for stimulating the growth and development of the root system, which in turn contributes to better uptake and translocation of macro- and microelements, and it is associated with a significant increase in crops [48].

Biostimulants based on protein products and protein hydrolysates consist of a mixture of peptides, animal or vegetal amino acids, and single amino acids. Amino acids are a building material for proteins, but they are also precursors of phytohormones. They are involved in the synthesis of, e.g., vitamins, enzymes, terpenes, amines, purines, pyrimidines, and alkaloids [49,50]. They also play an important role in the process of pollination and fruit formation [51]. The application of exogenous amino acids, which are active in metabolic signaling (glutamate, histidine, proline, glycine, betaine), induces plant defense mechanisms by increasing their resistance to abiotic stress factors [8,52]. Due to the presence of specific peptides and precursors of phytohormone biosynthesis (tryptophan, which is the main precursor of IAA (Indole-3-acetic acid) biosynthesis and bioactive peptides), protein hydrolysates affect the hormonal balance of plants, which is related to stimulating plant growth [53]. The use of these biopreparations positively impacts the quality of agricultural produce, increasing the content of carotenoids, flavonoids, polyphenols, and ascorbic acid [54–57], and reducing the amount of undesirable compounds, e.g., nitrates [56].

Terra Sorb Complex (Bioiberica, S.A.U.) is a biostimulant containing 20% free vegetal amino acids and 5.5% total N (including 5% organic N), 0.8% MgO, 1.5% B, 1% Fe, 0.1% Mn, 0.001% Mo, 0.1% Zn, and 25% organic matter.

The beneficial effect of biostimulants with seaweed extracts or amino acids on plant growth and development, and on the quantity and quality of the yield, regardless of its developmental stage, was confirmed in numerous studies [13,34,35,39–41,57–62]. An interesting solution is to combine these two components in a single preparation, as in the case of the Fylloton biostimulant (Biolchim Poland), which contains the extract of *Ascophyllum nodosum* (L.) Le Jolis, as well as vegetal amino acids. The composition of this preparation includes *Ascophyllum nodosum* extract, amino acid complexes of vegetal origin 37.5%, organic nitrogen 6%, organic carbon of biological origin 11%, and organic substance 35%.

Bean is an economically important legume that is sensitive to low temperatures in the early stages of its development and flowering. The use of biostimulants that positively affect the metabolic processes occurring in the plant, especially in the time of climate change, which causes stress factors for this sensitive plant, can be one of the elements contributing to the improvement in the quantity and quality of bean yield. Plant response to the biostimulant often depends on the variety, as demonstrated in earlier studies [15,39,59,60]. There are also no reports regarding the reaction of two-colored coat seed of bean to treatment with biostimulants. In view of the above and based on the importance being given to the improving crop yields, the research objective was to investigate the effect of the use of Kelpak, Terra Sorb Complex, and Fylloton biostimulants on improving the yield and nutritional properties of common bean (*Phaseolus vulgaris* L.) variety Oczko.

2. Materials and Methods

2.1. Plant Materials and Growth Conditions

The research material came from field studies carried out in the year 2015–2017 in Perespa (50°66′ north (N); 23°63′ east (E)), Poland, on common beans (*Phaseolus vulgaris* L.), variety Oczko. The experiment was established in a random block system, in four replications, on an area of 10 m^2. The experiment was established on an alkaline (pH in 1 M KCl-7.4) soil of the brown rendzina subtype. Soil fertility level was as follows: phosphorus medium (12.6–14.2 mg P_2O_5 in 100 g of soil), potassium medium (15.3–17.1 mg K_2O in 100 g of soil), and magnesium medium (6.2–6.8 mg Mg in 100 g of soil). In each year of research, the forecrop for common beans was winter wheat. Tillage for bean was carried out in accordance with good agricultural practice [63]. Pre-winter plowing was performed in the first week of November. In the spring, soil treatment combined with mineral fertilization was performed. Mineral fertilizers were in the following doses: 30 kg N·ha^{-1}, 60 kg P_2O_5·ha^{-1}, and 120 kg K_2O·ha^{-1}.

Mineral fertilization was used at a constant level throughout all experimental combinations. Beans of common bean, variety Oczko (with a red and white bean coat), were sown with a mechanical precision bean drill in the first week of May (2 May in 2015, 2016, and 2017) at a depth of 3–4 cm, in rows 45 cm apart, using 30 plants per 1 m^2. Biostimulants Terra Sorb Complex, Kelpak, and Fylloton were used during the growing season, according to the experiment design (Table 1), and the obtained results were compared with the control, in which pure water was used for double spraying the plants. In individual years of research, biostimulants were used in time frames dependent on the development phase of plants, as shown in Table 1. Plants single sprayed with the biostimulant in BBCH 12-13, in the second period (plant stage of BBCH 61), were sprayed with pure water.

Table 1. Overview of biostimulant application in bean variety Oczko cultivation.

Biostimulant	Concentration	Number of Sprays and Plant Developmental Stages (BBCH)	Volume of Working Solution/ Working Pressure	Date		
				2015	2016	2017
Fylloton (F)	1%	Single spraying: BBCH 12–13 (SS)	300 L·ha^{-1}/0.30 MPa	June 5	June 7	June 9
		Double spraying: BBCH 12–13, BBCH 61 (DS)		June 5, June 20	June 7, June 23	June 9, June 26
Terra Sorb Complex (TS)	0.5%	Single spraying: BBCH 12–13 (SS)	300 L·ha^{-1}/0.30 MPa	June 5	June 7	June 9
		Double spraying: BBCH 12–13, BBCH 61 (DS)		June 5, June 20	June 7, June 23	June 9, June 26
Kelpak (K)	1%	Single spraying: BBCH 12–13 (SS)	300 L·ha^{-1}/0.30 MPa	June 5	June 7	June 9
		Double spraying: BBCH 12–13, BBCH 61 (DS)		June 5, June 20	June 7, June 23	June 9, June 26

Abbreviations: single spraying BBCH 12-13—single spraying at the 2–3-leaf stage; double spraying BBCH 12-13, BBCH 61—double spraying first at the 2–3-leaf stage and second at the beginning of bean blooming.

The plants were sprayed using a GARLAND FUM 12B backpack sprayer. The Lechler LU 120–03 atomizer was used, at a working pressure of 0.30 MPa, using 300 L of the working liquid per 1 ha. All variants of treatment with biostimulants were compared with the control, where plants were treated with the same volume of water (no biostimulant was applied). No pesticides were used in the cultivation, as pathogens, pests and weeds did not exceed the damage threshold. Plants were weeded manually. The average temperature and rainfall during the bean growing season are shown in Table 2. The weather station (W200P, Vector Instruments Ltd., Rhyl, UK) was located in the experimental field, in which the experiment was carried out, at 210 m above sea level.

After harvesting plants in the third week of August (22 August 2015; 27 August 2016; 24 August 2017), 20 plants were randomly selected from each plot, and the number of pods, number of seeds, seed yield, and weight of one thousand seeds was determined. The beans obtained from each plot were dried, ground in a laboratory mill, and sieved with a 0.310-mm sieve. The flours were stored at −20 °C and used for further chemical analysis.

Table 2. Conditions during the growing seasons in bean variety Oczko cultivation in 2015–2017.

Month	Year						Average from 2002–2015	
	2015		2016		2017			
	T (°C) Average (min/max)	Rainfall (mm)	T (°C) Average (min/max)	Rainfall (mm)	T (°C) Average (min/max)	Rainfall (mm)	T (°C)	Rainfall (mm)
IV	8.2 (−1.7/24.3)	30.1	9.2 (−1.2/22.6)	68.4	7.7 (−1.6/23.3)	37.2	8.6	41.9
V	12.7 (1.5/24.9)	108.6	13.8 (2.6/26.7)	61.3	13.7 (−1.4/26.9)	100.0	12.6	64.1
VI	17.4 (6.6/30.5)	14.1	18.1 (4.2/31.5)	97.1	18.3 (5.7/30.2)	38.6	17.8	68.3
VII	19.6 (8.4/33.4)	59.2	19.5 (8.8/31.2)	107.6	18.5 (5.3/32.9)	61.1	18.8	79.4
VIII	21.6 (5.6/35.5)	23.4	18.2 (7.1/30.7)	95.3	19.5 (4.3/34.4)	25.5	19.5	71.5
IX	15.1 (4.2/34.5)	137.6	15.2 (1.6/28.7)	41.2	13.2 (−0.3/27.3)	100.4	14.0	69.6
Average/Total	15.8	373.0	17.1	470.9	15.2	362.8	15.2	394.8

Abbreviation: T—temperature.

2.2. Determination of Polyphenols

A ground sample of bean seeds of 0.25 g was weighed, to which 4 cm^3 of extraction solution (acetone:water:hydrochloric acid 70:29:1) was added. The solutions were shaken for 1 h. Then, 100 μL of distilled water and 0.4 mL of Folina–Ciocalteu reagent were added to 100 μL of extract, and, after 10 min, 2 mL of 10% Na_2O_3 solution was added. After 30 min, the absorbance of methanol was measured at $\lambda = 725$ nm. The polyphenol content was calculated in mg·100 g^{-1}, from the gallic acid calibration curve (1 mg·mL^{-1}).

2.3. Determination of Flavonoids

A ground sample of bean seeds of 0.25 g was weighed, to which 4 cm^3 of extraction solution (acetic acid:methanol 1:19) was added. The solutions were shaken for 1 h. Then, 0.1 mL of a 2% $AlCl_3 \cdot 6H_2O$ methanolic solution was added to 1 mL of the extract, together with 1.4 mL of CH_3COOH methanolic solution (1:19). The sample was then incubated at 20 °C for 30 min. Absorbance was measured at $\lambda = 425$ nm against methanol. The flavonoid content was calculated in mg·100 g^{-1}, from the calibration curve for quercetin (0.2 mg·mL^{-1}).

2.4. Determination of Proteins

The protein content of bean extracts was determined using the Bradford reagent, according to the method of Redmile-Gordon et al. [57] with modifications. The Bradford reagent (150 μL) was applied on a microplate and 50 μL of assay or standard protein (BSA, bovine serum albumin) was added. The samples were shaken at room temperature for 15 min. Absorbance at 595 nm was measured using an Epoch Microplate Spectrophotometer (BioTek-USA). The resulting protein was expressed in mg·g^{-1} of dry weight (DW).

2.5. Statistical Analysis

The statistical analysis was performed using the Statistica 10PL program by StatSoft®. The normal distribution of variables was tested using the Shapiro–Wilk test. The one-way (for 2015, 2016, and 2017) and the two-way (for average 2015–2017) analysis of variance was used. The significance of the mean was determined using the Tukey test, at a significance level of $p < 0.05$.

3. Results

Based on the two-way ANOVA analysis, the effect of the number of applications and biostimulant treatment on seed yield, number of seeds, and phenol content in bean seeds was found (Table 3). The effect of biostimulant treatment on the number of pods, the weight of one thousand seeds, and the content of proteins and flavonoids in bean seeds were demonstrated. The interaction of biostimulant treatment with its number of applications regarding the impact on the number of seeds, as well as the content of protein and phenols in bean seeds, was found.

Table 3. Two-way ANOVA of number of applications and treatment of bean variety Oczko (average 2015–2017).

Effect	Sum of Squares	Degrees of Freedom	Mean Squares	F Ratio	*p*-Values
Seed yield					
Intercept	5,439,549	1	5,439,549	23,099.18	0.000000
Number of applications	3181	1	3181	13.51	0.000697
Treatment	34,486	3	11,495	48.82	0.000000
Number of applications × treatment	1510	3	503	2.14	0.110568
Error	9419	40	235		
1000 seed weight					
Intercept	10,599,621	1	10,599,621	44,665.77	0.000000
Number of applications	314	1	314	1.32	0.257138
Treatment	11,114	3	3705	15.61	0.000001
Number of applications × treatment	281	3	94	0.40	0.757166
Error	9492	40	237		
Number of pods					
Intercept	2,570,576	1	2,570,576	10,667.40	0.000000
Number of applications	290	1	290	1.20	0.279127
Treatment	15,025	3	5008	20.78	0.000000
Number of applications × treatment	1506	3	502	2.08	0.117750
Error	9639	40	241		
Number of seeds					
Intercept	24,743,716	1	24,743,716	33,040.26	0.000000
Number of applications	22,838	1	22,838	30.50	0.000002
Treatment	209,494	3	69,831	93.25	0.000000
Number of applications × treatment	12,377	3	4126	5.51	0.002905
Error	29,956	40	749		
Protein					
Intercept	20,792.66	1	20,792.66	88,921.99	0.000000
Number of applications	0.39	1	0.39	1.65	0.206357
Treatment	31.28	3	10.43	44.59	0.000000
Number of applications × treatment	2.36	3	0.79	3.37	0.027788
Error	9.35	40	0.23		
Total phenols					
Intercept	731.7033	1	731.7033	3095.779	0.000000
Number of applications	1.3763	1	1.3763	5.823	0.020490
Treatment	6.8737	3	2.2912	9.694	0.000062
Number of applications × treatment	2.5409	3	0.8470	3.583	0.021918
Error	9.4542	40	0.2364		
Total flavonoids					
Intercept	0.159506	1	0.159506	7312.597	0.000000
Number of applications	0.000001	1	0.000001	0.024	0.877973
Treatment	0.006708	3	0.002236	102.506	0.000000
Number of applications × treatment	0.000174	3	0.000058	2.663	0.060961
Error	0.000873	40	0.000022		

Number of applications (1 or 2); treatment (Fylloton; Terra Sorb Complex, Kelpak, control).

Treating plants with the biostimulants had a positive effect on increasing the number of pods (Table 4). In 2015, no significant effect of the biostimulants on the studied trait was identified; however, a tendency to increase the number of pods was observed after foliar application of biostimulants, especially Kelpak. On the other hand, in 2016, after treating the plants with biostimulants based on seaweed extract, i.e., both after a single application of Kelpak (*Ecklonia maxima* extract) and after a single or double application of Fylloton (*Ascophyllum nodosum* extract, amino acids), a significant increase in this trait was identified (increases by 36%, 28%, and 33%, respectively, as compared to the control). Treatment of plants with Kelpak in 2017 significantly increased the number of pods, regardless of the number of applications (increase by 18% for a single application, and by 19% for a double application, as compared to the control). A synthesis of the three years of research (2015–2017) confirmed that a single spraying of plants with the Kelpak biostimulant in the BBCH 12-13 phase significantly increased the number of pods (by 26%), as compared to the control.

Table 4. Effect of Fylloton, Terra Sorb Complex, and Kelpak biostimulants treatment on number of pods and seeds of bean variety Oczko.

Parameters	Biostimulant Treatment	Season			Average 2015–2017
		2015	2016	2017	
Number of pods (per m^2)	F_1	224 ± 7.8 n.s.	240 ± 8.5 a	231 ± 11.3 ab	232 ± 4.0 b
	F_2	234 ± 9.9 n.s.	249 ± 8.5 a	245 ± 4.9 ab	243 ± 7.8 ab
	TS_1	223 ± 29.7 n.s.	207± 7.8 bc	249 ± 7.8 ab	226 ± 4.7 b
	TS_2	248 ± 12.0 n.s.	234 ± 9.9 ab	254 ± 9.8 ab	245 ± 2.6 ab
	K_1	257 ± 10.6 n.s.	254 ± 7.8 a	258 ± 12.0 a	256 ± 3.1 a
	K_2	252 ±18.4 n.s.	225 ± 6.4 ab	260 ± 6.3 a	245 ± 6.1 ab
	C	203 ±17.7 n.s.	187 ± 9.2 c	219 ± 9.2 b	203 ± 5.9 c
Number of seeds (per m^2)	F_1	776 ± 9.2 a	649 ± 9.2 d	686 ± 24.0 bc	703 ± 1.9 c
	F_2	809 ±17.0 a	774 ± 17.7 b	799 ± 23.3 a	794 ± 3.8 a
	TS_1	713 ±19.1 b	720 ± 4.9 c	693 ± 21.9 bc	708 ± 2.6 c
	TS_2	762 ±6.4 ab	752 ± 6.4 bc	751 ± 21.8 ab	755 ± 3.1 b
	K_1	764 ± 22.6 ab	777 ± 9.9 b	754 ± 20.5 ab	765 ± 2.6 b
	K_2	793 ± 15.6 a	835 ± 5.7 a	780 ± 25.5 a	803 ± 5.2 a
	C	602 ± 14.8 c	611 ± 19.1 d	613 ± 5.7 c	608 ± 9.4 d

Abbreviations: F_1, single spraying of Fylloton; F_2, double spraying of Fylloton; TS_1, single spraying of Terra Sorb Complex; TS_2, double spraying of Terra Sorb Complex; K_1, single spraying of Kelpak; K_2, double spraying of Kelpak; C, control; n.s., not significant. Means in the columns, concerning the selected traits, followed by different small letters are significantly different at $p < 0.05$.

Analysis of variance showed that, regardless of the number of applications, the treatment of plants with Fylloton significantly increased the number of seeds in 2015, as did the double spraying with Kelpak (increases by 29%, 34%, and 32% respectively, as compared to the control) (Table 4). In the second year of research, the best results were obtained after double spraying with biostimulant based on the extract of *Ecklonia maxima*, with a significant increase of 37%, as compared to the control. In 2017, it was found that spraying plants with biostimulants containing seaweed extracts, i.e., Fylloton and Kelpak, significantly increased the number of seeds, by 30% and 27%, respectively, as compared to the control. The average of three years of research demonstrated that an increase of this trait, by 31% and 32%, respectively, as compared to the control, was obtained after a double foliar application of Fylloton and Kelpak.

Double application of Kelpak, in 2015 and 2016, had the most beneficial effect on seed yield, increasing this trait by 22% and 38%, respectively, as compared to the control (Figure 1). In 2017, the best effects in increasing the bean crop were obtained after double treatment of plants with Fylloton and Kelpak biostimulants, when a 25% increase of this trait was obtained, as compared to the control. A synthesis of the three years of research demonstrated that double spraying plants with Kelpak biostimulant was most beneficial for increasing the bean yield (increase by 28% as compared to the control).

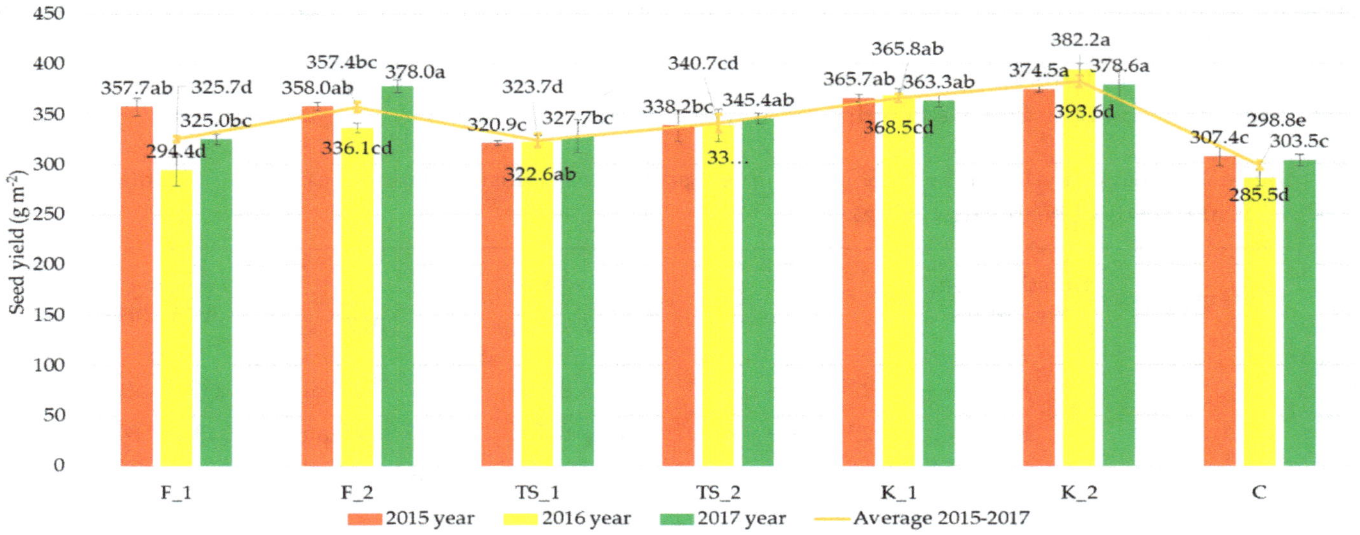

Figure 1. Effect of Fylloton, Terra Sorb Complex, and Kelpak biostimulant treatment on seed yield of bean variety Oczko. Abbreviations: F_1, single spraying of Fylloton; F_2, double spraying of Fylloton; TS_1, single spraying of Terra Sorb Complex; TS_2, double spraying of Terra Sorb Complex; K_1, single spraying of Kelpak; K_2, double spraying of Kelpak; C, control. Means over the study years followed by different small letters are significantly different at $p < 0.05$.

Analysis of variance showed that the use of biostimulants in bean cultivation resulted in a reduction of the weight of one thousand seeds (Figure 2). In the first year of research, a significant increase in the weight of one thousand seeds was obtained in the control, by 11%–15%, as compared to the combination with Fylloton or Terra Sorb Complex. However, in 2016 there were no significant differences in the weight of one thousand seeds between the combinations that included the biostimulants and the control. Plants in the control plot in 2017 were characterized by a higher weight of one thousand seeds by 8%, as compared to the double application of the Terra Sorb Complex. In turn, the average of three years of research confirmed that the highest weight of one thousand seeds (an increase of 9%, as compared to the double use of Fylloton) was obtained in the control.

Foliar application of biostimulants increased the protein content in the seeds (Table 5). In 2015 and 2017, the best effects were obtained after applying Terra Sorb Complex as a single or double spraying of plants, which increased this trait by 12% and 13% (in 2015), and by 9% and 10% (in 2017), respectively, as compared to the control. Double treatment of plants with the Terra Sorb Complex biostimulant in 2016 had the most beneficial effect on increasing the protein content by 13%, as compared to the control. A synthesis of the three years of research showed that, regardless of the number of applications, the use of an amino acid-based biostimulant significantly increased the protein content, as did the double application of the biostimulant containing the of the *Ecklonia maxima* extract (11%, 12%, and 10% increases, respectively, as compared to the control).

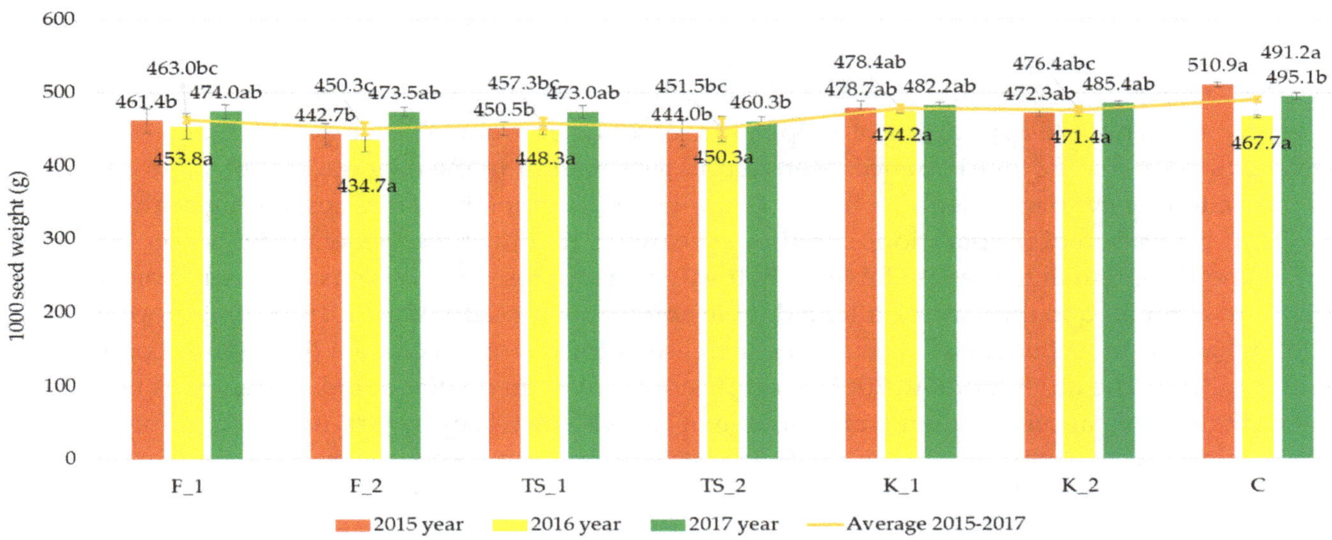

Figure 2. Effect of Fylloton, Terra Sorb Complex, and Kelpak biostimulant treatment on the weight of one thousand seeds of bean variety Oczko. Abbreviations: F_1, single spraying of Fylloton; F_2, double spraying of Fylloton; TS_1, single spraying of Terra Sorb Complex; TS_2, double spraying of Terra Sorb Complex; K_1, single spraying of Kelpak; K_2, double spraying of Kelpak; C, control. Means over the study years followed by different small letters are significantly different at $p < 0.05$.

Table 5. Effect of Fylloton, Terra Sorb Complex, and Kelpak biostimulant treatment on protein, phenol, and flavonoid content of bean seeds.

Parameters	Biostimulant Treatment	Season			Average 2015–2017
		2015	2016	2017	
Protein (% DM)	F_1	21.19 ± 0.23 bc	20.59 ± 0.29 bc	21.30 ± 0.14 ab	21.03 ± 0.38 a
	F_2	20.82 ± 0.10 c	20.19 ± 0.14 c	21.01 ± 0.10 ab	20.67 ± 0.60 ab
	TS_1	21.81 ± 0.02 a	21.20 ± 0.05 ab	22.04 ± 0.02 a	21.68 ± 0.43 a
	TS_2	21.99 ± 0.04 a	21.38 ± 0.10 a	22.27 ± 0.18 a	21.88 ± 0.46 a
	K_1	20.74 ± 0.27 c	20.05 ± 0.27 c	21.14 ± 0.43 ab	20.64 ± 0.55 ab
	K_2	21.79 ± 0.08 ab	21.15 ± 0.16 ab	21.61 ± 0.59 ab	21.52 ± 0.33 a
	C	19.53 ± 0.15 d	18.94 ± 0.17 d	20.18 ± 0.58 b	19.55 ± 0.62 b
Total phenols ($mg \cdot g^{-1}$ DM)	F_1	3.76 ± 0.0 b	4.02 ± 0.02 c	4.13 ± 0.03 c	3.97 ± 0.19 n.s.
	F_2	5.96 ± 0.05 a	3.56 ± 0.01 d	4.11 ± 0.02 c	4.54 ± 1.29 n.s.
	TS_1	4.01 ±0.01 b	4.43 ± 0.01 b	4.36 ± 0.03 b	4.26 ± 0.22 n.s.
	TS_2	4.01 ± 0.03 b	4.48 ± 0.12 ab	3.74 ± 0.06 d	4.07 ± 0.37 n.s.
	K_1	2.82 ± 0.22 d	3.89 ± 0.09 c	3.54 ±0.08 e	3.42 ± 0.54 n.s.
	K_2	3.83 ± 0.01 b	4.70 ± 0.04 a	4.63 ± 0.04 a	4.39 ± 0.48 n.s.
	C	3.27 ± 0.08 c	3.30 ± 0.05 e	3.29 ± 0.01 f	3.29 ± 0.02 n.s.
Total flavonoids ($mg \cdot g^{-1}$ DM)	F_1	0.061± 0.001 bc	0.075 ± 0.001 a	0.078 ± 0.001 a	0.071 ± 0.009 a
	F_2	0.058 ± 0.001 c	0.066 ± 0.001 ab	0.068 ± 0.001 b	0.064 ± 0.005 a
	TS_1	0.059 ± 0.001 c	0.060 ± 0.008 b	0.069 ± 0.001 b	0.063 ± 0.006 a
	TS_2	0.063 ± 0.001 ab	0.061 ± 0.001 ab	0.065 ± 0.001 b	0.063 ± 0.002 a
	K_1	0.064 ± 0.002 ab	0.062 ± 0.003 ab	0.066 ± 0.001 b	0.064 ± 0.002 a
	K_2	0.067 ± 0.001 a	0.073 ± 0.001 ab	0.075 ± 0.001 a	0.072 ± 0.004 a
	C	0.036 ± 0.001 d	0.038 ± 0.001 c	0.041 ± 0.001 c	0.038 ± 0.003 b

Abbreviation: F_1, single spraying of Fylloton; F_2, double spraying of Fylloton; TS_1, single spraying of Terra Sorb Complex; TS_2, double spraying of Terra Sorb Complex; K_1, single spraying of Kelpak; K_2, double spraying of Kelpak; C, control; n.s., not significant; DM, dry matter. Means in the columns, concerning the selected traits, followed by different small letters are significantly different at $p < 0.05$.

Double application of biostimulants containing seaweed extracts most favorably impacted the content of phenols in the seeds (Table 5). In 2015, an increase in the phenolic compound content in beans was noted after a double application of Fylloton (an increase of this trait by 82%, as compared to the control). However, after a single application of Kelpak, this trait was found to be reduced, as compared to the control. In turn, in 2016 and 2017, the best effects in increasing the phenol content were observed after a double application of Kelpak (increases by 42% and 41%, respectively, as compared to the control). No significant differences were found for this trait in the average for the years 2015–2017, but only a tendency to increase after a double application of biostimulants containing seaweed.

The application of preparations containing seaweed had the most beneficial effect on increasing the flavonoid content in the seeds (Table 5). In the first year of research, this characteristic was increased by 87%, as compared to the control, after a double application of Kelpak. In 2016, a single application of Fylloton had the most beneficial effect on the increase of the flavonoid content (by 98%, as compared to the control). On the other hand, in 2017, a significant increase in the flavonoid content was noted both with a single application of Fylloton and double of Kelpak (increases by 90% and 82%, respectively, as compared to the control). The average of three years of research showed that the application of all biostimulants significantly increased the flavonoid content, by 64%–88%, as compared to the control.

4. Discussion

The results of our research show a positive effect of natural biostimulants on bean yield, as well as on its quality. A more beneficial effect in modifying the yield components (number of pods and seeds, seed yield) was obtained after using biostimulants based on seaweed extract, especially upon double application of the *Ecklonia maxima* (Kelpak) extract. Only in the case of the weight of one thousand seeds was a reduction of this trait observed as a result of using biostimulants. The use of other biostimulants, i.e., Fylloton, which contains an *Ascophyllum nodosum* extract and amino acids, as well as Terra Sorb Complex containing amino acids, also had a positive effect on the yield components. In turn, the conditions favorable for setting pods and seeds, and increasing the weight of seeds were the most beneficial in 2016, when the double use of Kelpak resulted in the largest increase in these characteristics compared to control. This year, there were also favorable temperature and humidity conditions conducive to flowering, and setting pods and seeds. The significant increase in protein content was most positively influenced by weather conditions in 2015 and 2016, when the highest value of this feature was obtained after double application of Terra Sorb Complex. The application of biostimulants based on marine algae had a positive effect on the flavonoid content in all years of research. In 2015, however, a significant increase in polyphenols was found after double application of Fylloton.

Numerous studies conducted on arable crops confirmed the beneficial effect of seaweed extracts on increasing yield components [11,12,47,64–66]. Previous research on other bean and soybean varieties confirmed the stimulating effect of the *Ecklonia maxima* or *Ascophyllum nodosum* extracts on the number of pods and seeds, and the weight of beans [39–41,67]. The lack of effect of Kelpak on the one-thousand-seed weight of bean was also confirmed by previous studies on common bean [66]. Use of extracts from *Kappaphycus alvarezii* and *A. nodosum* increased the number of pods, seeds, and yield in soybean [68,69], even under stress conditions (reduced NPK fertilization) [70]. Bean plants reacted favorably to the foliar application of *Caulerpa racemosa* and *A. nodosum* extract by increasing the number of pods and seeds, one thousand weight of bean, and seed yield in common beans [71,72], mung beans [73], and broad beans [74].

The foliar application of amino acids positively affects the yield of many plants, even growing under stress [11,75,76]. In previous studies, the studied bean responded positively to foliar application of the Terra Sorb Complex; however, the yielding effect depended on the variety, concentration, and number of applications of the biostimulant, as well as climatic conditions prevailing in a given study year [59]. The Aura variety (with white seeds) responded more favorably to a single application of a 0.5% concentration, and the Toska variety (with red seeds) responded more favorably to a single application of a 0.3% concentration of this biostimulant. The plants increased the number of pods and seeds, as well as the seed yield; however, no effect of this biostimulant was found for the weight of one thousand seeds. Foliar application of biostimulants containing amino acids increased the number of pods and seeds, weight of one thousand seeds, and yield in seeds of beans [71,77], peas [78], and broad beans [79,80].

The positive effect of seaweed extracts on plant growth and development and, as a result, on the increase in yield is undoubtedly associated with the presence of phytohormones, especially cytokinins [81]. Together with auxins, cytokinins regulate many physiological processes, including those affecting plant growth and development [82–84]. Aremu et al. [85] and Masondo et al. [86] also observed a positive effect of Kelpak on increasing the content of cytokinins. In addition, Kulkarni et al. [87] found an increase in the content of *cis*-zeatin, dihydrozeatin, and isopentenyladenine after using the *Ecklonia maxima* extract (Kelpak). The many active substances and compounds included in Kelpak suggest that it is not only cytokinin that is responsible for the growth and development of plants, but also probably the cross-reactions of these compounds with other bioactive molecules included in biostimulants that are based on seaweed extracts [70].

Thanks to the content of endogenous auxins, seaweed extracts have a positive effect on the growth and development of the root system [88]. This improves the uptake of water and nutrients and, in effect, stimulates the growth and development of plants, contributing to the improvement of yield quantity and quality [81]. The application of biostimulants based on seaweed extracts also has a positive effect on plant growth and development due to the content of gibberellins (GA1, GA3, GA4, GA5, GA6, GA7, GA13) [14], which affect seed germination, stem elongation, leaf expansion, and flower and seed development [89–91], as well as of gibberellin-like substances, e.g., terpenoids and tocopherol [92,93]. In the seaweed extract (Kelpak), the presence of brassinosteroids, brassinolide and castasterone [10], was identified. As phytohormones, brassinosteroids promote cell division and elongation, stimulate stem and root growth, and initiate flowering and flower development, as well as fruit development and increases in seed yield. Under stress conditions, they protect plants against abiotic and biotic stress [94,95].

The positive effect of amino acid-based preparations on plant growth and development probably results from the fact that, at the molecular level, they stimulate the plant's defense response to biotic and abiotic stress factors [96]. The amino acids contained in them are easily absorbed by plants. They participate in the synthesis of a number of organic compounds and affect the uptake of macro- and microelements [97]. Garcia et al. [98] showed that foliar application of amino acids and peptides together with nutrients increases the content of potassium, calcium, magnesium, iron, copper, and zinc in leaves, affecting their nutritional condition and promoting improved growth and development of plants. Applied on a leaf, preparations of this type exhibit phytohormone-like effects, comparable to that of auxin and gibberellin [54]. They also contribute to increasing the content of phytohormones (gibberellins, cytokinins, auxins) [76]. In addition, the use of protein hydrolysates in plant cultivation has a beneficial effect on the uptake of water and nutrients, resulting in increasing crop yielding thanks to the increased microbial and enzymatic activity of the soil, improved mobility and solubility of microelements (iron, zinc, manganese, copper), modified structure of the root system (its length, compaction, and number of lateral roots), or the increased synthesis of nitrate and glutamine reductase, as well as the activity of iron reductase [11,54,99–102]. Numerous reports confirmed that protein hydrolysates, such as auxins and gibberellins, have hormone-like effects, stimulating root and shoot growth. This, in turn, has a positive effect on crop productivity [11,53,54,100,102–106].

In our research, the use of biostimulants based on amino acids and seaweed extracts had a positive effect on the nutritional value of beans by increasing their protein content. Application of seaweed extract and amino acids had a positive effect on increasing the protein content in bean, pea, and faba bean seeds [59,71,74,78,80]. This was confirmed by Rouphael [62]; in their research, the increase in protein content in spinach plants was obtained after using biostimulants containing an extract of *Ecklonia maxima* and *Ascophyllum nodosum* and legume-derived protein hydrolysate. The protein content of legumes leaves was also determined by seaweed extracts. Numerous authors found an increase in this trait after the application of *Ulva rigida, Fucus spiralis, Hypnea musciformis,* and

Colpomenia sinuosa extracts in bean leaves [16,107,108]. However, the use of biostimulants is not always beneficial for the protein content in beans [15,109]. Schubert and Mengel [110] demonstrated that amino-acid uptake is an important mechanism for recovering carbon and nitrogen that was lost in the rhizosphere. Ertani et al. [100] found that the stimulation of nitrogen assimilation is responsible for accelerating the growth and metabolism of nitrogen in plants treated with protein hydrolysates. This is due to an increase in the activity of two key enzymes, nitrate reductase and glutamine synthetase, thereby contributing to increasing protein firmness.

After the application of biostimulants containing seaweed extracts, researchers observed an increase in the phenolic content. Ertani [56] showed that the treatment of plants with biostimulants stimulates numerous metabolic pathways in plants. The pathways are also associated with the synthesis of secondary metabolites, including phenolic compounds, which play an important role in protecting plants against stress factors. In turn, a frequent indicator of plant resistance to biotic factors is the content of phenolic compounds, which are precursors to more complex phenolic structures, such as flavonoids and lignins [60,111]. The presence of bioactive compounds in biostimulants, including phytohormones, amino acids, protein, and phenols, is responsible for the physiological response of plants treated with these preparations [56,112,113]. Eckol (phenolic compound) found in seaweed extracts affects the phenylpropanoid pathway in the biochemical synthesis of phenolic acids [114]. Aremu et al. [85,115] showed that the timing of eckol plants increases the content of phenolic compounds, such as *p*-hydroxybenzoic and ferulic acids. In turn, the Kelpak application increases the content of caffeic acid, ferulic acid [85,116], protocatechuic acid, *p*-hydroxybenzoic acid, gentisic acid, *p*-coumaric acid, and *trans*-cinnamic acid in *Eucomis autumnalis* [85]; however, the content of phenolic compounds depended on the biostimulant concentration. In turn, Rouphael et al. [62] found an increased content of phenolic compounds after the treatment of plants with biostimulants containing alginians, fucoidans, and laminarins that affect endogenic hormonal homeostasis [117]. Treatment of plants with eckol also had a positive effect on the flavonoid content, increasing the amount of kempferol in plants several times. However, foliar application of Kelpak increased the content of kaempferol in tubers and entire plants, as well as taxifolin in leaves of *Eucomis autumnalis* [85,115]. Increasing the content of bioactive compounds in plants treated with biostimulants is associated with a mechanism that includes the stimulation of the chalcone isomerase enzyme, involved in the biosynthesis of flavanone precursors [117].

Paul et al. [118] found that tomato plants treated with protein hydrolysates were characterized by a higher content of, e.g., low-molecular-weight phenolic compounds, phytohormones (polyamines), hydroxy-carotenoids, poly-hydroxy fatty acids, and membrane lipids (glycoand phospholipids). They suggest that the metabolic changes caused by treating plants with protein hydrolysates can be correlated with a relatively small number of processes that converge toward the ROS-related (reactive oxygen species-related) plant signaling network. Increasing the content of secondary metabolites, such as phenols and carotenoids, which play a key role in protecting plants against oxidative stress [62,102,119], suggests fine-tuning of ROS signaling in plants after the application of protein hydrolysates [118]. In addition, the use of animal protein hydrolysates had a positive effect on the content of protein, phenols, and flavonoids in bananas [120], and vegetal protein hydrolysates stimulated an increase in the content of phenolic compounds and anthocyanins in grape [55].

So far, there are few reports confirming the beneficial effect of a preparation consisting of a combination of seaweed extract and amino acids. In our research, combining the Fylloton biostimulant, which induces the extract's effect, with *Ascophyllum nodosum* and amino acids increased the yield components, particularly the number of seeds after its double application. Moreover, previous studies, conducted on three soybean varieties, confirmed that this preparation has a positive effect on the number of pods and seeds, as well as seed yield [40]. The application of Fylloton in the cultivation of winter oilseed rape positively influenced the increase in the number of pods, the yield of seeds, and the weight of one thousand seeds, especially after applying the biostimulant together with the Perfektmikro micronutrient fertilizer containing EDTA-chelated (ethylenediaminetetraacetic acid-chelated) manganese, copper, iron, and zinc, as well as molybdenum, boron, and nitrogen [121].

5. Conclusions

All studied biostimulants had a positive effect on quantity and quality of bean yield. Double application of Kelpak biostimulant (*Ecklonia maxima* extract) stimulated morphological features and seed yield, as well as the content of polyphenols and flavonoids in seeds to a greater extent. In turn, the biostimulant containing amino acids (Terra Sorb Complex) significantly increased the protein content in beans. In contrast, Fylloton containing *Ascophyllum nodosum* extract and amino acids had also a more favorable effect on the number of seeds. In 2015 and 2017, biostimulants containing seaweed extracts had the most beneficial effect on bean yield and its quality. On the other hand, in 2016, treatment of plants with Kelpak had a more beneficial effect on the studied features.

Author Contributions: All authors read and agreed to the published version of the manuscript. A.K. and J.L. conceptualized and designed the research; R.K. and M.T. performed the field experiments; J.L, N.N., and K.K. analyzed the plant material; S.T. carried out statistical analysis of the data; A.K., R.K., and D.L. wrote the paper; M.T. and S.T. revised the manuscript.

References

1. Caruso, G.; De Pascale, S.; Cozzolino, E.; Cuciniello, A.; Cenvinzo, V.; Bonini, P.; Colla, G.; Rouphael, Y. Yield and nutritional quality of Vesuvian Piennolo tomato PDO as affected by farming system and biostimulant application. *Agronomy* **2019**, *9*, 505. [CrossRef]
2. Caruso, G.; De Pascale, S.; Cozzolino, E.; Giordano, M.; El-Nakhel, C.; Cuciniello, A.; Cenvinzo, V.; Colla, G.; Rouphael, Y. Protein hydrolysate or plant extract-based biostimulants enhanced yield and quality performances of greenhouse perennial wall rocket grown in different seasons. *Plants* **2019**, *8*, 208. [CrossRef] [PubMed]
3. Du Jardin, P. Plant biostimulants: Definition, concept, main categories and regulation. *Sci. Hortic.* **2015**, *196*, 3–14. [CrossRef]
4. Matyjaszczyk, E. The introduction of biostimulants on the Polish market. The present situation and legal requirements. *Przem. Chem.* **2015**, *10*, 1841–1844. [CrossRef]
5. Matyjaszczyk, E. Problems of implementing compulsory integrated pest management. *Pest. Manag. Sci.* **2019**, *75*, 2063–2067. [CrossRef]
6. Traon, D.; Amat, L.; Zotz, F.; Du Jardin, P. *A Legal Framework for Plant Biostimulants and Agronomic Fertiliser Additives in the EU. Report for the European Commission Enterprise & Industry Directorate—General*; Publications Office of the European Union: Luxembourg, 2014.
7. Matyjaszczyk, E. "Biorationals" in integrated pest management strategies. *J. Plant Dis. Prot.* **2018**, *125*, 523–527. [CrossRef]
8. Calvo, P.; Nelson, L.; Kloepper, J.W. Agricultural uses of plant biostimulants. *Plant Soil* **2014**, *383*, 3–41. [CrossRef]
9. Rouphael, Y.; Cardarelli, M.; Bonini, P.; Colla, G. Synergistic action of a microbial-based biostimulant and a plant derived-protein hydrolysate enhances lettuce tolerance to alkalinity and salinity. *Front. Plant Sci.* **2017**, *8*, 131. [CrossRef]
10. Colla, G.; Rouphael, Y. Biostimulants in horticulture. *Sci. Hortic.* **2015**, *196*, 1–2. [CrossRef]
11. Colla, G.; Hoagland, L.; Ruzzi, M.; Cardarelli, M.; Bonini, P.; Canaguier, R.; Rouphael, Y. Biostimulant action of protein hydrolysates: unraveling their effects on plant physiology and microbiome. *Front. Plant Sci.* **2017**, *8*, 2202. [CrossRef]
12. Battacharyya, D.; Babgohari, M.Z.; Rathor, P.; Prithiviraj, B. Seaweed extracts as biostimulants in horticulture. *Sci. Hortic.* **2015**, *196*, 39–48. [CrossRef]
13. Basak, A. Biostimulators. Definitions, classification and legislation. In *Biostimulators in Modern Agriculture, General Aspects*; Gawrońska, H., Ed.; Editorial House Wieś Jutra: Warszawa, Poland, 2008; pp. 7–17.

14. Stirk, W.A.; Tarkowská, D.; Turecová, V.; Strnad, M.; van Staden, J. Abscisic acid, gibberellins and brassinosteroids in Kelpak®, a commercial seaweed extract made from *Ecklonia maxima*. *J. Appl. Phycol.* **2014**, *26*, 561–567. [CrossRef]
15. Kocira, A.; Świeca, M.; Kocira, S.; Złotek, U.; Jakubczyk, A. Enhancement of yield, nutritional and nutraceutical properties of two common bean cultivars following the application of seaweed extract (*Ecklonia maxima*). *Saudi J. Biol. Sci.* **2018**, *25*, 563–571. [CrossRef]
16. Latique, S.; Chernane, H.; Mansori, M.; El Kaoua, M. Seaweed liquid fertilizer effect on physiological and biochemical parameters of bean plant (*Phaesolus vulgaris* variety Paulista) under hydroponic system. *Eu. Sci. J.* **2013**, *9*, 174–191.
17. Sharma, S.H.S.; Lyons, G.; McRoberts, C.; McCall, D.; Carmichael, E.; Andrews, F.; Swan, R.; McCormack, R.; Mellon, R. Biostimulant activity of brown seaweed species from Strangford Lough: compositional analyses of polysaccharides and bioassay of extracts using mung bean (*Vigno mungo* L.) and pak choi (*Brassica rapa chinensis* L.). *J. Appl. Phycol.* **2012**, *24*, 1081–1091. [CrossRef]
18. Stirk, W.A.; Van Standen, J. Flow of cytokinins through the environment. *Plant. Growth Regul.* **2010**, *62*, 101–116. [CrossRef]
19. Tarakhovskaya, E.R.; Maslov, Y.I.; Shishova, M.F. Phytohormones in algae. *Russ. J. Plant. Physiol.* **2007**, *54*, 163–170. [CrossRef]
20. Klarzynski, O.; Descamps, V.; Plesse, B.; Yvin, J.C.; Fritig, B. Sulfated fucan oligosaccharides elicit defense responses in tobacco and local and systemic resistance against Tobacco Mosaic Virus. *Mol. Plant-Microbe Interact.* **2003**, *16*, 115–122. [CrossRef]
21. Zhang, Q.; Zhang, J.; Shen, J.; Silva, A.; Dennis, D.A. A simple 96-well microplate method for estimation of total polyphenol content in seaweeds. *J. Appl. Phycol.* **2006**, *18*, 445–450. [CrossRef]
22. Rioux, L.E.; Turgeon, S.L.; Beaulieu, M. Characterization of polysaccharides extracted from brown seaweeds. *Carbohydr. Polym.* **2007**, *69*, 530–537. [CrossRef]
23. Stirk, W.A.; Novak, O.; Hrandecka, V.; Pencik, A.; Rolcik, J.; Strnad, M.; Van Staden, J. Endogenous cytokinins, auxins and abscisic acid in *Ulva fasciata* (Chlorophyta) and *Dictyota humifusa* (Phaeophyta): towards understanding their biosynthesis and homoeostasis. *Eur. J. Phycol.* **2009**, *44*, 231–240. [CrossRef]
24. Yokoya, N.S.; Stirk, W.A.; Van Staden, J.; Novak, O.; Tureckova, V.; Pencik, A.; Strnad, M. Endogenous cytokinins, auxins, and abscisic acid in red algae from Brazil. *J. Phycol.* **2010**, *46*, 1198–1205. [CrossRef]
25. Matysiak, K.; Kaczmarek, S.; Leszczyńska, D. Influence of liquid seaweed extract of *Ecklonia maxima* on winter wheat cv Tonacja. *J. Res. Appl. Agric. Engin.* **2012**, *57*, 44–47.
26. Papenfus, H.B.; Stirk, W.A.; Finnie, J.F.; Van Staden, J. Seasonal variation in the polyamines of *Ecklonia maxima*. *Botanica Marina* **2012**, *55*, 539–546. [CrossRef]
27. Rengasamy, K.R.R.; Kulkarni, M.G.; Stirk, W.A.; Van Staden, J. Eckol—a new plant growth stimulant from the brown seaweed *Ecklonia maxima*. *J. Appl. Phycol.* **2015**, *27*, 581–587. [CrossRef]
28. Rengasamy, K.R.R.; Kulkarni, M.G.; Stirk, W.A.; Van Staden, J. Eckol improves growth, enzyme activities, and secondary metabolite content in maize (*Zea mays* cv. Border King). *J. Plant Growth Regul.* **2015**, *34*, 410–416. [CrossRef]
29. Panda, D.; Pramanik, K.; Nayak, B.R. Use of sea weed extracts as plant growth regulators for sustainable agriculture. *Int. J. Bioresour. Stress Manag.* **2012**, *3*, 404–411.
30. Crough, I.J.; Van Staden, J. Effect of seaweed concentrate on the establishment and yield of greenhouse tomato plants. *J. Appl. Phycol.* **1992**, *4*, 291–296. [CrossRef]
31. Stirk, W.A.; Bálint, P.; Tarkowská, D.; Strnad, M.; van Staden, J.; Ördög, V. Endogenous brassinosteroids in microalgae exposed to salt and low temperature stress, *Eur. J. Phycol.* **2018**. [CrossRef]
32. Ecoforce, A. Extracto de Algas para Agricultura Ecológica—Fertilizantes Ecoforce. Available online: http://fertilizantesecoforce.es/es/blog/index/list/cat/agricultura-ecologica/?p=19 (accessed on 1 December 2019).
33. Van Oosten, M.J.; Pepe, O.; De Pascale, S.; Silletti, S.; Maggio, A. The role of biostimulants and bioeffectors as alleviators of abiotic stress in crop plants. *Technol. Agric.* **2017**, *4*, 5. [CrossRef]
34. Sosnowski, J.; Malinowska, E.; Jankowski, K.; Redzik, P. Morpho-chemical diversity in *Festuca pratensis* and *Lolium perenne* depending on concentrations of *Ecklonia maxima* extract. *Appl. Ecol. Env. Res.* **2016**, *14*, 369–379. [CrossRef]
35. Drobek, M.; Frąc, M.; Cybulska, J. Plant biostimulants: importance of the quality and yield of horticultural crops and the improvement of plant tolerance to abiotic stress—A review. *Agronomy* **2019**, *9*, 335. [CrossRef]

36. Ali, N.; Farrell, A.; Ramsubhag, A.; Jayaraman, J. The effect of *Ascophyllum nodosum* extract on the growth, yield and fruit quality of tomato grown under tropical conditions. *J. Appl. Phycol.* **2016**, *28*, 1353–1362. [CrossRef]
37. Mikiciuk, M.; Dobromilska, R. Assessment of yield and physiological indices of small sized tomato cv. 'Bianka F1' under the influence of biostimulators of marine algae origin. *Acta Sci. Pol., Hortorum Cultus* **2014**, *13*, 31–41.
38. Chanthini, K.-P.; Stanley-Raja, V.; Thanigaivel, A.; Karthi, S.; Palanikani, R.; Shyam Sundar, N.; Sivanesh, H.; Soranam, R.; Senthil-Nathan, S. Sustainable agronomic strategies for enhancing the yield and nutritional quality of wild tomato, *Solanum lycopersicum* (l) var *cerasiforme* Mill. *Agronomy* **2019**, *9*, 311. [CrossRef]
39. Kocira, S.; Kocira, A.; Kornas, R.; Koszel, M.; Szmigielski, M.; Krajewska, M.; Szparaga, A.; Krzysiak, Z. Effect of seaweed extract on yield and protein content of two common bean (*Phaseolus vulgaris* L.) cultivars. *Legume Res.* **2017**, LR-383. [CrossRef]
40. Kocira, S.; Szparaga, A.; Kocira, A.; Czerwińska, E.; Wójtowicz, A.; Bronowicka-Mielniczuk, U.; Koszel, M.; Findura, P. Modelling biometric traits, yield and nutritional and antioxidant properties of seeds of three soybean cultivars through the application of biostimulant containing seaweed and amino acids. *Front. Plant Sci.* **2018**, *9*. [CrossRef]
41. Kocira, S.; Szparaga, A.; Kuboń, M.; Czerwińska, E.; Piskier, T. Morphological and biochemical responses of *Glycine max* (L.) Merr. to the use of seaweed extract. *Agronomy* **2019**, *9*, 93. [CrossRef]
42. Szczepanek, M.; Siwik-Ziomek, A. P and K Accumulation by rapeseed as affected by biostimulant under different NPK and S fertilization doses. *Agronomy* **2019**, *9*, 477. [CrossRef]
43. Abd El-Gawad, H.G.; Osman, H.S. Effect of exogenous application of boric acid and seaweed extract on growth, biochemical content and yield of eggplant. *J. Hortic. Sci. Ornam. Plants* **2014**, *6*, 133–143. [CrossRef]
44. Manna, D.; Sarkar, A.; Maity, T.K. Impact of Biozyme on growth, yield and quality of chilli (*Capsicum annuum* L.). *J. Crop Weed* **2012**, *8*, 40–43.
45. Kumar, N.A.; Vanlalzarzova, B.; Sridhar, S.; Baluswami, M. Effect of liquid seaweed fertilizer of *Sargassum wightii* Grev. on the growth and biochemical content of green gram (*Vigna radiata* (L.) R. Wilczek). *Recent Res. Sci. Technol.* **2012**, *4*, 40–45.
46. Di Mola, I.; Cozzolino, E.; Ottaiano, L.; Giordano, M.; Rouphael, Y.; Colla, G.; Mori, M. Effect of vegetal- and seaweed extract-based biostimulants on agronomical and leaf quality traits of plastic tunnel-grown baby lettuce under four regimes of nitrogen fertilization. *Agronomy* **2019**, *9*, 571. [CrossRef]
47. Colla, G.; Cardarelli, M.; Bonini, P.; Rouphael, Y. Foliar applications of protein hydrolysate, plant and seaweed extracts increase yield but differentially modulate fruit quality of greenhouse tomato. *HortScience* **2017**, *52*, 1214–1220. [CrossRef]
48. Lötze, E.; Hoffman, E.W. Nutrient composition and content of various biological active compounds of three south African-based commercial seaweed biostimulants. *J. Appl. Phycol.* **2016**, *28*, 1379–1386. [CrossRef]
49. Moreira Sisalema, J.M. Comportamiento Agronómico del Cultivo de Soya (*Glycine max* L.), a la Aplicación de Tres Extractos de Algas Marinas, en la Zona de Puebloviejo. Available online: http://dspace.utb.edu.ec/bitstream/49000/5048/1/TE-UTB-FACIAG-ING%20AGROP-000018.pdf (accessed on 1 December 2019).
50. Marschner, H. *Mineral Nutrition of Higher Plants*; Academic Press: London, UK, 1995.
51. Pratelli, R.; Pilot, G. Altered amino acid metabolism in glutamine dumper1 plants. *Plant. Signal. Behav.* **2007**, *2*, 182–184. [CrossRef]
52. Stitt, M.; Müller, C.; Matt, P.; Gibon, Y.; Carillo, P.; Morcuende, R.; Scheible, W.R.; Krapp, A. Steps towards an integrated view of nitrogen metabolism. *J. Exp. Bot.* **2002**, *53*, 959–970. [CrossRef]
53. Colla, G.; Nardi, S.; Cardarelli, M.; Ertani, A.; Lucini, L.; Canaguiere, R.; Rouphael, J. Protein hydrolysates as biostimulants in horticulture. *Sci. Hortic.* **2015**, *196*, 28–38. [CrossRef]
54. Colla, G.; Rouphael, Y.; Canaguier, R.; Svecova, E.; Cadarelli, M. Biostimulant action of a plant-derived protein hydrolysate produced through enzymatic hydrolysis. *Front. Plant Sci.* **2014**, *5*, 448. [CrossRef]
55. Parrado, J.; Escudero-Gilete, M.L.; Frriaza, V.; Garcia-Martinez, A.; González-Miret, M.L.; Bautista, J.D.; Heredia, F.J. Enzymatic vegetable extract with bioactive components: influence of fertilizer on the colour and anthocyanins of red grapes. *J. Sci. Food Agric.* **2007**, *87*, 2310–2318. [CrossRef]
56. Ertani, A.; Pizzeghello, D.; Francioso, O.; Sambo, P.; Sanchez-Cortes, S.; Nardi, S. *Capsicum chinensis* L. growth and nutraceutical properties are enhanced by biostimulants in a long-term period: Chemical and metabolomic approaches. *Front. Plant Sci.* **2014**, *5*, 375. [CrossRef] [PubMed]

57. Di Mola, I.; Ottaiano, L.; Cozzolino, E.; Senatore, M.; Giordano, M.; El-Nakhel, C.; Sacco, A.; Rouphael, Y.; Colla, G.; Mori, M. Plant-based biostimulants influence the agronomical, physiological, and qualitative responses of baby rocket leaves under diverse nitrogen conditions. *Plants* **2019**, *8*, 522. [CrossRef] [PubMed]
58. Liu, X.Q.; Ko, K.Y.; Kim, S.H.; Lee, K.S. Effect of amino acid fertilization on nitrate assimilation of leafy radish and soil chemical properties in high nitrate soil. *Commun. Soil Sci. Plant Anal.* **2008**, *39*, 269–281. [CrossRef]
59. Kocira, S.; Kocira, A.; Szmigielski, M.; Piecak, A.; Sagan, A.; Malaga-Toboła, U. Effect of an amino acids-containing biostimulator on common bean crop. *Przem. Chem.* **2015**, *94*, 1732–1736. [CrossRef]
60. Kocira, S.; Szparaga, A.; Kocira, A.; Czerwińska, E.; Depo, K.; Erlichowska, B.; Deszcz, E. Effect of applying a biostimulant containing seaweed and amino acids on the content of fiber fractions in three soybean cultivars. *Legume Res.* **2018**, LR-412. [CrossRef]
61. Rouphael, Y.; De Micco, V.; Arena, C.; Raimondi, G.; Colla, G.; De Pascale, S. Effect of *Ecklonia maxima* seaweed extract on yield, mineral composition, gas exchange and leaf anatomy of zucchini squash grown under saline conditions. *J. Appl. Phycol.* **2017**, *29*, 459–470. [CrossRef]
62. Rouphael, Y.; Giordano, M.; Cardarelli, M.; Cozzolino, E.; Mori, M.; Kyriacou, M.C.; Bonini, P.; Colla, G. Plant and seaweed-based extracts increase yield but differentially modulate nutritional quality of greenhouse spinach through biostimulant action. *Agronomy* **2018**, *8*, 126. [CrossRef]
63. Duer, I.; Fotyma, M.; Madej, A. *Kodeks Dobrej Praktyki Rolniczej*, 3rd ed.; Fundacja Programów Pomocy dla Rolnictwa: Warszawa, Poland, 2004. (In Polish)
64. Craigie, J.S. Seaweed extract stimuli in plant science and agriculture. *J. Appl. Phycol.* **2011**, *23*, 371–393. [CrossRef]
65. Sharma, H.S.S.; Fleming, C.; Selby, C.; Rao, J.R.; Martin, T. Plant biostimulants: A review on the processing of microalgae and use of extracts for crop management to reduce abiotic and biotic stresses. *J. Appl. Phycol.* **2014**, *26*, 465–490. [CrossRef]
66. Bulgari, R.; Cocetta, G.; Trivellini, A.; Vernieri, P.; Ferrante, A. Biostimulants and crop responses: A review. *Biol. Agric. Hortic.* **2015**, *31*, 1–17. [CrossRef]
67. Kocira, A.; Kornas, R.; Kocira, S. Effect assessment of Kelpak on the bean yield (*Phaseolus vulgaris* L.). *J. Cent. Eur. Agric.* **2013**, *14*, 545–554. [CrossRef]
68. Rathore, S.S.; Chaudhary, D.R.; Boricha, G.N.; Gosh, A.; Bhatt, B.P.; Zodape, S.T.; Patolia, J.S. Effect of seaweed extract on the growth, yield and nutrient uptake of soybean (*Glycine max*) under rainfed conditions. *S. Afr. J. Bot.* **2009**, *75*, 351–355. [CrossRef]
69. Karthikeyan, K.; Shanmugam, M. Development of a protocol for the application of commercial bio-stimulant manufactured from *Kappaphycus alvarezii* in selected vegetable crops. *J. Exp. Biol. Agric. Sci.* **2016**, *4*, 92–102.
70. Tandon, S.; Dubey, A. Effects of Biozyme (*Ascophyllum nodosum*) biostimulant on growth and development of soybean [*Glycine max* (L.) Merill]. *Commun. Soil Sci. Plant Anal.* **2015**, *46*, 845–858. [CrossRef]
71. Zewail, R.M.Y. Effect of seaweed extract and amino acids on growth and productivity and some biocostituents of common bean (*Phaseolus vulgaris* L.) plants. *J. Plant. Production Mansoura Univ.* **2014**, *5*, 1441–1453. [CrossRef]
72. Abo-Sedera, F.A.; Shams, A.S.; Mohamed, M.H.M.; Hamoda, A.H.M. Effect of organic fertilizer and foliar spray with some safety compounds on growth and productivity of snap bean. *Ann. Agric. Sci. Moshtohor* **2016**, *54*, 105–118.
73. Sujatha, K.; Vijayalakshmi, V. Foliar application of *Caulerpa racemosa* seaweed extract as bio-stimulant for enhancement of growth and yield of blackgram (*Vigna mungo* L.). *IJOART* **2013**, *2*, 216–230.
74. Jasim, A.H.; Obaid, A.S. Effect of foliar fertilizers spray, boron and their interaction on broad bean (*Vicia faba* L.) yield. *Sci. Pap. B Hortic.* **2014**, *58*, 271–276.
75. Delchev, G.; Zhelyazkova, T.; Stoyanova, A. Stability valuation of some mixtures between foliar fertilizers and combined herbicides for the grain yield of durum wheat. *Turk. J. Agric. Natur. Sci., Special Issue* **2014**, *1*, 1128–1133.
76. El-Gamal, I.S.; Abd El-Aal, M.M.M.; El-Desouky, S.A.; Khedr, Z.M.; Abo Shady, K.A. Effect of some growth substances on growth, chemical compositions and root yield productivity of sugar beet (*Beta vulgaris* L.) plant. *Middle East. J. Agric. Res.* **2016**, *5*, 171–185.
77. Abdel-Mawgoud, A.M.R.; El-Bassiouny, A.M.; Ghoname, A.; Abou-Hussein, S.D. Foliar application of amino acids and micronutrients enhance performance of green bean crop under newly reclaimed land conditions. *Aust. J. Basic & Appl. Sci.* **2011**, *5*, 51–55.

78. Shafeek, M.R.; Hafez, M.M.; Mahmoud, A.R.; Ali, A.R. Comparative effect on N-fixing bacterial with foliar application of amino acid mixed on growth and yield of pea plants (*Pisum sativum* L.). *Middle East. J. Appl. Sci.* **2014**, *4*, 755–761.
79. El-Ghamry, A.M.; Abd El-Hai, K.M.; Ghoneem, K.M. Amino and humic acids promote growth, yield and disease resistance of faba bean cultivated in clayey soil. *Aust. J. Basic Appl. Sci.* **2009**, *3*, 731–739.
80. Sadak, M.S.H.; Abdelhamid, M.T.; Schmidhalter, U. Effect of foliar application of aminoacids on plant yield and some physiological parameters in bean plants irrigated with seawater. *Acta Biol. Colomb.* **2015**, *20*, 141–152.
81. Khan, W.; Rayirath, U.P.; Subramanian, S.; Jithesh, M.N.; Rayorath, P.; Hodges, D.M.; Critchley, A.T.; Craigie, J.S.; Norrie, J.; Prithiviraj, B. Seaweed extracts as biostimulants of plant growth and development. *J. Plant Growth Regul.* **2009**, *28*, 386–399. [CrossRef]
82. Santner, A.; Calderon-Villalobos, L.I.A.; Estelle, M. Plant hormones are versatile chemical regulators of plant growth. *Nat. Chem. Biol.* **2009**, *5*, 301–307. [CrossRef] [PubMed]
83. Depuydt, S.; Hardtke, C.S. Hormone signalling crosstalk in plant growth regulation. *Curr. Biol.* **2011**, *21*, 365–373. [CrossRef]
84. Sakakibara, H. Cytokinins: Activity, biosynthesis, and translocation. *Ann. Rev. Plant Biol.* **2006**, *57*, 431–449. [CrossRef]
85. Aremu, A.O.; Plačková, L.; Grúz, J.; Biba, O.; Novák, O.; Stirk, W.A.; Doležal, K.; Van Staden, J. Seaweed-derived biostimulant (Kelpak) influences endogenous cytokinins and bioactive compounds in hydroponically grown *Eucomis autumnalis*. *J. Plant Growth Regul.* **2016**, *35*, 151–162. [CrossRef]
86. Masondo, N.A.; Aremu, A.O.; Kulkarni, M.G.; Petřík, I.; Plačková, L.; Šubrtová, M.; Novák, O.; Grúz, J.; Doležal, K.; Strnad, M.; et al. Elucidating the role of Kelpak® on the growth, phytohormone composition, and phenolic acids in macronutrient-stressed *Ceratotheca triloba*. *J. Appl. Phycol.* **2019**, *31*, 2687–2697. [CrossRef]
87. Kulkarni, M.G.; Rengasamy, K.R.R.; Pendota, S.C.; Grúz, J.; Plačková, L.; Novák, O.; Doležal, K.; Van Staden, J. Bioactive molecules derived from smoke and seaweed *Ecklonia maxima* showing phytohormone-like activity in *Spinacia oleracea* L. *New Biotechnol.* **2019**, *48*, 83–89. [CrossRef] [PubMed]
88. Crouch, I.J.; Smith, M.T.; Van Staden, J.; Lewis, M.J.; Hoad, G.V. Identification of auxins in a commercial seaweed concentrate. *J. Plant Physiol.* **1992**, *139*, 590–594. [CrossRef]
89. Jennings, R.C. Gibberellins as endogenous growth regulators in green and brown algae. *Planta* **1968**, *80*, 34–42. [CrossRef]
90. Yamaguchi, S. Gibberellin metabolism and its regulation. *Annu. Rev. Plant Biol.* **2008**, *59*, 225–251. [CrossRef]
91. Tanimoto, E. Gibberellins. In *Plant Roots—The Hidden Half*; Waisel, Y., Eshel, A., Kafkafi, U., Eds.; Marcel Dekker: New York, NY, USA, 2002; pp. 405–416.
92. Gopala, R.P. Gibberellin-like behaviour of α-tocopherol in green gram *Vigna radiata*. *Geobios* **1984**, *11*, 21–25.
93. Jensen, A. Tocopherol content of seaweed and seaweed meal. I. Analytical methods and distribution of tocopherols in benthic algae. *J. Sci. Food Agric.* **1969**, *20*, 449–453.
94. Bajguz, A.; Hayat, S. Effects of brassinosteroids on the plant responses to environmental stresses. *Plant Physiol. Biochem.* **2009**, *47*, 1–8. [CrossRef]
95. Divi, U.K.; Krishna, P. Overexpression of the brassinosteroid biosynthetic gene AtDWF4 in *Arabidopsis* seeds overcomes abscisic acid-induced inhibition of germination and increases cold tolerance in transgenic seedlings. *J. Plant Growth Regul.* **2010**, *29*, 385–393. [CrossRef]
96. Cambri, D.; Filippini, L.; Apone, F.; Arciello, S.; Colucci, G.; Portoso, D. Effect of Aminoplant® on expression of selected genes in *Arabidopsis thaliana* L. plants. In *Biostimulators in Modern Agriculture: General Aspects*; Gawrońska, H., Ed.; Wieś Jutra: Warszawa, Poland, 2008; pp. 77–82.
97. Maini, P. The experience of the first biostimulant, based on amino acids and peptides: A short retrospective review on the laboratory researches and the practical results. *Fertilitas Agrorum* **2000**, *1*, 29–43.
98. Garcia, A.L.; Madrid, R.; Gimeno, V.; Rodriguez-Ortega, W.M.; Nicolas, N.; Garcia-Sanchez, F. The effects of amino acids fertilization incorporated to the nutrient solution on mineral composition and growth in tomato seedlings. *Span. J. Agric. Res.* **2011**, *9*, 852–861. [CrossRef]
99. Cerdán, M.; Sánchez-Sánchez, A.; Oliver, M.; Juárez, M.; Sánchez-Andreu, J.J. Effect of foliar and root applications of amino acids on iron uptake by tomato plants. *Acta Hortic.* **2009**, *830*, 481–488. [CrossRef]
100. Ertani, A.; Cavani, L.; Pizzeghello, D.; Brandellero, E.; Altissimo, A.; Ciavatta, C.; Nardi, S. Biostimulant activities of two protein hydrolysates on the growth and nitrogen metabolism in maize seedlings. *J. Plant Nutr. Soil Sci.* **2009**, *172*, 237–244. [CrossRef]

101. García-Martínez, A.M.; Díaz, A.; Tejada, M.; Bautista, J.; Rodríguez, B.; María, C.S.; Revilla, E.; Parrado, J. Enzymatic production of an organic soil biostimulant from wheat condensed distiller solubles: effects on soil biochemistry and biodiversity. *Process. Biochem.* **2010**, *45*, 1127–1133. [CrossRef]
102. Lucini, L.; Rouphael, Y.; Cardarelli, M.; Canaguier, R.; Kumar, P.; Colla, G. The effect of a plant-derived protein hydrolysate on metabolic profiling and crop performance of lettuce grown under saline conditions. *Sci. Hortic.* **2015**, *182*, 124–133. [CrossRef]
103. Matsumiya, Y.; Kubo, M. Soybean peptide: Novel plant growth promoting peptide from soybean. In *Soybean and Nutrition*; El-Shemy, H., Ed.; InTech, 2011. Available online: http://www.intechopen.com/books/soybean-and-nutrition/soybean-peptide-novel-plant-growth-promoting-peptide-from-soybean (accessed on 1 December 2019).
104. Schiavon, M.; Ertani, A.; Nardi, S. Effects of an alfalfa protein hydrolysate on the gene expression and activity of enzymes of TCA cycle and N metabolism in *Zea mays* L. *J. Agric. Food Chem.* **2008**, *56*, 11800–11808. [CrossRef]
105. Halpern, M.; Bar-Tal, A.; Ofek, M.; Minz, D.; Muller, T.; Yermiyahu, U. The use of biostimulants for enhancing nutrient uptake. *Adv. Agron.* **2015**, *130*, 141–174. [CrossRef]
106. Nardi, S.; Pizzeghello, D.; Schiavon, M.; Ertani, A. Plant biostimulants: physiological responses induced by protein hydrolyzed-based products and humic substances in plant metabolism. *Sci. Agric.* **2016**, *73*, 18–23. [CrossRef]
107. Selvam, G.G.; Sivakumar, K. Micromorphological study of *Vigna mungo* L. using seaweed liquid fertilizer from *Hypnea musciformis* (Wulf.) Lamouroux. *Indian J. Mar. Sci.* **2016**, *45*, 1199–1207.
108. John, P.P.J.; Yuvaraj, P. Effect of Seaweed Liquid Fertilizer of *Colpomenia sinuosa* (Mert. ex Roth) Derbes & Solier (Brown Seaweed) on *Vigna radiata* (L.) R. Wilczek. in Koothankuzhi, Tirunelveli district, Tamil Nadu, India. *Int. J. Pure Appl. Biosci.* **2014**, *2*, 177–184.
109. Gugała, M.; Sikorska, A.; Zarzecka, K.; Findura, P.; Malaga–Toboła, U. Chemical composition of winter rape seeds depending on the biostimulators used. *Agronomy* **2019**, *9*, 716. [CrossRef]
110. Schubert, S.; Mengel, K. Important factors in nutrient availability: root morphology and physiology. *Z. Pflanzenernähr. Bodenkd.* **1989**, *152*, 169–174. [CrossRef]
111. Czewińska, E.; Szparaga, A. Antibacterial and antifungal activity of plant extracts. *Rocz. Ochr. Sr.* **2015**, *17*, 209–229.
112. Jindo, K.; Martim, S.A.; Navarro, E.C.; Pérez-Alfocea, F.; Hernandez, T.; Garcia, C.; Aguiar, N.O.; Canellas, L.P. Root growth promoting by humic acids from composted and non-composted urban organic wastes. *Plant Soil* **2012**, *353*, 209–220. [CrossRef]
113. Ertani, A.; Schiavon, M.; Muscolo, A.; Nardi, S. Alfalfa plant-derived biostimulant stimulate short-term growth of salt stressed *Zea mays* L. plants. *Plant. Soil* **2013**, *364*, 145–158. [CrossRef]
114. Cheynier, V.; Comte, G.; Davies, K.M.; Lattanzio, V.; Martens, S. Plant phenolics: Recent advances on their biosynthesis, genetics, and ecophysiology. *Plant. Physiol. Biochem.* **2013**, *72*, 1–20. [CrossRef]
115. Aremu, A.O.; Masondo, N.A.; Rengasamy, K.R.R.; Amoo, S.O.; Grúz, J.; Biba, O.; Šubrtová, M.; Pencık, A.; Novák, O.; Doležal, K.; et al. Physiological role of phenolic biostimulants isolated from brown seaweed *Ecklonia maxima* on plant growth and development. *Planta* **2015**, *241*, 1313–1324. [CrossRef]
116. Świeca, M.; Gawlik-Dziki, U.; Kowalczyk, D.; Złotek, U. Impact of germination time and type of illumination on the antioxidant compounds and antioxidant capacity of *Lens culinaris* sprouts. *Sci. Hortic.* **2012**, *140*, 87–95. [CrossRef]
117. Fan, D.; Hodges, D.M.; Critchley, A.T.; Prithiviraj, B. A commercial extract of brown macroalga (*Ascophyllum nodosum*) affects yield and the nutritional quality of spinach in vitro. *Commun. Soil Sci. Plant Anal.* **2013**, *44*, 1873–1884. [CrossRef]
118. Paul, K.; Sorrentino, M.; Lucini, L.; Rouphael, Y.; Cardarelli, M.; Bonini, P.; Reynaud, H.; Canaguier, R.; Trtílek, M.; Panzarová, K.; et al. Understanding the biostimulant action of vegetal-derived protein hydrolysates by high-throughput plant phenotyping and metabolomics: a case study on tomato. *Front. Plant Sci.* **2019**, *10*, 47. [CrossRef]
119. Shalaby, S.; Horwitz, B.A. Plant phenolic compounds and oxidative stress: integrated signals in fungal–plant interactions. *Curr. Genet.* **2015**, *61*, 347–357. [CrossRef]

120. Gurav, R.G.; Jadhav, J.P. A novel source of biofertilizer from feather biomass for banana cultivation. *Environ. Sci. Pollut. Res. Int.* **2013**, *20*, 4532–4539. [CrossRef] [PubMed]
121. Wenda-Piesik, A.; Kazek, M.; Ropińska, P. Impact of amino acid biostimulation and microelements fertilization in foliar application on productivity of winter oilseed rape. *Fragm. Agron.* **2017**, *34*, 119–129.

6

Appraisal of Biodegradable Mulching Films and Vegetal-Derived Biostimulant Application as Eco-Sustainable Practices for Enhancing Lettuce Crop Performance and Nutritive Value

Eugenio Cozzolino [1,†], Maria Giordano [2,†], Nunzio Fiorentino [2], Christophe El-Nakhel [2], Antonio Pannico [2], Ida Di Mola [2], Mauro Mori [2], Marios C. Kyriacou [3], Giuseppe Colla [4] and Youssef Rouphael [2,*]

1 Council for Agricultural Research and Economics (CREA)—Research Center for Cereal and Industrial Crops, 81100 Caserta, Italy; eugenio.cozzolino@crea.gov.it

2 Department of Agricultural Sciences, University of Naples Federico II, 80055 Portici, Italy; maria.giordano@unina.it (M.G.); nunzio.fiorentino@unina.it (N.F.); nakhel_christophe@hotmail.com (C.E.N.); antonio.pannico@unina.it (A.P.); ida.dimola@unina.it (I.D.M.); mori@unina.it (M.M.)

3 Department of Vegetable Crops, Agricultural Research Institute, 1516 Nicosia, Cyprus; m.kyriacou@ari.gov.cy

4 Department of Agriculture and Forest Sciences, University of Tuscia, 01100 Viterbo, Italy; giucolla@unitus.it

* Correspondence: youssef.rouphael@unina.it

† Co-first authors (E.C. and M.G.).

Abstract: Scientists, extensions specialists, and growers are seeking sustainable agricultural practices that are able to cope with these objectives in order to ensure global food security and minimize environmental damage. The use of mulching films and plant biostimulants in agriculture seems to be a valid solution for tackling these rising concerns. A greenhouse experiment was conducted in order to elucidate the morpho-physiological and nutritive characteristics of lettuce (*Lactuca sativa* L.) in response to foliar application of a tropical plant extract (PE) biostimulant and the use of plastic mulches. Two biodegradable mulch treatments (Mater-Bi® 1 and Mater-Bi® 2) were compared to black polyethylene (LDPE) and bare soil. Biodegradable mulch film Mater-Bi® 1 produced a comparable marketable fresh yield to the commercial standard polyethylene (LDPE), whereas Mater-Bi® 2 exhibited the highest crop productivity. When averaged over biostimulant application, lettuce plants grown with biodegradable film Mater-Bi® 2 exhibited superior quality traits in terms of K, Ca, total ascorbic acid, and carotenoids content. The combination of film mulching (LDPE, Mater-Bi® 1 or Mater-Bi® 2) with the tropical plant extract biostimulant exhibited a positive and significant synergistic effect (+30%) on yield. The PE-biostimulant induced higher values of SPAD index and total chlorophyll content when compared to untreated greenhouse lettuce. The mineral content of leaf tissues was greater by 10% and 17% (for P and Ca, respectively) when compared to the untreated lettuce (no PE application). Nitrate content was significantly reduced by 23% in greenhouse lettuce plants receiving PE as compared to the untreated control. The positive effect of Mater-Bi® 2 film on the ascorbic acid content has also been highlighted when combined with the biostimulant application, where a major amplification of total ascorbic acid (+168%) was recorded in comparison to the untreated lettuce. Overall, our work can assist leafy vegetables growers in adopting good agricultural practices, such as biodegradable plastic mulches and vegetal-derived biostimulants, to improve the sustainability of greenhouse production.

Keywords: eco-friendly practices; *Lactuca sativa* L.; total ascorbic acid; tropical plant extract; Mater-Bi®; nitrate; mineral composition; SPAD index; functional quality

1. Introduction

A widespread agricultural practice across the world consists of covering the soil around plants with plastic films. The introduction of this technique in agriculture dates back to the 1970s, and its success is still linked to multiple benefits. In fact, plastic films can: (i) increase soil temperature and keep it constant throughout the first 20–30 cm layer, so that plants' roots develop faster [1,2]; (ii) reduce soil evapotranspiration and preserve moisture; (iii) prevent soil erosion and excessive leaching of nutrients from plants' rhizosphere; and, (iv) improve the performance of plants in a quantitative and qualitative manner [1,3–5]. In addition, mulching films suppress weeds growth, protect crops against pests and various diseases, and reduce the use of pesticides and herbicides. Based on their color (black, clear or white), they absorb and/or reflect sunlight, differently varying soil temperature, thus affecting crop growth and productivity [5]. Plastic films are widely used for growing vegetables under both open-field and greenhouse conditions [6]. Moreover, these films are mainly made by low-density polyethylene (LDPE) [3], having a strong resistance and high durability, even though, like all petroleum products, they are non-compostable and non-biodegradable. The presence of LDPE residues in the soil beyond the duration of a crop cycle is associated to soil contamination with phthalate and phthalic acid esters due to thermal degradation [7]. Therefore, farmers must manually or mechanically collect from the field and recycle or dispose them to comply with the legislative directives of each country. Unfortunately, the frequent illegal burning of plastic mulches by farmers is becoming a common practice, with the aim of reducing production costs by avoiding disposal expenses, which results in a consequent emission of toxic and harmful substances for humans and the environment [1,3]. In such a way, plastic mulching films increase plastic wastes that are used in agriculture, such as pipes and fittings; agricultural packaging, such as bags, liners, and containers [3]. Therefore, there is an urgent need to use compostable and biodegradable materials in modern agriculture. Nowadays, research is projected towards the creation of films made of biopolymers, such as starch, polylactic acid, and cellulose. These materials are derived from renewable resources, such as corn, potato, and rice [1,6]. Their degradation is in compliance with the European laws and Italian ones (UNI 10785, 1999) on biodegradability (EN 13432, 2000). In fact, these materials are entirely degraded by soil microorganisms and they are mineralized in carbon dioxide and methane, water, and biomass, without the production of toxic substances. Any biodegradable material is designed to disappear within the soil in 5–6 months after the end of the crop [2].

Efficient management of natural resources, such as water and soil, is needed in a scenario where the world population is growing, and agriculture must meet an increasing food demand. On the other hand, the use of plant biostimulants in agriculture has been recognized during the last two decades as an efficient tool to boost yield under optimal and sub-optimal conditions, to improve quality as well as increase nutrient uptake and use efficiency of field and horticultural crops [8–11]. Under the new European Union Regulation 2019/1009, plant biostimulants are specified based on their agronomical effects on crops (i.e., claims), and they include humic substances, protein hydrolysates, algae and plant extracts, inorganic compounds (e.g., silicon), growth-promoting bacteria, and mycorrhizal fungi. Many recent studies on vegetal-based biostimulants have shown to increase the tolerance of crops to abiotic stress (extreme temperature, drought, and salinity), and improve the quality of the produce, in terms of organoleptic and nutraceutical characteristics [11]. They have also contributed to the reduction of unwanted substances content, such as nitrates and heavy metals, in crops [12]. Among these, plant extracts that mainly contain signaling molecules (i.e., small peptides and free amino acids) can influence both primary and secondary metabolism in plants, by stimulating glycolysis enzymes' activity, Krebs' cycle, and nitrates' assimilation [13,14]. Moreover, it has been shown that vegetal-derived plant biostimulants effects involve the size modifications of roots by increasing the length and the number of root hairs, as well as the intake of both macroelements and microelements, leading to better crop performance and the nutritive value of the final produce [8,13,14].

Lettuce (*Lactuca sativa* L.) belongs to the *Asteraceae* family and it is one of the most intensively produced leafy vegetables being widespread all over the world. It is valued for its organoleptic properties and is considered an important source for health-promoting metabolites (carotenoids, chlorophylls, macro and trace elements, phenolics, and vitamins), which are crucial in human nutrition [15,16]. Lettuce has a high water and low fat content, which makes it ideal for dietary plans [15]. Italy dedicates vast areas to lettuce production, and has a broad market, which places it as a European leader in this sector [10]. More importantly, production systems and agronomic practices are pre-harvest factors that can determine the quantitative and qualitative variations in lettuce bioactive compounds [17].

On the basis of the above-mentioned considerations, the aim of our work was to combine two eco-sustainable agricultural practices, such as the use of biodegradable films and plant-based biostimulant (tropical plant extract), and test their effect on the morpho-physiological performance, mineral composition, and nutritive value of greenhouse lettuce plants. The films used were two biodegradable mulching films, namely Mater-Bi® with different composition, which effect was compared with that of a polyethylene film and bare soil. The findings of the study will elucidate the biostimulant × mulch interaction to select the best combination (s) able to improve crop performance and nutritive value of this important leafy vegetable. We also believe that these results will be of great interest for horticulturists, extension specialists, and scientists.

2. Materials and Methods

2.1. Greenhouse Growth Conditions, Treatments and Experimental Design

The experimental test was implemented in a protected environment made of an unheated greenhouse, which was located at the experimental farm of the Department of Agriculture, University of Naples Federico II, Portici—Naples (lat. 40°49′ N; long 14°20′ E, 37 a.s.l). The main physical and chemical characteristics of the soil at the experimental site were: sandy loam texture (74% sand, 20% silt, 6% clay), electrical conductivity of 0.5 dS m^{-1}, neutral pH-7.0, total nitrogen (N) of 0.12%, and organic matter of 1.20% (w/w). The nitrate N, ammoniacal N, Olsen phosphorus, and exchangeable potassium were 105, 12, 40, and 936 mg kg^{-1}, respectively. The butterhead lettuce F1 hybrid SINTIA RZ (42–160; Rijk Zwaan, Der Lier, The Netherlands) was used in this test. This lettuce is very resistant to tip burn and bolting and it is characterized by bright green leaves. SINTIA RZ was selected as the most representative commercial cultivar that was used in Italy during the autumn and winter growing seasons under protected environment. On 16 September 2017 three mulching films (M) were installed, two black biodegradable films, namely Mater-Bi® PC 17 N1 (15 µm thick, commercial; Novamont S.p.A, Novara, Italy) and Mater-Bi® PC 17 N2 (15 µm thick, experimental; Novamont S.p.A, Novara, Italy), and one traditional black low-density polyethylene (LDPE) plastic film (50 µm thick, Idroland s.r.l., Bari, Italy). The compositions of the two biodegradable films are composed of thermoplastic starch and copolyester. The two Mater-Bi® mulching films differ in the presence of Masterbatch (PC 17 N2), a solid additive that is used for imparting color or other properties to plastics, with innovative characteristics to improve the color of mulches with low impact on the original polymer. The soil additive is a concentrated mixture of pigments that was made through a heating process and it includes a carrier resin (e.g., wax) that is cut into granules and then added to plastics.

The greenhouse consisted of a galvanized steel frame with plastic covering material, two non-automated side openings, and a mechanized roof opening. The total greenhouse surface corresponded to an area of 162 square meters (27 m × 6 m). The soil was prepared with low energy inputs consisting of a manual grubbing-up of weeds and then a shallow hoeing (20–25 cm) to allow for a leveling of the soil in a single pass. Water was not a limiting factor, the crop evapotranspiration was calculated with the Hargreaves method, and the water deficit was fully restored by using a drip irrigation system. The irrigation system consisted of a main polyethylene pipeline with a diameter of 32 mm with a low operating pressure of 2 atm, while a series of semi-compensating dripping wings (16 mm diameter and 10 cm interpolation) were laterally attached.

The lettuce seedlings were transplanted in the greenhouse on September 25th on raised furrows. On each furrow, the lettuce seedlings were arranged in double rows, at a plant density of 12.3 plants per m^2. The antiperonosporic protection was performed with Metalaxil seven days after transplantation in order to limit the development of fungal pathogens.

Figure 1 presents the trend of minimum, maximum, and mean daily air temperature inside the greenhouse during the cropping cycle. The soil temperature measurements (minimum, maximum, and mean temperature) were also recorded with microchip sensors (0.5 °C sensitivity) that were placed at 10 cm depth. All of the measurements were collected on a data logger (Davis Vantage Pro2, CA, USA). Nitrogen fertilization was applied by fertigation with ammonium nitrate at eight and 16 days after transplantation (DAT). Half of the plots were treated with Auxym® (Italpollina USA Inc., Anderson, IN, USA) product in order to assess the action of the biostimulant (B). This biostimulant is obtained from fermented tropical plant biomass. It contains phytohormones, amino acids, vitamins, phytochelatins, and enzymes. Auxym® contains as well micro and macroelements (g/kg): N 8.3, P 4.0, K 25.0, Ca 0.9, Mg 1.2, Fe 6.6, Mn 6.4, B 4.4, Zn 0.4, and Cu 0.2 [18]. The biostimulant was applied -at a concentration of 2 mL per liter of water- on plants by a sprayer shoulder pump and application took place five times at seven days' intervals starting 10 DAT (i.e., foliar application).

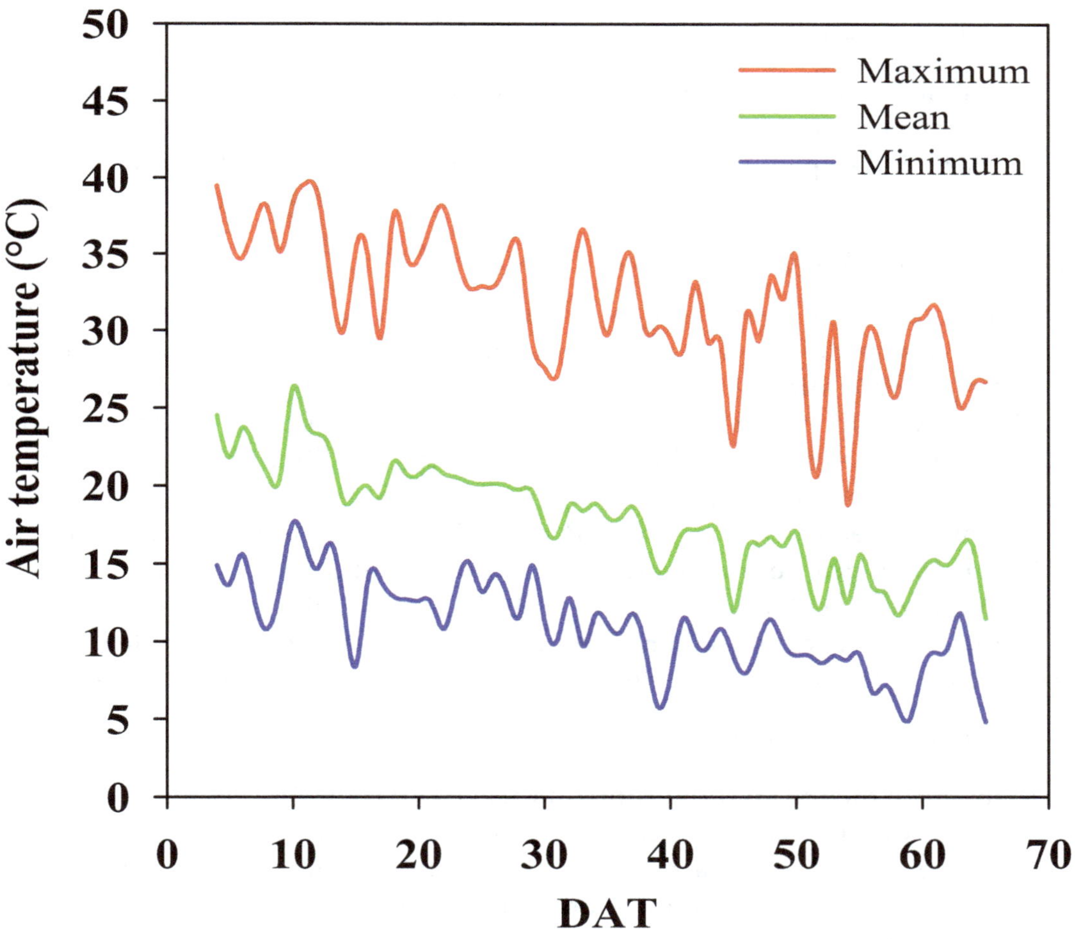

Figure 1. Daily maximum, mean, and minimum values of air temperature recorded inside the greenhouse during the growing period of lettuce.

The experimental scheme provided a two factors factorial combination that resulted in eight treatments in which the factors were mulching (M; three mulching films and bare soil) and biostimulant application (B; control treatment and foliar application of biostimulant). Each treatment was replicated three times and all of the treatments were organized in a randomized complete-block design, resulting in a total of 24 experimental plots.

2.2. *Growth Analysis, Yield, Harvest and Quality Analysis Sampling*

The harvest was manually carried out by cutting the plants at the crown area, just when commercial weight was attained. For each replicate, a total of 15 representative plants were collected. Each plant was first weighed as a whole (leaves and stem) in order to determine the total yield, while the commercial yield was estimated after separation and weighing of leaves. In both cases, yield was expressed in g plant^{-1}. Finally, the leaves were counted and the leaf area (cm^2 plant^{-1}) was determined using a LiCor 3100C leaf area meter (LI-COR Biosciences, Lincoln, NE, USA). Five fresh plants from each treatment were randomly sampled, and then stored at −80 °C until the determination of bioactive compounds content.

2.3. *Soil Plant Analysis Development (SPAD) Index and Color Measurements*

Fifteen SPAD index measurements were performed by a chlorophyll meter (Minolta SPAD-502, Tokyo, Japan) and averaged to a single value on five fully expanded lettuce leaves per replicate. Leaf color (space parameters L*, a* and b*) was recorded with a Minolta chroma meter (CR-300, Minolta Camera Co. Ltd., Tokyo, Japan), on the center of the upper leaf surface with special care to avoid the central vein.

2.4. *Total Chlorophyll and Carotenoid Content Determination*

On 1 g of fresh leaf samples, the total chlorophylls and carotenoids content was determined following the Lichtenhaler and Buschman [19] method. Fresh sample was extracted in pure acetone, for 15 min in the dark. Subsequently, the absorbance of the extracted solutions was measured at 662, 645, and 470 nm, while using a Hach DR 2000 spectrophotometer (Hach Co., Loveland, CO, USA).

2.5. *Dry Matter, Nitrate and Macromineral Content Analysis*

After the determination of fresh yield, the leaves were put to a ventilated stove at a temperature of 65 °C for 72 h until constant weight for dry weight determination. The dry mater content was expressed as percentage (%). Mineral analysis was carried out in 250 mg of dry ground leaves (IKA, MF 10.1, Staufen, Germany), which were sieved and diluted in 50 mL of ultrapure water (Milli-Q, Merck Millipore, Darmstadt, Germany). A syringe with a 0.45 µm pore filter (Phenomenex, Torrance, CA, USA) was used to inject each sample into an ion chromatography (ICS-3000, Dionex, Sunnyvale, CA, USA). For macrocations determination, an IonPac CG12A (4 × 250 mm) guard column and IonPac CS12A (4 × 250 mm) analytical column were used. While, for macroanions determination, an IonPac AG11-HC (4 × 50 mm) guard column and IonPac AS11-HC analytical column (4 × 250 mm) were used. All of the macrominerals were expressed on a dry weight (dw) basis (g kg^{-1}), while the nitrate content was expressed as mg kg^{-1} fw based on the respective leaf sample dry matter content.

2.6. *Hydrophilic Antioxidant Activity Determination*

In order to measure the hydrophilic antioxidant activity (HAA), 200 mg of lyophilized sample were extracted twice with distilled water, following the *N*,*N*-dimethyl-p-phenylenediamine (DMPD) method [20]. An aliquot of 20 µL of extract was combined with 2 mL of DMPD + solution. The bleaching of solution was proportional to the amount of antioxidant compounds concentration. The reduction in absorbance, as measured by UV Vis spectrophotometry at 505 nm, allows for determining the antioxidant activity. For this purpose, an ascorbate external standard calibration curve was used.

2.7. *Total Phenols and Total Ascorbic Acid Content Determination*

The total phenols content was assessed with the Folin–Ciocalteau procedure [21]. 250 mg of lyophilized sample were extracted with 10 mL of methanol/water (60:40 *v*/*v*). After an incubation of 90 min., the absorption was measured at 765 nm while employing a UV-Vis spectrophotometer. The results were calculated using an external gallic acid calibration curve (Sigma Aldrich Inc., St. Louis,

MO, USA). Total ascorbic acid content was measured according to the method of Kampfenkel et al. [22], and it was quantified by a spectrophotometer at 525 nm against an external ascorbate standard calibration curve.

2.8. Statistical Analysis

The normal distribution of the data was verified through the Shapiro–Wilk's and Kolmogorov–Smirnov's procedures. All of the data were subjected to Two-way ANOVA using SPSS 20 software package. For mulching factor, the treatment means were confronted utilizing Duncan's Multiple Range Test that was performed at $p \leq 0.05$, while, for the biostimulant effect, the means were compared using the t-test.

3. Results

3.1. Soil Temperature Trends

The minimum, maximum and mean soil temperatures under the three tested mulches were influenced by the composition of the utilized mulching material (Figure 2). The differences between the minimum and mean soil temperatures between LDPE and the two biodegradable mulches (Mater-Bi® 1 and Mater-Bi® 2) were notable during the first 15–20 days after transplanting, whereas the differences became narrowerer towards the end of the growing cycle (Figure 2). Concerning the maximum soil temperatures, the Mater-Bi® 1 film had similar maximum soil temperature values to LDPE and slightly higher ones than Mater-Bi® 2. However, the soil minimum, maximum, and mean temperatures trends that were recorded in bare soil were regularly lower than those reached among the three tested mulch materials (Figure 2). The soil temperature trends were similar under the three mulching films in all cases, since they had the highest values at the beginning of the crop cycle and underwent a gradual decrease afterwards, especially towards the end of the growing period. The average minimum soil temperature varied between 13.2–23.9 °C in LDPE, 12.7–22.2 °C in Mater-Bi® 1, 12.7–22.2 °C in Mater-Bi® 2, and 10.6–20.8 °C in bare soil. Finally, the average maximum soil temperature fluctuated between 15.1–30.0 °C in LDPE, 15.1–29.5 °C in Mater-Bi® 1, 14.7–28.3 °C in Mater-Bi® 2, and 12.0–25.2 °C in bare soil.

3.2. Yield and Biometric Parameters

The combination of LDPE or biodegradable mulching materials (Mater-Bi® 1 or Mater-Bi® 2) with the PE-based biostimulant positively affected the total and marketable yields of greenhouse lettuce when compared to the untreated plants, although the beneficial effect of biostimulant application was not apparent for bare soil treatment (Table 1). According to the average effect of the mulching films, a tendency to higher total yield values was recorded for Mater-Bi® 2 (319.5 g plant^{-1}), with a 22% increase as compared to bare soil (261.3 g plant^{-1}), even though no significant differences were recorded between the three mulching treatments. However, this trend became apparent for marketable yield with significantly higher values for Mater-Bi® 2 when compared to Mater-Bi® 1 or LDPE and especially to bare soil (Table 1). The positive effect of PE-treated lettuce plants that were cultivated under the three mulching materials (LDPE, Mater-Bi® 1, or Mater-Bi® 2) was mainly attributed to an increment in the total leaf area and not to an increase in the plant leaf number based on the M $\times$ B interaction (Table 1). Moreover, the effect of PE foliar application, when averaged over all mulching treatments, was shown to affect leaf number, which was higher by 10% in PE-treated than in untreated greenhouse lettuce plants. Finally, our findings demonstrated that lettuce plants that were grown under LDPE or Mater-Bi® 2 elicited a significant increment in the number of leaves confronted to the bare soil treatment, whereas the plants cultivated under Mater-Bi® 1 exhibited intermediate values (Table 1).

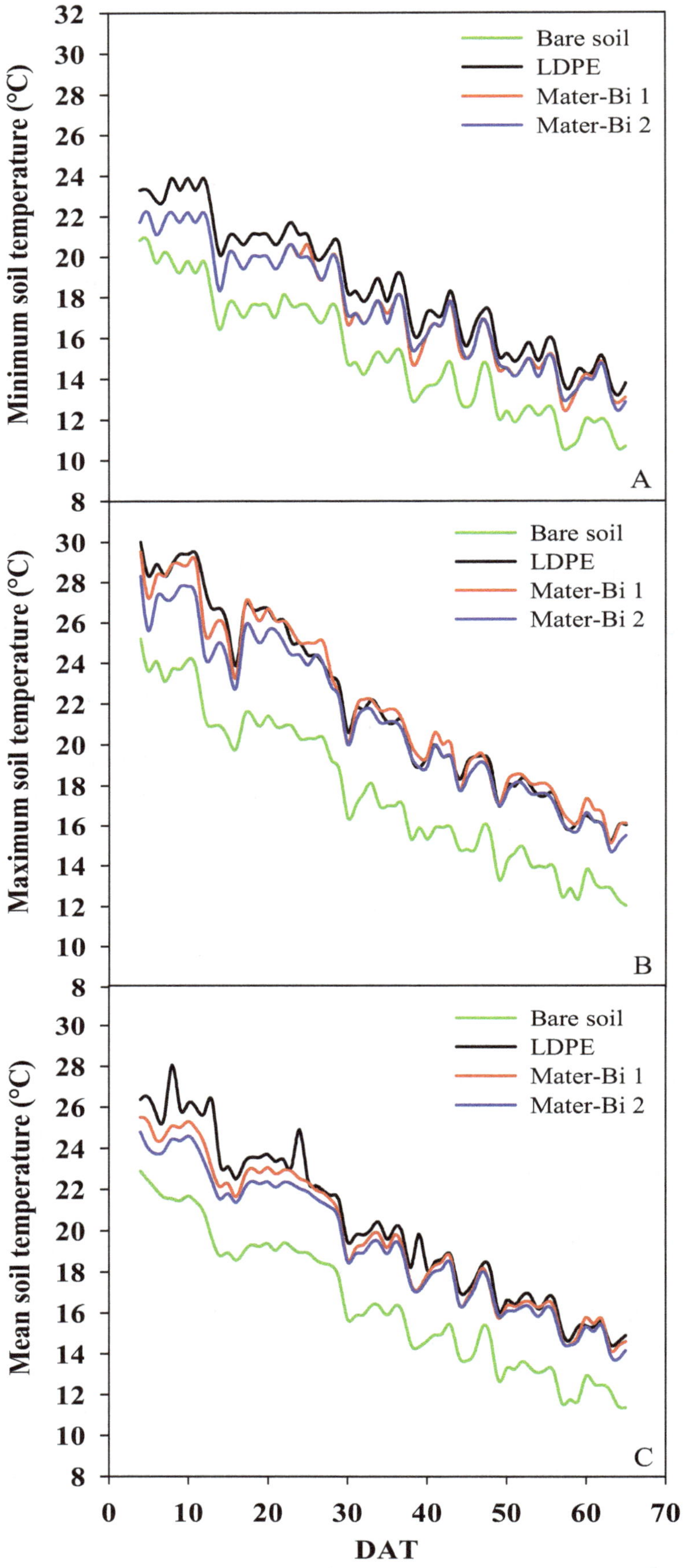

Figure 2. Daily minimum (**A**), maximum (**B**), and mean (**C**) values of soil temperature recorded at a depth of 10 cm in bare soil, LDPE, Mater-Bi® 1 and Mater-Bi® 2 mulches.

Table 1. Mean comparison and analysis of variance for total and marketable yield, leaf number, and total leaf area of untreated and biostimulant-treated greenhouse lettuce grown under low-density polyethylene (LDPE) mulch and biodegradable (Mater-Bi® 1 and Mater-Bi® 2) mulching materials in relation to bare soil.

Source of Variance	Total Yield (g plant^{-1})	Marketable Yield (g plant^{-1})	Leaf Number (no. plant^{-1})	Leaf Area (cm^2 plant^{-1})
Mulch (M)				
Bare soil	261.3 ± 7 b	243.9 ± 7 c	36.5 ± 0.4 c	3414 ± 58 c
LDPE	298.0 ± 18 a	274.3 ± 17 b	44.5 ± 1.2 a	3885 ± 193 a
Mater-Bi® 1	302.8 ± 21 a	274.7 ± 19 b	40.6 ± 1.4 b	3566 ± 168 b
Mater-Bi® 2	319.5 ± 17 a	296.5 ± 16 a	41.9 ± 1.0 a	3667 ± 168 ab
	***	***	***	*
Biostimulant (B)				
Control	266.4 ± 6	243.3 ± 5	38.9 ± 0.8	3375 ± 73
Tropical plant extract (PE)	324.5 ± 12	301.4 ± 10	42.8 ± 1.1	3891 ± 100
t-test	0.000	0.000	0.009	0.000
M × B				
Bare soil without biostimulant	260.6 ± 14 b	238.0 ± 13 b	35.5 ± 0.1	3433 ± 109 b
LDPE without biostimulant	263.9 ± 15 b	240.3 ± 12 b	42.7 ± 1.6	3545 ± 241 b
Mater-Bi® 1 without biostimulant	256.8 ± 5 b	232.9 ± 4 b	38.6 ± 0.5	3198 ± 47 b
Mater-Bi® 2 without biostimulant	284.8 ± 11 b	262.0 ± 8 b	38.9 ± 0.6	3323 ± 103 b
Bare soil + PE	262.0 ± 8 b	249.9 ± 6 b	37.4 ± 0.2	3395 ± 67 b
LDPE + PE	332.2 ± 13 a	308.4 ± 13 a	46.4 ± 1.0	4226 ± 114 a
Mater-Bi® 1 + PE	348.7 ± 9 a	316.5 ± 7 a	44.9 ± 0.3	3934 ± 57 a
Mater-Bi® 2 + PE	354.9 ± 4 a	331.0 ± 4 a	42.6 ± 0.8	4011 ± 112 a
	**	*	NS	*

NS, *, **, *** Non-significant or significant at $p \leq 0.05$, 0.01, and 0.001, respectively. Different letters in the same column indicate significant differences according to DMR test ($p = 0.05$). Means of biostimulant effect are compared according to Student's *t*-test ($p = 0.05$). All data are expressed as mean ± SE.

3.3. SPAD index, Chlorophyll Content and Colorimetric Indices

The non-destructive (SPAD index) and destructive measurement of chlorophylls content were significantly affected by mulching materials and biostimulant applications, with no significant effects from the M × B interaction (Table 2). The PE-based biostimulant provoked greater values of SPAD index and chlorophyll content (+6% and 30%, respectively) in comparison to the untreated control, irrespective of the mulching materials (Table 2). Moreover, when averaged over biostimulant applications, the total chlorophyll content was enhanced by 33% in mulched lettuce plants (avg. 51.3 mg 100 g^{-1} fw) when compared to bare soil (avg. 38.6 mg 100 g^{-1} fw), with no significant differences being observed among the three mulching materials (Table 2).

Concerning the Hunter color parameters, the ANOVA highlighted no significant M x B interaction for all of the examined color parameters (Table 2). In general, neither mulching nor biostimulant application had a significant effect on leaf yellowness (+b*; avg. 33.4) of greenhouse lettuce. Moreover, the use of LDPE as a mulching material resulted in greater lightness (i.e., lowest L* values) of greenhouse lettuce leaves (Table 2). Finally, the foliar application of PE-based biostimulant provoked greater values of brightness and greenness, in comparison to the untreated control, irrespective of the mulching materials (Table 2).

Table 2. Mean comparison and analysis of variance for Soil Plant Analysis Development (SPAD) index, total chlorophyll content, and Hunter color parameters L* (brightness), a* (−a* = green) and b* (+b* = yellow) of untreated and biostimulant-treated greenhouse lettuce grown under LDPE mulch and biodegradable (Mater-Bi® 1 and Mater-Bi® 2) mulching materials in relation to bare soil.

Source of Variance	SPAD Index	Total Chlorophyll (mg 100 g^{-1} fw)	L*	a*	b*
Mulch (M)					
Bare soil	27.3 ± 0.4 c	38.6 ± 0.0 b	54.4 ± 1.7 b	−19.4 ± 0.4	33.6 ± 0.4
LDPE	29.4 ± 0.4 b	50.3 ± 0.1 a	51.8 ± 2.1 c	−19.8 ± 0.4	33.3 ± 0.6
Mater-Bi® 1	30.3 ± 0.6 a	51.8 ± 0.1 a	60.2 ± 1.1 a	−18.6 ± 0.6	33.9 ± 0.9
Mater-Bi® 2	29.2 ± 0.4 b ***	51.7 ± 0.1 a *	56.5 ± 2.0 b ***	−18.9 ± 2.0 NS	32.8 ± 0.4 NS
Biostimulant (B)					
Control	28.3 ± 0.4	41.9 ± 0.0	52.1 ± 1.2	−19.9 ± 0.2	33.9 ± 0.3
Tropical plant extract (PE)	30.0 ± 0.5	54.4 ± 0.0	59.3 ± 0.9	−18.3 ± 0.3	32.9 ± 0.5
t-test	0.043	0.039	0.000	0.000	0.084
M × B					
Bare soil without biostimulant	26.7 ± 0.2	34.5 ± 0.0	50.6 ± 0.3	−20.4 ± 0.2	34.3 ± 0.4
LDPE without biostimulant	28.7 ± 0.9	39.3 ± 0.1	47.7 ± 0.5	−20.6 ± 0.3	33.6 ± 0.6
Mater-Bi® 1 without biostimulant	29.5 ± 0.7	40.8 ± 0.1	58.0 ± 1.0	−19.3 ± 0.3	34.8 ± 0.4
Mater-Bi® 2 without biostimulant	28.4 ± 0.1	52.9 ± 0.0	52.1 ± 1.2	−19.7 ± 0.5	33.2 ± 0.7
Bare soil + PE	27.9 ± 0.1	42.7 ± 0.1	58.1 ± 0.7	−18.5 ± 0.2	33.0 ± 0.5
LDPE + PE	30.0 ± 0.4	61.4 ± 0.1	56.0 ± 2.1	−19.0 ± 0.4	33.2 ± 1.1
Mater-Bi® 1 + PE	31.1 ± 0.6	62.9 ± 0.1	62.3 ± 0.5	−17.9 ± 1.0	32.9 ± 1.9
Mater-Bi® 2 + PE	30.0 ± 0.2	50.5 ± 0.1	60.9 ± 0.4	−18.0 ± 0.2	32.6 ± 0.3
	NS	NS	NS	NS	NS

NS, *, *** Non-significant or significant at $p \leq 0.05$ or 0.001, respectively. Different letters in the same column indicate significant differences according to DMR test ($p = 0.05$). Means of biostimulant effect are compared according to Student's *t*-test ($p = 0.05$). All data are expressed as mean ± SE.

3.4. Dry Matter Percentage and Leaf Mineral Profile

The leaf dry matter percentage and nitrate content were significantly influenced by M × B interaction (Table 3). The recorded leaf dry matter percentage across the eight experimental treatments ranged from 3.5 to 4.2%, with the lowest values being recorded in bare soil without biostimulant application (Table 3). The recorded nitrate content across the eight experimental treatments (836–2685 mg kg^{-1} fw) was within the limits set by the EU Commission Regulation No 1258/2011 for the commercialization of fresh lettuce (3000–5000 mg kg^{-1} fw). Our results also demonstrated that the presence of mulching materials, in particular, the use of Mater-Bi® 2, evoked a significant increment in nitrate content confronted to bare soil in both untreated and biostimulant-treated lettuce plants. Interestingly, the nitrate content was significantly reduced by 23% in greenhouse lettuce plants receiving foliar application with tropical plant extract (1566 mg kg^{-1} fw) confronted to the control (2037 mg kg^{-1} fw) (Table 3).

Table 3. Mean comparison and analysis of variance for leaf dry matter percentage and mineral composition of untreated and biostimulant-treated greenhouse lettuce grown under LDPE mulch and biodegradable (Mater-Bi® 1 and Mater-Bi® 2) mulching materials in relation to bare soil.

Source of Variance	Dry Matter (%)	Nitrate (mg kg^{-1} fw)	P (g kg^{-1} dw)	K (g kg^{-1} dw)	Ca (g kg^{-1} dw)	Mg (g kg^{-1} dw)	S (g kg^{-1} dw)	Na (g kg^{-1} dw)
Mulch (M)								
Bare soil	3.7 ± 0.1 c	955 ± 54 d	7.9 ± 0.3 b	90.2 ± 2.1 a	6.4 ± 0.3 b	3.5 ± 0.2	0.8 ± 0.1	2.2 ± 0.1 b
LDPE	4.2 ± 0.0 a	1898 ± 74 c	9.4 ± 0.4a	77.3 ± 2.7 b	4.8 ± 0.4 c	2.7 ± 0.2	1.0 ± 0.1	2.3 ± 0.2 b
Mater-Bi® 1	4.1 ± 0.1 ab	1984 ± 152 b	9.0 ± 0.2 a	79.6 ± 1.2 b	5.9 ± 0.4 bc	3.3 ± 0.1	1.1 ± 0.1	2.3 ± 0.1 b
Mater-Bi® 2	3.9 ± 0.1 bc	2368 ± 142 a	7.5 ± 0.3 b	88.7 ± 1.2 a	8.3 ± 0.6 a	4.0 ± 0.1	1.1 ± 0.1	2.7 ± 0.0 a
	***	***	***	***	***	NS	NS	*
Biostimulant (B)								
Control	3.9 ± 0.1	2037 ± 180	8.1 ± 0.3	82.1 ± 2.3	5.8 ± 0.5	3.3 ± 0.2	0.9 ± 0.1	2.5 ± 0.1
Tropical plant extract (PE)	4.0 ± 0.1	1566 ± 135	8.9 ± 0.3	85.9 ± 1.8	6.8 ± 0.4	3.5 ± 0.2	1.1 ± 0.1	2.2 ± 0.1
t-test	0.758	0.048	0.050	0.206	0.049	0.320	0.157	0.024
M × B								
Bare soil without biostimulant	3.5 ± 0.1 c	1075 ± 2 f	7.3 ± 0.3	88.6 ± 2.7	6.0 ± 0.6	3.4 ± 0.3	0.8 ± 0.1	2.5 ± 0.1
LDPE without biostimulant	4.2 ± 0.1 a	2063 ± 10 c	8.7 ± 0.5	72.8 ± 2.7	4.1 ± 0.0	2.5 ± 0.1	1.0 ± 0.1	2.5 ± 0.1
Mater-Bi® 1 without biostimulant	4.2 ± 0.1 a	2325 ± 8 b	8.9 ± 0.2	78.9 ± 2.4	5.7 ± 0.4	3.3 ± 0.2	1.0 ± 0.1	2.4 ± 0.2
Mater-Bi® 2 without biostimulant	3.9 ± 0.2 b	2685 ± 3 a	7.3 ± 0.2	88.0 ± 2.4	7.9 ± 1.2	3.9 ± 0.2	0.9 ± 0.1	2.7 ± 0.1
Bare soil + PE	3.9 ± 0.1 b	836 ± 9 g	8.5 ± 0.2	91.9 ± 3.6	6.9 ± 0.4	3.6 ± 0.2	0.9 ± 0.1	2.0 ± 0.1
LDPE + PE	4.2 ± 0.1 a	1733 ± 10 d	10.0 ± 0.3	81.8 ± 3.1	5.5 ± 0.8	2.9 ± 0.3	1.1 ± 0.1	1.9 ± 0.3
Mater-Bi® 1 + PE	3.9 ± 0.1 b	1643 ± 13 e	9.0 ± 0.4	80.4 ± 0.5	6.2 ± 0.7	3.3 ± 0.2	1.1 ± 0.1	2.2 ± 0.2
Mater-Bi® 2 + PE	3.8 ± 0.0 b	2051 ± 3 c	7.9 ± 0.5	89.4 ± 1.0	8.7 ± 0.4	4.2 ± 0.1	1.1 ± 0.1	2.7 ± 0.1
	*	***	NS	NS	NS	NS	NS	NS

NS, *, *** Non-significant or significant at $p \leq 0.05$ and 0.001, respectively. Different letters in the same column indicate significant differences according to DMR test ($p = 0.05$). Means of biostimulant effect are compared according to Student's *t*-test ($p = 0.05$). All data are expressed as mean ± SE.

Neither mulching materials nor PE-application had a significant influence on Mg and S concentrations in greenhouse lettuce leaves (avg. 3.4 and 1.0 g kg^{-1} dw, respectively). The concentrations of target macronutrients and sodium in leaf tissues were significantly affected by mulching materials, with the highest values of K, Ca, and Na being recorded in lettuce plants that were grown under Mater-Bi® 2 mulching material (Table 3). Interestingly, PE biostimulant treatment, as averaged over mulching materials (M × B interaction = ns), affected P, Ca, and Na leaf tissues concentrations, which were greater by 10% and 17% (for P and Ca, respectively) and lower by 12% (for Na) when compared to the untreated lettuce (Table 3).

3.5. Antioxidant Activity and Bioactive Compounds

The hydrophilic antioxidant fraction of lettuce ranged from 5.6 to 7.5 mmol ascorbate eq. 100 g^{-1} dw (Table 4). Regardless of mulching materials, the antioxidant capacity in lettuce that was treated with the commercial biostimulant Auxym® was significantly higher (+9%) as compared to the untreated control (Table 4). Neither mulching materials nor PE-application had a significant influence on total phenols content in lettuce leaves (avg. 3.4 mg gallic acid eq. 100 g^{-1} dw). Moreover, phytochemicals with antioxidant properties, such as total ascorbic acid and carotenoids, were affected by both the tested factors (mulching materials, biostimulant application, and their combination). When averaged over the biostimulant application, the use of the Mater-Bi® 2 film evoked a significant increase in the biosynthesis and the accumulation of carotenoids (Table 4). The positive effect of Mater-Bi® 2 film on total ascorbic acid content has also been highlighted in the interaction with the biostimulant, where a major increase of total ascorbic acid (+168%) was recorded in comparison to the untreated and PE-treated lettuce grown in bare soil (Table 4).

Table 4. Mean comparison and analysis of variance for hydrophilic antioxidant activity, total phenols, total ascorbic acid and carotenoid contents of untreated and biostimulant-treated greenhouse lettuce grown under LDPE mulch and biodegradable (Mater-Bi® 1 and Mater-Bi® 2) mulching materials in relation to bare soil.

Source of Variance	Hydrophilic Antioxidant Activity (mmol ascorbate eq. 100 g^{-1} dw)	Total Phenols (mg gallic acid eq. 100 g^{-1} dw)	Ascorbic Acid (mg 100 g^{-1} fw)	Carotenoids (mg g^{-1} fw)
Mulch (M)				
Bare soil	7.3 ± 0.1	3.9 ± 0.2	7.2 ± 0.6 b	18.4 ± 2.2 ab
LDPE	6.4 ± 0.3	3.1 ± 0.3	8.1 ± 1.1 b	13.1 ± 1.4 b
Mater-Bi® 1	6.8 ± 0.4	3.4 ± 0.3	7.7 ± 2.3 b	17.4 ± 1.4 ab
Mater-Bi® 2	7.2 ± 0.3	3.3 ± 0.1	13.7 ± 3.1 a	21.0 ± 2.1 a
	NS	NS	*	*
Biostimulant (B)				
Control	6.6 ± 0.2	3.3 ± 0.2	6.0 ± 0.7	16.0 ± 1.3
Tropical plant extract (PE)	7.2 ± 0.2	3.6 ± 0.2	12.3 ± 1.6	18.9 ± 1.6
t-test	0.036	0.241	0.002	0.158
M × B				
Bare soil without biostimulant	7.1 ± 0.2	3.7 ± 0.5	6.1 ± 0.3 c	14.3 ± 1.9
LDPE without biostimulant	5.6 ± 0.1	3.0 ± 0.3	7.1 ± 0.8 bc	12.7 ± 3.1
Mater-Bi® 1 without biostimulant	6.7 ± 0.3	3.3 ± 0.6	3.0 ± 0.4 c	16.2 ± 1.1
Mater-Bi® 2 without biostimulant	6.9 ± 0.2	3.2 ± 0.2	8.0 ± 1.4 bc	20.8 ± 1.5
Bare soil + PE	7.5 ± 0.1	4.1 ± 0.0	8.3 ± 0.6 bc	22.5 ± 2.0
LDPE + PE	7.0 ± 0.1	3.1 ± 0.6	9.1 ± 1.9 bc	13.4 ± 0.6
Mater-Bi® 1 + PE	6.9 ± 0.8	3.6 ± 0.1	12.4 ± 2.0 b	18.6 ± 2.6
Mater-Bi® 2 + PE	7.5 ± 0.4	3.5 ± 0.1	19.3 ± 3.7 a	21.2 ± 4.3
	NS	NS	*	NS

NS, * Non-significant or significant at $p \leq 0.05$, respectively. Different letters in the same column indicate significant differences according to DMR test ($p = 0.05$). Means of biostimulant effect are compared according to Student's *t*-test ($p = 0.05$). All data are expressed as mean ± SE.

4. Discussion

The use of biodegradable mulching films and plant-based biostimulants has revolutionized modern agriculture in the last two decades. Nevertheless, no scientific studies have assessed the combinatorial effect of these two agricultural practices on crop performance and nutritional value of an important greenhouse leafy vegetable, such as lettuce. Our findings indicated that biodegradable mulching film Mater-Bi® 1 produced comparable marketable fresh yield to the commercial standard polyethylene (LDPE), while Mater-Bi® 2 exhibited the highest crop productivity. It is well established that plastic mulching films increase soil temperature in comparison to bare ground. This was the case in the current experiment, since the soil minimum, maximum, and mean temperature trends that were recorded in bare soil were always lower by 2.3–3.3 °C, 3.5–4.2 °C, and 2.8–3.8 °C, respectively, than those that were observed among the three tested mulching materials. The differences in fresh yield could be also associated to differences in soil temperatures, when temperature is a limiting factor (autumn–winter growing season; [23]). The results that were recorded in this greenhouse experiment endorse the previous study, where the span of soil temperature under the different mulching materials had a pronounced effect on marketable lettuce yield [24–27]. Our findings concerning the beneficial effect of mulching versus bare soil were also reported in previous studies on open-field and greenhouse vegetables. For instance, melon plants had more fruits and higher fruit mean weight when grown with biodegradable films and LDPE, as compared to bare soil [2]. An increase in marketable yield in the presence of polyethylene and biodegradable (Mater-Bi®) films when compared to bare soil was also observed in pumpkin [24], tomato [1,4,25], strawberry [3,26], garlic chives [5], as well as lettuce [27]. The use of plastic films may have preserved soil moisture and prevented water evaporation and the excessive leaching of nutrients in the rhizosphere [1].

Interestingly, the combination of film mulching (LDPE, Mater-Bi® 1, or Mater-Bi® 2) with the tropical plant extract biostimulant exhibited a positive and important synergistic effect (+30%) on both total and marketable yield. Particularly, the higher marketable production that was observed in greenhouse lettuce plants that were grown under mulching films and treated with PE-biostimulant, was due to an increase in the leaf area and not to the number of leaves per plant. The increase in crop productivity and biometric parameters of lettuce plants grown under protected cultivation has been previously reported in several research studies testing the action of this tropical plant extract biostimulant on leafy and fruit vegetables, such as tomato, jute, wall rocket, and lettuce [18,28,29]. The biostimulant action of the commercial product Auxym® on PE-treated lettuce plants could be associated to the presence of signaling molecules, such as carbohydrates, vitamins, but especially free amino acids and soluble peptides [14,18,30]. The hormone-like activity of plant-derived peptides that are contained in Auxym® has been proposed in many scientific papers, where the foliar application of vegetal-based biostimulants elicited auxin- and gibberellin-like activities and, thus, boosted yield [31,32]. Since many other signaling peptides have been identified in plant cells controlling growth, development, and stress responses of plants [33], it is expected that more signaling-peptide based PE will be developed in the near future. Furthermore, some indirect effects of amino acids can be postulated. The amino acid L-tryptophan is a precursor of indole compounds (thus including auxins), while L-methionine is known as the precursor of ethylene [30]. Finally, these bioactive compounds that are present in the plant-based biostimulants can act on the primary metabolism, increasing the photosynthetic activity of the plants, and it can act as well on root growth, which might increase water and nutrient absorption efficiency, thus resulting in a yield increase [18]. This was the case in the current study, where plants that were grown under plastic mulching films and treated with tropical plant extract were characterized by better physiological and biochemical status. The greater SPAD index and chlorophyll content of lettuce leaves corroborated this, thus confirming the better photosynthetic efficiency that leads to better plant performance. Similar results on the stimulation of the physiological and biochemical status of biostimulant-treated plants were also previously observed in greenhouse tomato [18], spinach [34], lettuce [35], and jute [28].

The leaf appearance in peculiar color is among the visual characteristics of leafy vegetables that steadily govern consumer preference and selection choice [36]. Lettuce green color is directly dependent upon chlorophyll synthesis in leaf tissue. Plant extract-biostimulant application affected lettuce greenness color ($-a^*$) to the extent it affected chlorophyll content, as observed earlier in a broad span of leafy greens, such as spinach, lamb's lettuce, and baby lettuce [29,37].

A negative aspect in the quality of leafy vegetables is, certainly, the high content of nitrates, as they are involved in the onset of different diseases [38]. Generally, vegetables that belong to the *Brassicaceae*, *Chenopodiaceae*, and *Asteraceae* families [18] may accumulate nitrates in their leaves. Significant genotypic variations in nitrate accumulation are shown for lettuce [39–41]. The nitrate concentration in plants is closely related to nitrate reductase activities [42]. This reality has prompted the European Commission to regulate the nitrate limits for lettuce. In our experiment, nitrate concentrations for plants cultivated with LDPE films, Mater Bi® 1, and Mater Bi® 2 films (1898–2368 mg kg^{-1} fw), were within the set limits for fresh lettuce according to Commission regulation (EU) No 1258/2011 (3000–5000 mg kg^{-1} fw) [43]. The PE-biostimulant application decreased nitrates concentration in lettuce leaves by 23% (avg. 1566 mg kg^{-1}), as compared to the control (avg. 2037 mg kg^{-1}). This positive effect could be linked to the presence of a high content of free amino acids in the biostimulant product, which, once absorbed by the plant, might exert the inhibition of the nitric ion transporters that are present in the root. On the other hand, the ability of the plant-based biostimulant Auxym® to reduce nitrates accumulation could be associated with the regulation of nitrogen metabolism in plants, which involves the activity of nitrate and nitrite reductase, glutamate synthase, as well as glutamine synthetase [14,18,44]. Various studies confirmed our results, such as that of Bulgari [45] performed on iceberg lettuce, which showed that a biostimulant of vegetal origin enriched with micro-elements (one), kept nitrate levels well under the limit required by the EC. Similar results were also obtained in spinach, on which the effect of amino acid-based biostimulant (Aminoplant) was evaluated [46]. Other studies on corn, soy, and wheat also showed that exogenous amino acids application can significantly reduce nitrate absorption [18].

Scientists recommend that people should consume fruits and vegetables daily, because they satisfy 11%, 35%, 7%, and 24% of the daily intake of P, K, Ca, and Mg, respectively [10]. These macronutrients help against certain diseases, such as blood pressure imbalances, hypertension (K), and osteoporosis (P, Ca, and Mg) [15]. For lettuce, several authors reported a potassium content between 48–72 mg g^{-1}, phosphorus 4–6 mg g^{-1}, magnesium 1.4–2.8 mg g^{-1}, and calcium 4–10 mg^{-1} on a dry weight basis [15,16,47]. In our work, the use of biodegradable films influenced the biofortification of macronutrinets in lettuce leaves. In particular, the use of polyethylene film Mater-Bi® 1 increased P content, confronted to the control, whereas lettuce plants that were grown under Mater-Bi® 2 exhibited higher values of K and Ca when compared to the bare soil treatment. Our results match with previous studies on the 'nutrient acquisition response' of plant-based biostimulant application on tomato [18], jute [28], and spinach [29]. In addition to the accumulation of macronutrients in leaf tissues of biostimulant-treated plants, the use of PE reduced sodium concentration in lettuce leaves by 12%, confronted to the control, which is in harmony with Carillo et al. [28] findings. This is a very important aspect, because Na causes hypertension and cardiovascular diseases [48].

Furthermore, lettuce is considered to be a good source of nutraceutical molecules, such as vitamin C and carotenoids [15]. These molecules represent the radical scavenging power that protects plants from the oxidative damage caused by free radicals. In our work, when averaged over biostimulant application, lettuce plants that were grown under Mater-Bi® 2 had the highest total ascorbic acid and carotenoids content. Similarly, Morra et al. [3] recorded a higher antioxidant activity, total polyphenols, and anthocyanins in two strawberry cultivars grown under biodegradable Mater-Bi film as compared to those cultivated with LDPE or in bare ground. These results are also confirmed for melon plants that are grown with biodegradable mulching films [27,49]. The Mater Bi® 2 behavior could be related to the fact that below this film there is a greater evaporation of the soil, which results in a lower accumulation of water in plants. Therefore, this mild condition of stress might trigger the plant to synthesize defensive molecules [27,49]. More compelling, these secondary metabolites are also crucial to human well-being [50,51].

In our work, the foliar application of PE on greenhouse lettuce also influenced antioxidant activity and health-promoting secondary metabolites, since the antioxidant potential increased by 10% when compared to the untreated control plants. The latter is a notable qualitative functional parameter in leafy vegetables, since it is correlated to the synergetic effect of low-molecular weight biologically active compounds, such as phenolic compounds and carotenoids [50]. Ertani et al. [52] showed an increase in antioxidant activity, lycopene, phenols, and ascorbic acid of *Capsicum chinense* L., in response to the application of plant extract based biostimulants. The synergistic action of Mater-Bi® 2 with tropical plant extract is of significant interest for scientists and nutritionists, because it resulted in the production of superior greenhouse lettuce leaves in terms of vitamin C content (+168% as compared to the control). In fact, as also shown by Carillo et al. [28], signaling compounds that are present in the tropical plant extract Auxym®, like glutamic and aspartic acids are involved in the stimulation of primary and secondary metabolism, thus, leading to a greater synthesis of antioxidant molecules, such as vitamin C [28].

5. Conclusions

In recent years, horticultural research has focused on improving farming practices in the framework of a more sustainable agricultural, including the use of biodegradable mulching films and vegetal-based plant biostimulants to improve the crop performance and nutritive quality of the produced commodities. Our greenhouse experiment on lettuce confirmed that the use of biodegradable plastic mulching materials, especially Mater-Bi® 2, could be considered as an alternative to LDPE and bare soil cultivation. This biodegradable mulching material increased marketable yield irrespective of the biostimulant application, due to many agronomic benefits, in particular, the better microclimate (minimum and maximum soil temperatures) in the rhizosphere. Our results also demonstrated, that lettuce plants grown under biodegradable film especially Mater-Bi® 2 exhibited superior quality traits in terms of K, Ca, total ascorbic acid, and carotenoids. Interestingly, the foliar application of PE-biostimulant in the presence of mulching materials was able to improve the total and marketable yield and biometric traits. The synergistic effect of mulching with plant-based biostimulant was linked to better physiological and biochemical status (higher SPAD index and chlorophyll content) and a higher nutrient acquisition response (higher P and Ca and lower Na content). The PE-biostimulant treated lettuce had a lower nitrate content and higher antioxidant scavenging capacity than the non-treated control, while the combination of Mater-Bi® 2 and PE-biostimulant resulted in the production of premium greenhouse lettuce leaves in terms of vitamin C content. The outcomes of the current study can encourage leafy vegetables producers to replace LDPE films with biodegradable ones in combination with plant-based biostimulants in order to attain high productivity and reach consumer expectations for high quality produce. In addition, the substitution of plastic mulching with biodegradable ones can significantly tackle the environmental issues that are related to the disposal of mulching materials at the end of the cropping cycles. The absence of dumping costs for farmers could likely offset the higher costs due to biodegradable mulching, favoring the application of biodegradable mulching materials on a wide scale.

Author Contributions: Conceptualization, E.C. and Y.R. methodology, E.C. software, C.E.-N., A.P., M.G., I.D.M., and M.M. formal analysis, C.E.-N., A.P., M.G., I.D.M., and M.M. investigation, E.C. resources, Y.R. data curation, C.E.-N., A.P., M.G., I.D.M., and M.M. writing—original draft preparation, M.G. writing-review and editing, M.G., N.F., C.E.-N., M.C.K., G.C. and Y.R. visualization, E.C. and Y.R. supervision, E.C., N.F. and Y.R. project administration, Y.R. funding acquisition, Y.R. All authors have read and agreed to the published version of the manuscript.

Acknowledgments: The authors are grateful to Vincenzo Esposito for his technical assistance in the greenhouse experiment.

References

1. Moreno, M.M.; Moreno, A. Effect of different biodegradable and polyethylene mulches on soil properties and production in a tomato crop. *Sci. Hortic.* **2008**, *116*, 256–263. [CrossRef]
2. Filippi, F.; Magnani, G.; Guerrini, S.; Ranghino, F. Agronomic evaluation of green biodegradable mulch on melon crop. *Ital. J. Agron.* **2011**, *6*, e18. [CrossRef]
3. Morra, L.; Bilotto, M.; Cerrato, D.; Coppola, R.; Leone, V.; Mignoli, E.; Pasquariello, M.S.; Petriccione, M.; Cozzolino, E. The Mater-Bi® biodegradable film for strawberry (*Fragaria* x *ananassa Duch.*) mulching: Effects on fruit yield and quality. *Ital. J. Agron.* **2016**, *11*, 203–206. [CrossRef]
4. Cowan, J.S.; Miles, C.; Andrews, K.P.; Inglis, D.A. Biodegradable mulch performed comparably to polyethylene in high tunnel tomato (*Solanum lycopersicum* L.) production. *J. Sci. Food Agric.* **2013**, *94*, 1854–1864. [CrossRef] [PubMed]
5. Adamczewska-Sowinska, K.; Turczuk, J. Yielding and biological value of garlic chives (*Allium tuberosum Rottl.* ex *Spreng.*) depending to the type of mulch. *J. Elem.* **2016**, *21*, 7–19. [CrossRef]
6. Scaringelli, M.A.; Giannoccaro, G.M.; Lopolito, A. Adoption of biodegradable mulching films in agriculture: Is there a negative prejudice towards materials derived from organic wastes? *Ital. J. Agron.* **2016**, *11*, 92–99. [CrossRef]
7. Thompson, A.A.; Samuelson, M.B.; Ignatius Kadoma, I.; Erick Soto-Cantu, E.; Rhae Drijber, R.; Wortman, S.E. Degradation rate of bio-based agricultural mulch is influenced by mulch composition and biostimulant application. *J. Polym. Environ.* **2019**, *27*, 498–509. [CrossRef]
8. Du Jardin, P. Plant biostimulants: Definition, concept, main categories and regulation. *Sci. Hortic.* **2015**, *196*, 3–14. [CrossRef]
9. Colla, G.; Rouphael, Y. Biostimulants in horticulture. *Sci. Hortic.* **2015**, *196*, 1–2. [CrossRef]
10. Rouphael, Y.; Kyriacou, M.C.; Vitaglione, P.; Giordano, M.; Pannico, A.; Colantuono, A.; De Pascale, S. Genotypic variation in nutritional and antioxidant profile among iceberg lettuce cultivars. *Acta Sci. Pol. Hortorum Cultus* **2017**, *16*, 37–45. [CrossRef]
11. Rouphael, Y.; Colla, G. Editorial: Biostimulants in agriculture. *Front. Plant Sci.* **2020**, *11*, 40. [CrossRef]
12. Rouphael, Y.; Colla, G. Synergistic biostimulatory action: Designing the next generation of plant biostimulants for sustainable agriculture. *Front. Plant Sci.* **2018**, *9*. [CrossRef]
13. Battacharyya, D.; Babgohari, M.Z.; Rathor, P.; Prithiviraj, B. Seaweed extracts as biostimulants in horticulture. *Sci. Hortic.* **2015**, *196*, 39–48. [CrossRef]
14. Colla, G.; Nardi, S.; Cardarelli, M.; Ertani, A.; Lucini, L.; Canaguier, R.; Rouphael, Y. Protein hydrolysates as biostimulants in horticulture. *Sci. Hortic.* **2015**, *196*, 28–38. [CrossRef]
15. Kim, M.J.; Moon, Y.; Tou, J.C.; Mou, B.; Waterland, N.L. Nutritional value, bioactive compounds and health benefits of lettuce (*Lactuca sativa* L.). *J. Food Comp. Anal.* **2016**, *49*, 19–34. [CrossRef]
16. Baslam, M.; Idoia, G.; Nieves, G. The arbuscular mycorrhizal symbiosis can overcome reductions inyield and nutritional quality in greenhouse-lettuces cultivated at inappropriate growing seasons. *Sci. Hortic.* **2013**, *164*, 145–154. [CrossRef]
17. Rouphael, Y.; Cardarelli, M.; Bassal, A.; Leonardi, C.; Giufrida, F.; Colla, G. Vegetable quality as affected by genetic, agronomic and environmental factors. *J. Food Agric. Environ.* **2012**, *10*, 680–688.
18. Colla, G.; Cardarelli, M.; Bonini, P.; Rouphael, Y. Foliar applications of protein hydrolysate, plant and seaweed extracts increase yield but di_erentially modulate fruit quality of greenhouse tomato. *HortScience* **2017**, *52*, 1214. [CrossRef]
19. Lichtentahler, H.K.; Buschmann, C. Chlorophylls and carotenoids; Measurement and characterization by UV-VIS spectroscopy. *Curr. Protocols Food Anal. Chem.* **2001**, *1*. [CrossRef]
20. Fogliano, V.; Verde, V.; Randazzo, G.; Ritieni, A. Method for measuring antioxidant activity and its application to monitoring the antioxidant capacity of wines. *J. Agric. Food Chem.* **1999**, *47*, 1035–1040. [CrossRef]
21. Singleton, V.L.; Orthofer, R.; Lamuela-Raventós, R.M. Analysis of Total Phenols and Other Oxidation Substrates and Antioxidants by Means of Folin-Ciocalteu Reagent. In *Methods in Enzymology*; Academic Press: Cambridge, MA, USA, 1999; Volume 299, pp. 152–178.
22. Kampfenkel, K.; Van Montagu, M.; Inzé, D. Extraction and determination of ascorbate and dehydroascorbate from plant tissue. *Anal. Biochem.* **1995**, *225*, 165–167. [CrossRef] [PubMed]

23. Brown, J.E.; Goff, W.D.; Dangler, J.M.; Hogue, W.; West, M.S. Plastic mulch color inconsistently affects yield and earliness of tomato. *HortScience* **1992**, *27*, 1135. [CrossRef]
24. Miles, C.; Shuresh Ghimire, S.; Wszelaki, A.; Moore, J. Biodegradable Plastic Mulch in Organic Vegetable Production Systems. In Proceedings of the New England Fruit and Vegetable Conference, Manchester, NH, USA, 17 December 2015.
25. Cirujeda, A.; Aibar, J.; Anzalone, A.; Martín-Closas, L.; Meco, R.; Moreno, M.M.; Pardo, A.; Pelacho, A.M.; Rojo, F.; Royo-Esnal, A.; et al. Biodegradable mulch instead of polyethylene for weed control of processing tomato production. *Agron. Sustain. Dev.* **2012**, *2*, 889–897. [CrossRef]
26. Cozzolino, E.; Leone, V.; Lombardi, P.; Piro, F. Telo biodegradabile su fragola, buoni effetti su resa e qualità. *L'Informatore Agrario* **2011**, *29*, 46–48.
27. Cozzolino, E.; Leone, V.; Lombardi, P. Nuovi teli pacciamanti in prova su lattuga sotto terra. *L'Informatore Agrario* **2014**, *24*, 37.
28. Carillo, P.; Colla, G.; El-Nakhel, C.; Bonini, P.; D'Amelia, L.; Dell'Aversana, E.; Pannico, A.; Giordano, M.; Sifola, M.I.; Kyriacou, M.C.; et al. Biostimulant application with a tropical plant extract enhances *Corchorus olitorius* adaptation to sub-optimal nutrient regimens by improving physiological parameters. *Agronomy* **2019**, *9*, 249. [CrossRef]
29. Di Mola, I.; Cozzolino, E.; Ottaiano, L.; Giordano, M.; Rouphael, Y.; Colla, G.; Mori, M. Effect of vegetal- and seaweed extract-based biostimulants on agronomical and leaf quality traits of plastic tunnel-grown baby lettuce under four regimes of nitrogen fertilization. *Agronomy* **2019**, *9*, 571. [CrossRef]
30. Colla, G.; Hoagland, L.; Ruzzi, M.; Cardarelli, M.; Bonini, P.; Canaguier, R.; Rouphael, Y. Biostimulant action of protein hydrolysates: Unraveling their effects on plant physiology and microbiome. *Front. Plant Sci.* **2017**, *8*, 2202. [CrossRef]
31. Ertani, A.; Schiavon, M.; Muscolo, A.; Nardi, S. Alfalfa plant-derived biostimulant stimulate short-term growth of salt stressed *Zea mays* L. plants. *Plant Soil* **2013**, *64*, 145–158. [CrossRef]
32. Colla, G.; Rouphael, Y.; Canaguier, R.; Svecova, E.; Cardarelli, M. Biostimulant action of a plant-derived protein hydrolysate produced through enzymatic hydrolysis. *Front. Plant Sci.* **2014**, *5*, 448. [CrossRef]
33. Tavormina, P.; De Coninck, B.; Nikonorova, N.; De Smet, I.; Cammue, P.A.B. The plant peptidome: An expanding repertoire of structural features and biological functions. *Plant Cell* **2015**, *27*, 2095–2118. [CrossRef] [PubMed]
34. Rouphael, Y.; Giordano, M.; Cardarelli, M.; Cozzolino, E.; Mori, M.; Kyriacou, M.C.; Bonini, P.; Colla, G. Plant and seaweed-based extracts increase yield but differentially modulate nutritional quality of greenhouse spinach through biostimulant action. *Agronomy* **2018**, *8*, 126. [CrossRef]
35. Luziatelli, F.; Ficca, A.G.; Colla, G.; Baldassarre Švecová, E.; Ruzzi, M. Foliar application of vegetal-derived bioactive compounds stimulates the growth of beneficial bacteria and enhances microbiome biodiversity in lettuce. *Front. Plant Sci.* **2019**, *10*, 60. [CrossRef]
36. Kyriacou, M.C.; Rouphael, Y.; Colla, G.; Zrenner, R.M.; Schwarz, D. Vegetable grafting: The implications of a growing agronomic imperative for vegetable fruit quality and nutritive value. *Front. Plant Sci.* **2017**, *8*, 741. [CrossRef]
37. Di Mola, I.; Ottaiano, L.; Cozzolino, E.; Senatore, M.; Giordano, M.; El-Nakhel, C.; Sacco, A.; Rouphael, Y.; Colla, G.; Mori, M.. Plant-based biostimulants influence the agronomical, physiological, and qualitative responses of baby rocket leaves under diverse nitrogen conditions. *Plants* **2019**, *8*, 522. [CrossRef] [PubMed]
38. Santamaria, P. Nitrate in vegetables: Toxicity, content, intake and EC regulation. *J. Sci. Food Agric.* **2006**, *86*, 10–17. [CrossRef]
39. Burns, I.G.; Zhang, K.; Turner, M.K.; Meacham, M.; Al-Redhiman, K.; Lynn, J.; Broadley, M.R.; Hand, P.; Pink, D. Screening for genotype and environment effects on nitrate accumulation in 24 species of young lettuce. *J. Sci. Food Agric.* **2011**, *91*, 553–562. [CrossRef] [PubMed]
40. Burns, I.G.; Durnford, J.; Lynn, J.; McClement, S.; Hand, P.; Pink, D. The influence of genetic variation and nitrogen source on nitrate accumulation and iso-osmotic regulation by lettuce. *Plant Soil* **2012**, *352*, 321–339. [CrossRef]
41. López, A.; Javier, G.; Fenoll, J.; Pilar, H.; Pilar, F. Chemical composition and antioxidant capacity of lettuce: Comparative study of regular-sized (*Romaine*) and baby-sized (*Little Gem* and *Mini Romaine*) types. *J. Food Comp. Anal.* **2014**, *33*, 39–48. [CrossRef]

42. Pinto, E.; Fidalgo, F.; Teixeira, J.; Aguiarc, A.A.; Ferreira, I.M.P.L.V.O. Influence of the temporal and spatial variation of nitrate reductase, glutamine synthetase and soil composition in the N species content inlettuce (*Lactuca sativa*). *Plant Sci.* **2014**, *219–220*, 35–41. [CrossRef] [PubMed]
43. Commission Regulation (EU). Commission Regulation (EU) n° 1258/2011 of 2 of December 2011 amending Regulation (EC) n° 1881/2006 as regards maximum level for nitrates in foodstuffs. *Off. J. Eur. Union* **2011**.
44. Calvo, P.; Nelson, L.; Kloepper, J.W. Agricultural uses of plant biostimulants. *Plant Soil* **2014**, *383*, 3–41. [CrossRef]
45. Bulgari, R.; Cocetta, G.; Trivellin, A.; Martinetti, L.; Ferrante, A. Prodotti biostimolanti ed effetti sulle colture ortofloricole. *Acta Ital. Horts* **2015**, *15*, 55–63.
46. Kunicki, E.; Grabowska, A.; Sekara, A.; Wojciechowska, R. The effect of cultivar type, time of cultivation and biostimulant treatment on the yield of spinach (*Spinacia oleracea* L.). *Folia Hortic.* **2010**, *22*, 9–13. [CrossRef]
47. Giordano, M.; El-Nakhel, C.; Pannico, A.; Kyriacou, M.C.; Stazi, S.R.; De Pascale, S.; Rouphael, Y. Iron biofortification of red and green pigmented lettuce in closed soilless cultivation impacts crop performance and modulates mineral and bioactive composition. *Agronomy* **2019**, *9*, 290. [CrossRef]
48. Krummel, R.D. Nutricíon en la Hypertension. In *10th Nutricíon y Dietoterapia*; De Krause Mahan, L.K., Escott-Stump, S., Eds.; McGraw-Hill: Contadero, Mexico, 2000; pp. 649–664.
49. Cozzolino, E.; Bilotto, M.; Leone, V.; Zampella, L.; Petriccione, M.; Cerrato, D.; Morra, L. Produzione e qualità di melone retatosu pacciamatura in in Mater-Bi®. *Colt. Protette* **2015**, *6*, 66–71.
50. Khanam, U.K.; Oba, S.; Yanase, E.; Murakami, Y. Phenolic acids, flavonoids and total antioxidant capacity of selected leafy vegetables. *J. Func. Foods* **2012**, *4*, 979–987. [CrossRef]
51. Subhasree, B.; Baskar, R.; Keerthana, R.L.; Susan, R.L.; Rajasekaran, P. Evaluation of antioxidant potential in selected green leafy vegetables. *Food Chem.* **2009**, *115*, 1213–1220. [CrossRef]
52. Ertani, A.; Pizzeghello, D.; Francioso, O.; Sambo, P.; Sanchez-Cortes, S.; Nardi, S. *Capsicum chinensis* L. growth and nutraceutical properties are enhanced by biostimulants in a long-term period: Chemical and metabolomic approaches. *Front. Plant Sci.* **2014**, *5*, 375.

7

Saffron Cultivation in Marginal Alpine Environments: How AMF Inoculation Modulates Yield and Bioactive Compounds

Matteo Caser [1], Íris Marisa Maxaieie Victorino [2,3,4], Sonia Demasi [1], Andrea Berruti [2], Dario Donno [1], Erica Lumini [2], Valeria Bianciotto [2] and Valentina Scariot [1,*]

1 Department of Agricultural, Forest and Food Sciences, University of Torino, Largo Paolo Braccini 2, 10095 Grugliasco, Italy; matteo.caser@unito.it (M.C.); sonia.demasi@unito.it (S.D.); dario.donno@unito.it (D.D.)

2 Institute for Sustainable Plant Protection, National Research Council (CNR), Viale Mattioli 25, 10125 Torino, Italy; irisvictorino@gmail.com (Í.M.M.V.); berruti.andrea@gmail.com (A.B.); erica.lumini@unito.it (E.L.); valeria.bianciotto@ipsp.cnr.it (V.B.)

3 Department of Life Sciences and Systems Biology, University of Turin, Viale Mattioli 25, 10125 Torino, Italy

4 Biological Science Department, Science Faculty, Eduardo Mondlane University (UEM), Av. Julius Nyerere-Campus Universitário, Maputo 3453, Mozambique

* Correspondence: valentina.scariot@unito.it

Abstract: Arbuscular mycorrhizal fungi (AMF) establish mutualistic symbiotic associations with plant roots and act as biofertilizers by enhancing plant nutrient and water uptake. Moreover, AMF colonization may influence the biosynthesis of plant bioactive compounds in medicinal and aromatic plants. There is limited information on AMF associations with *Crocus sativus* L. (saffron) roots and their effect on crop performances and spice quality. In the present work we verified the efficiency of root mycorrhization in potted conditions, and then we evaluated the yield and quality of the saffron produced in two Alpine sites during two cultivation cycles with the application of AMF. Two inocula were applied, either a single-species (*Rhizophagus intraradices*) or a multispecies mixture (*R. intraradices* and *Funneliformis mosseae*). The trial conducted in potted conditions confirmed that both AMF commercial inocula established symbiotic relationships with saffron roots. The multispecies inoculation yielded the highest content of arbuscules in colonized portions of the root (100%), while the single-species was slightly less (82.9%) and no AMF were recorded in untreated control corms. In open-field conditions, AMF colonization of the root systems, flower production, and saffron yields were monitored, and bioactive compounds contents and antioxidant activity in the dried spice were analyzed using spectrophotometry and high performance liquid chromatography. Overall, the saffron produced was high quality (ISO category) and had high contents of bioactive compounds, with very high total polyphenol content and elevated antioxidant activity. The use of arbuscular mycorrhizal symbionts as biostimulants positively affected saffron cultivation, improving the crop performances and the content of important nutraceutical compounds. In particular, the inoculum composed by *R. intraradices* and *F. mosseae* increased flower production and the saffron yield. *R. intraradices* alone enhanced the spice antioxidant activity and the content of bioactive compounds such as picrocrocin, crocin II, and quercitrin. Since saffron is the world's highest priced spice, the increases in yield and quality obtained using AMF suggests that farms in marginal areas such as alpine sites can increase profitability by inoculating saffron fields with arbuscular mycorrhiza.

Keywords: *Crocus sativus* L.; biofertilization; arbuscular mycorrhizal fungi; antioxidant activity; crocin; picrocrocin; polyphenols; safranal

1. Introduction

Saffron (*Crocus sativus* L.) is a triploid herbaceous geophyte that is reproduced by means of replacement corms and is cultivated in environments with very different soil characteristics [1–3] for its red scarlet stigmas that are used worldwide as a spice and natural dye [4]. Origin, abiotic stresses, agronomical practices, and processing methods (stigma separation, drying, and storage) can influence both the plant and the saffron spice yield, composition, and quality [5,6]. The spice's organoleptic properties are ascribed to the relative percentage of peculiar secondary metabolites—crocin, picrocrocin, and safranal—which provide the unique color, bitter taste, and aroma, respectively. The concentrations of these constituents combine to determine the saffron spice quality, as defined by the International Organization for Standardization [7]. Studies related to saffron quality are expanding mainly due to the antioxidant properties of this spice and their positive influence on human health [8]. Antitumor and cancer-preventive properties are mainly attributed to the high carotenoids content [9].

Reproductive, vegetative, and dormancy are the main phenological stages [10]. Saffron flower induction is a very complicated mechanism directly related to ecological conditions and field management [11,12]. As in most geophyte plants, both seasonal and daily thermoperiodism are involved as the main environmental factors [11]. Flower induction requires an incubation of the corms at high temperature (23–27 °C), followed by a period of exposure at moderately low temperature (17 °C) for flower emergence. In Mediterranean environments, flower induction occurs from early spring to mid summer, while flower emergence occurs from early- to late-autumn. Differences in the time required for flower initiation have mostly been attributed to the corm size [13]. In addition, Molina et al. [14] reported that air and soil temperatures might be responsible for differential flower induction and duration of up to two months. Flowering is followed by a vegetative stage throughout the winter and formation of replacement corms at the base of shoots. At the end of spring, the leaves reach the highest length, start to senesce, and wither, and the bulbs go into dormancy [14].

Due to its unique biological, physiological, and agronomic traits, saffron is able to exploit marginal land and is included in low-input cropping systems, even if high amount of skilled labour is required [11]. In Italy, saffron cultivation is gaining increasing attention as an alternative crop for sustainable agriculture systems [11,15], where it could represent a valid mean for increasing incomes of multifunctional farms, with a positive impact on the recovery and economy of these areas [15,16]. Since saffron is the world's highest-priced spice due to the intensive hand labour required for daily flower picking and stigma separation [14], small increases in the yield and/or quality can connote a large increase in profitability. In this context, the adoption of sustainable cultivation techniques such as the use of biostimulants may represent further help in both the increase in spice yield and active ingredients accumulation [17].

Recent research has focused on the benefits of soil organisms to crops, especially to promote plant nutrient uptake and assimilation [18,19]. Indeed the soil is not only the location of plant life cycle stages, but also the main reservoir for a wide range of plant biostimulants (PBs), including arbuscular mycorrhizal fungi (AMF) [19–21]. Ubiquitous and abundant, AMF are obligate endosymbionts living inside most plant roots present in diverse environments, including productive agricultural systems [22–25]. When colonizing roots, hyphae extend root limits, improving water and inorganic nutrient acquisition from the soil, mainly phosphorus (P) and other minerals, in exchange for photosynthetic products. The use of AMF has a demonstrated economic impact on agriculture and horticulture and they may also confer pathogen protection by altering plant physiological parameters, and improving soil nutrition and aggregation under different growing conditions [26–28].

Mounting evidence indicates that AMF may induce changes in primary and secondary metabolism of host plants, increasing polyphenols, flavonoids, and phytohormone dynamics [29,30]. Such metabolic changes may be ascribed to a transient activation of host defence reactions in colonized roots [20,31]. The role of AMF symbiosis in flowering date and flower production is fragmented [32].

In medicinal and aromatic plants (MAPs), such as *Arnica montana* L., *Coriandrum sativum* L., and *Anethum graveolens* L., AMF colonization influenced bioactive compound biosynthesis such as ascorbic acid, flavonoids, polyphenols, carotenoids, and vitamins [33–36]. Inoculation with *Funneliformis mosseae* Gerd. & Trappe and *G. versiforme* P. Karst. improved plant growth and enhanced the glycyrrhizin concentration in *Glycyrrhiza uralensis* Fisch plants [26]. Moreover, under low P availability, a mix of AMF increased the production of root biomass and of pseudohypericin and hypericin content in flowers of *Hypericum perforatum* L. [32]. Although widely applied, evidence for AMF symbiosis efficacy and persistence is scant, incomplete, or lacking [37,38] and the use of AMF in crop production is facing some limitations due to product costs, producer awareness levels, and variability in mycorrhizal inoculum quality [21,27]. Many factors can affect the success of inoculation and AMF persistence, including environmental and cultivation conditions, species compatibility, degree of spatial competition with other soil organisms, and the time of inoculation. However, once AMF inoculation is restored and well established in soil, the AMF community will persist through time. If detrimental practices are minimized before and after cultivation, biodiverse mycorrhizal hyphal networks will remain unaltered and infective in the field [27]. Hence, it is important to assess the effects of AMF on crop traits both as early application and as residual persistence in the following crop cultivation seasons.

Incidence of AMF, alone or in combination with plant growth promoting bacteria (PGPB), was reported in corms of *C. sativus* [39–44]. Different authors report that well-established AMF colonization of saffron roots results in increased corm P content, chlorophyll, fresh and dry corm mass, and leaf matter, and greater soil P and nitrogen assimilation [43–45]. Shajari et al. [44] indicated a significant effect of AMF in corm growth and mineral assimilation during the second cultivation season, supporting their effective residual effects in saffron cultivation. However, little is known about the effects of AMF on spice yield, and phytochemical profiles in open field cultivation [46,47].

The possibility that AMF can enhance the economic value of saffron by increasing yield and quality is even more interesting if we consider the worldwide increase in use of biocompounds in the food and pharmaceutical industries. Thus, the aims of the present study were (1) to preliminarily verify the constitutive association of AMF with saffron roots in sterile pot conditions, and (2) to assess the AMF symbiosis in open field conditions and its effects on saffron plant growth, productivity, and bioactive compounds content in Alpine open field conditions.

2. Materials and Methods

2.1. AMF Inoculation in Pot

Saffron corms with horizontal diameters of 1.3 to 2.8 cm were sown in pots (4 L; 1 corm per pot) in the last ten days of August 2016. Pots were filled with sterile quartz sand (3 L per pot) on a layer of sterilized expanded clay (1 L per pot). Corms were treated with two inocula (MycAgro Lab, Breteniére, FR), one composed of a single fungus *Rhizophagus intraradices* (Ri) and one of *R. intraradices* and *Funneliformis mosseae* (Ri + Fm). Ten grams of each inoculum were placed under each corm in order to guarantee the contact between the inoculum and the roots and therefore to favor the symbiosis between AMF and roots. Saffron corms used as controls were not inoculated (AMF-). Corms were not treated against fungal pathogens. A randomized block design was used with a total of 48 pots displayed in two experimental plot units (24 pots per unit) and three treatments (8 pots per treatment). Cultivation lasted for one cycle (August 2016–April 2017) in a heated glasshouse of the Department of Agricultural Forest and Food Sciences (DISAFA) of the University of Torino (Italy, 45°06′23.21″ N Lat, 7°57′82.83″ E Long; 293 m a.s.l.), with an average temperature of 22 °C during the day and 16 °C in the night. Irrigation water (pH 7.4, EC 505 µS cm) was added weekly (250 mL per pot) with a drip system. The corms were fertilized by fertigation (VIGORFLOR, AL.FE. srl, MN, Italy) every two weeks starting from the emergence of the spate, in quantities of 1.5 g L^{-1} of water. No flowering occurred because of the small size of the corms.

2.2. AMF Inoculation in Open Field

Saffron corms with horizontal diameters of 2.5 to 3.5 cm were planted in the last ten days of August 2016 in two Alpine experimental sites located in the municipality of Morgex (45°45′35.1″ N; 7°02′37.3″ E; 1000 m a.s.l.) and Saint Cristophe (45°45′06.9″ N; 7°20′37.0″ E; 700 m a.s.l.) in Italy and cultivation lasted for two cycles (2016–2017 and 2017–2018). Both sites were cultivated with saffron for at least the previous three years. Before starting the experiment both fields were milled. To assess the effects of AMF inocula on saffron cultivation and production, the same treatments used in the pot trial were applied (Ri, Ri + Fm or AMF-). A randomized block design was used, with three experimental plot units (blocks). Each plot unit consisted of 56 corms, planted in a 1.44 m^2 area (39 corms m^{-2}). Inter-row planting distance was of 7 cm, while between-row distance was 25 cm. Plots were separated from each other with at least 4 m distance. Before planting, 10 g of inoculum was placed under the corms to ensure contact between plant and the treatment. Irrigation was provided when needed and hand weeding control was conducted during cultivation, while no preplanting fertilization, tillage, or treatments against pathogens were applied. The two Alpine sites were characterized by semicontinental climate, with a long and cold winter (Supplementary Figure S1). In general, both sites had a sandy-loam texture according to the USDA classification and similar chemical characteristics (Supplementary Table S1).

2.3. AMF Evaluation

At the end of the vegetative phase in both pot (February 2017) and open field experiments (April 2017 and 2018), saffron roots were harvested, rid of topsoil, cleaned and stained with 0.1% (*w/v*) cotton blue in 80% lactic acid overnight, then destained 3 times with lactic acid for 18 h, cut into 1-cm-long segments and placed on microscope slides for further morphological analysis. Approximately 25 fragments were observed under light microscope for each replicate for a total of 300 root fragments. Fungal colonization was determined and calculated as described by Trouvelot et al. [48].

2.4. Plant Performance and Saffron Yield in Open Field

The daily number of picked flowers per corm (Supplementary Figure S2) and the yield of spice (i.e., stigmas dried at 40 °C for 8 h in an oven) were measured at flowering (November 2016 and 2017). When leaves were fully expanded (April 2017 and 2018), 50 mg of fresh leaves per treatment were used to determine chlorophyll and carotenoids content as described by Caser et al. [49]. Simultaneous with leaf sampling, the Chlorophyll Meter SPAD-502 (Konica Minolta Sensing Inc., Osaka, Japan) was used to determine the relative quantity of chlorophyll present in 27 randomly selected plants per treatment in the field.

At the end of full plant development (April 2017 and 2018), the leaves length of all corms was measured. Then, 27 plants per treatment were lifted, and corms rid of topsoil, cleaned, and detunicated. The wilted rate as the ratio between the number of wilted corms and the total number of sown corms, the shoot caliber size, and the number, the size and the weight of replacement corms were determined.

2.5. Saffron Extract Preparation and Quality

The saffron aqueous extracts were prepared according to Gresta et al. [11]. Fifty mg of powdered saffron from each treatment and both cultivation years were put into 5 mL of deionized water. After stirring for 1 h at room temperature (circa 21 °C) in the dark, the solution was filtered with polytetrafluoroethylene (PTFE, VWR international, Milano, Italy) filters of 25 mm diameter and 0.45 μm pore size. The saffron extract obtained was diluted 1:10 with deionized water (1 mg mL^{-1}). Saffron extracts were analyzed with a spectrophotometer (Ultrospec 2100 Pro, GE Healthcare, UK Ltd., Little Chalfont, Buckinghamshire, UK) to determine the amount of picrocrocin, crocin, and safranal, according to ISO 3632 [7].

2.6. Total Phenols

The content of total phenols (TPC) was measured by using the Folin–Ciocalteu's phenolic method and determined as reported by Donno et al. [50]. Five hundred μl of saffron extract was added and mixed with 30 mL of deionized water, 2.5 mL of Folin–Ciocalteu's reagent (diluted 1:10), and, after eight minutes, 10 mL of 7.5% (*w/v*) saturated sodium carbonate solution. The solution was incubated at room temperature for 2 h in the dark and the absorbance was detected at 765 nm with a spectrophotometer (Ultrospec 2100 Pro, GE Healthcare, UK Ltd., Little Chalfont, Buckinghamshire,UK). The results were expressed as mg of gallic acid equivalents (GAE) per 100 g of fresh weight (FW).

2.7. Total Anthocyanins

The total anthocyanins content (TAC) was determined using the pH-differential method [50]. Saffron extracts were added to a pH 1 and pH 4.5 buffer solutions. Absorbance of samples was determined at 515 nm and 700 nm after a 15 min equilibration. The formula for calculating TAC is as follows

$$\text{TAC (mg L}^{-1}) = (\text{A} \times \text{sample dilution factor} \times 1000)/(\text{molar absorptivity} \times 1) \quad (1)$$

where A is (Absorbance 515 nm—Absorbance 700 nm) at pH 1.0—(Absorbance 515 nm Absorbance 700 nm) at pH 4.5. The results were expressed as milligrams of cyanidin 3-O-glucoside (C3G) per 100 gof fresh weight (mg of C3G 100 g^{-1} FW).

2.8. Antioxidant Activity

The antioxidant activity (AOA) was determined using the ferric reducing antioxidant power (FRAP) method as reported by Caser et al. [51] and the 2′-azinobis (3-ethylbenzothiazoline-6-sulfonic acid) (ABTS) method as described by Urbani et al. [52].

For the FRAP method, a total of 30 μL of saffron extract was added and mixed with 90 μL of deionized water and 900 μL of the FRAP reagent. After incubation at 37 °C for 30 min, the absorbance of the solutions was measured at 595 nm using a spectrophotometer (Ultrospec 2100 Pro, GE Healthcare, UK Ltd., Little Chalfont, Buckinghamshire, UK). Results were expressed as millimoles of ferrous iron (Fe^{2+}) equivalents per kilogram of fresh weight.

The ABTS radical cation (ABTS +) was obtained by the reaction of 7.0 mM ABTS stock solution with 2.45 mM potassium persulfate solution. After the incubation for 12–16 h before use in the dark and at room temperature, the solution was diluted with distilled water to obtain an absorbance of 0.70 (±0.02) at 734 nm. After addition of 0.6 mL of diluted ABTS·+ solution to 180 μL of saffron extract, the reaction was left in the dark at room temperature for six min. The absorbance was then measured at 734 nm using a spectrophotometer (Ultrospec 2100 Pro, GE Healthcare, UK Ltd., Little Chalfont, Buckinghamshire, UK). The antioxidant activity was expressed as μmol of Trolox equivalents per gram of dry weight (μmol TE g^{-1} DW).

All analyses were performed in three replicates.

2.9. Identification and Quantification of Bioactive Compounds

The chromatographic analysis of saffron extracts (Supplementary Table S2) was conducted with an Agilent 1200 high-performance liquid chromatograph coupled to a diode array detector (HPLC-DAD; Agilent Technologies, Santa Clara, CA, USA), according to established methods [53]. Different chromatographic methods were used for analysis: benzoic acids (ellagic and gallic acids), catechins ((+) catechin and (−) epicatechin), cinnamic acids (caffeic, chlorogenic, coumaric, and ferulic acids), flavonols (hyperoside, isoquercitrin, quercetin, quercitrin, and rutin), carotenoids (crocin I and II and safranal), and vitamin C (ascorbic + dehydroascorbic acids).

Four chromatographic methods were used to separate the bioactive molecules on a Kinetex C18 column (4.6 × 150 mm, 5 μm, Phenomenex, Torrance, CA, USA). Several mobile phases were used for bioactive compound identification and ultraviolet (UV) spectra were recorded at different wavelengths, based on HPLC methods, previously tested and validated [4], with some modifications. UV spectra were recorded at 330 nm (α), 280 nm (β), 310 and 441 nm (χ), and 261 and 348 nm (δ).

All single compounds were identified in samples by comparison and combination of their retention times and UV spectra with those of authentic standards analyzed with the same chromatographic conditions.

2.10. Chemicals and Reagents

All the chemicals and reagents used for the AMF evaluation, phenols, anthocyanins, FRAP, and ABTS assays and bioactive quantification were purchased from Sigma-Aldrich (Saint Louis, MO, USA).

2.11. Statistical Analysis

Arcsin transformation was performed on all percent incidence data before statistical analysis in order to improve homogeneity of variance (Levene test). All the analyzed data were checked for normality of variance. For all indices analyzed in the greenhouse assay, mean differences were computed using a one-way analysis of variance (ANOVA) with Tukey's post hoc test ($p \leq 0.05$). Data from open field were analyzed by means of a linear mixed effect models considering AMF treatments as a fixed factor, year as a repeated measure, and sites and blocks as random factors. The following interactions (year $\times$ AMF treatment) were included in the model. Pairwise comparisons (according to sequential Bonferroni post hoc tests) were used to separate means when a treatment was significantly affecting the variable at a $p \leq 0.05$. All presented values are means of untransformed data. All computations were conducted with SPSS statistical package (version 25.0; SPSS Inc., Chicago, IL, USA).

3. Results

3.1. Assessment of Saffron Mycorrhization at Pot and Open Field Scale

Values concerning intensity of colonization in the root system and abundance of arbuscules or coils in the saffron roots in potted conditions are shown in Table 1. Corms treated with Ri + Fm reached the highest level of mycorrhization (M%) (93.33%), however, high levels were also obtained with Ri inoculum (71.37%). The Ri + Fm treatment also had the highest occurrence of arbuscules (a%) in the mycorrhizal portions (100%), being significantly higher than Ri (82.99%) and AMF-(0%).

Table 1. AMF colonization indices (intensity in the whole root system, M; intensity of the mycorrhizal portions, m; presence of arbuscules in the whole root system, A; presence of arbuscules in the mycorrhizal portions, a) of *Crocus sativus* L. roots treated with the inoculum composed by *Rhizophagus intraradices* and *Funneliformis mosseae* (Ri + Fm), *R. intraradices* alone (Ri), or the control (AMF-) in the saffron pot cultivation.

	Index (%)			
Treatment	**M**	**m**	**A**	**a**
Ri + Fm	93.33 a	93.33 a	93.33 a	100.00 a
Ri	71.37 b	80.28 b	58.98 b	82.99 b
AMF-	0.07 c	0.33 c	0.00 c	0.00 c
p	***	***	***	***

Mean values with the same letter are not statistically different at $p \leq 0.05$ according to Tukey's post-hoc tests. The statistical relevance is provided (*** $p < 0.001$).

In open field conditions, the AMF root colonization measurements in *C. sativus* treated with Ri + Fm or with Ri alone during the two cultivation cycles are presented in Table 2. In general both the presence of arbuscules in the mycorrhizal portions (a%) and in the whole root system (A%) indices were affected by the inoculum composition only in the first cultivation year, while control plants

(AMF-) were not colonized. In the second year, low root colonization was observed and no differences among the treated and untreated corms were detected.

Table 2. AMF colonization intensity in open field conditions after the first and second cultivation year of the whole root system (M) and of the mycorrhizal portions (m), and presence of arbuscules in the whole root system (A) and in the mycorrhizal portions (a) of *Crocus sativus* roots treated with inoculum composed of *Rhizophagus intraradices* and *Funneliformis mosseae* (Ri + Fm), *R. intraradices* alone (Ri), or the control (AMF-).

Effect	Index (%)			
Year 1	**M**	**m**	**A**	**a**
Ri + Fm	11.6 a	11.7	4.0 a	26.6 a
Ri	13.8 a	14.2	6.9 a	38.1 a
AMF-	1.7 b	3.4	0.0 b	0.0 b
p	*	ns	***	***
Year 2				
Ri + Fm	7.0	8.5	0.8	12.5
Ri	16.1	16.5	1.6	8.31
AMF-	4.73	6.1	2.5	18.8
p	ns	ns	ns	ns
Year × Treatment (*p*)	*	ns	*	*

Values with the same letter denote no significant differences. The statistical relevance is provided (ns, not significant; * $p < 0.05$; *** $p < 0.001$).

3.2. Impact of AMF Symbiosis on Saffron Productivity and Qualitative Traits in Open Field

Significant differences between the two cultivation years emerged for several studied parameters. In general, the wilting rate, all the main productivity traits (number of flowers m^{-2}, number of flowers per corm, mg of saffron m^{-2}, saffron per flower, and the number of replacement corms), and the content of leaf chlorophyll and carotenoids significantly increased after the second year of cultivation (Table 3); a reduction in leaf length, SPAD unit, and shoot size was also observed.

Table 3. Effects of cultivation seasons (Year 1 and Year 2), AMF treatments (Ri + Fm was composed of *Rhizophagus intraradices* and *Funneliformis mosseae*, Ri of *R. intraradices* alone, and AMF-was the uninoculated control), and their interaction (Year × AMF treatment) on saffron plant growth and productivity based on linear mixed-effects models considering AMF treatments as a fixed factor, year as a repeated measure, and sites and blocks as random factors.

	Growing Seasons			AMF Treatments				Year × AMF
Traits	**Year 1**	**Year 2**	***p***	**Ri + Fm**	**Ri**	**AMF-**	***p***	***p***
Wilting rate (%)	39.0	54.4	***	44.3	50.0	45.8	ns	ns
Flower (n m^{-2})	49.8	101.7	***	91.8 a	61.9 b	66.4 b	*	*
Flower/corm (n)	1.5	4.2	***	5.1 a	3.8 b	3.9 b	*	*
Saffron yield (mg m^{-2})	278.0	700.0	***	645.3 a	377.4 b	477.2 b	*	*
Saffron/flower (mg)	6.0	7.0	**	7.1 a	5.8 b	7.3 a	*	*
Leaf length (cm)	36.8	24.1	***	31.4	30.3	29.9	ns	ns
SPAD unit	74.8	45.7	***	60.0	61.1	59.7	ns	ns
Shoot size (mm)	5.3	4.1	**	5.5 a	3.3 b	4.2 ab	**	*
Corm size (mm)	21.1	20.2	ns	19.8	20.0	22.2	ns	ns
Replacement corm (n)	2.2	3.7	*	2.8	3.4	2.7	ns	ns
Corm weight (g)	7.7	6.5	ns	7.8	7.4	6.3	ns	ns
Chlorophyll (µg mg^{-1})	1.6	4.1	***	2.9	2.9	2.7	ns	ns
Carotenoids (µg mg^{-1})	0.6	2.2	***	1.4	1.5	1.4	ns	ns

Values with the same letter denote no significant differences. The statistical relevance is provided (ns, not significant; * $p < 0.05$; ** $p < 0.01$; *** $p < 0.001$).

In Table 3, the productivity and growth traits influenced by the AMF treatments are also reported. Particularly, the inoculum composed of the mix of *R. intraradices* and *F. mosseae* significantly increased the number of flowers m^{-2} (+138.2%), the number of flowers $corm^{-1}$ (+130.8%), and the mg of saffron m^{-2} (+135.2%) in comparison to other treatments. In contrast, the mg of saffron $flower^{-1}$ and the shoot size were significantly reduced (−20% and −40%, respectively) by the inoculum of *R. intraradices* alone in comparison to Ri + Fm and AMF-. Significant interaction between cultivation year and AMF treatments resulted for the number of flowers m^{-2}, the number of flowers $corm^{-1}$, saffron yield, saffron $flower^{-1}$, and shoot size.

Regarding the synthesis of bioactive molecules in the studied saffron spice, differences between the two cultivation seasons occurred (Table 4). Overall, the saffron produced at the two experimental sites belonged to the quality category I for the picrocrocin, safranal, and crocins analysis [7] with a significant increase after the second cultivation year. On the contrary, different bioactive compounds (isoquercitirin, quercitrin, ellagic acid, safranal, and total vitamin C) were significantly reduced. Very few differences were observed among AMF treatments (Table 4). Both Ri + Fm and Ri positively affected the antioxidant activity (FRAP assay) of the saffron produced. While, the effect of the Ri inoculum significantly increased the absorbance value of picrocrocin (ISO 3632) and the content of quercitrin in comparison to Ri + Fm, and the content of crocin II compared to AMF- (Table 4). A significant interaction between cultivation seasons and AMF treatments resulted for picrocrocin (ISO), quercitrin, crocin II, and antioxidant activity (FRAP assay).

Table 4. Effects of cultivation seasons (Year 1 and Year 2), AMF treatments (Ri + Fm was composed of *Rhizophagus intraradices* and *Funneliformis mosseae*, Ri of *R. intraradices* alone, and AMF- was the uninoculated control), and their interaction (year × AMF treatment) on bioactive compounds, total polyphenol content (TPC), anthocyanins, quality traits as defined by ISO 3632 [7], and antioxidant activity of the produced saffron based on liner mixed-effects models considering AMF treatments as a fixed factor, year as a repeated measure, and sites and blocks as random factors.

	Growing Seasons			**AMF Treatments**				**Year × AMF**
Traits	**Year 1**	**Year 2**	***p***	**Ri + Fm**	**Ri**	**AMF-**	***p***	***p***
ISO 3632 [7]								
Picrocrocin	131.4	135.0	*	130.2 b	138.7 a	136.1 a	*	*
Safranal	38.8	44.2	**	39.9	43.8	40.8	ns	ns
Crocins	207.1	368.5	***	275.9	303.5	284.1	ns	ns
Bioactive compounds (mg 100 g^{-1} dry weight)								
Coumaric acid	23.6	23.5	ns	23.6	23.5	23.7	ns	ns
Isoquercitrin	2.6	2.5	*	2.5	2.5	2.6	ns	ns
Quercitrin	22.8	16.0	***	17.0 b	22.3 a	18.9 ab	*	*
Gallic acid	5.0	4.9	ns	4.9	4.9	5.1	ns	ns
Ellagic acid	2.7	0.8	***	2.0	2.1	1.3	ns	ns
Catechin	3.4	3.1	ns	2.7	3.0	4.3	ns	ns
Epicatechin	6.1	8.3	ns	6.4	6.3	9.0	ns	ns
Safranal	4.4	4.0	***	4.2	4.3	4.2	ns	ns
Crocin I	32.5	67.9	**	49.2	37.7	63.8	ns	ns
Crocin II	31.1	36.6	*	35.0 ab	38.8 a	27.7 b	*	*
Total Vitamin C	76.5	67.0	**	71.4	70.1	73.2	ns	ns
TPC (mg_{GAE} 100 g^{-1} DW)	1340.7	2355.5	ns	1906.1	1868.8	1819.5	ns	ns
Antocyanins (mg_{C3G} 100 g^{-1} DW)	1866.5	1633.6	ns	964.1	2418.8	1867.3	ns	ns
Antioxidant activity								
FRAP (mmol Fe^{2+} kg^{-1})	408.9	1937.1	***	424.8 a	463.8 a	338.2 b	***	***
ABTS (μmol_{TE} g^{-1})	4.2	4.6	ns	4.3	4.5	4.6	ns	ns

Values with the same letter denote no significant differences. The statistical relevance is provided (ns, not significant; * $p < 0.05$; ** $p < 0.01$; *** $p < 0.001$).

4. Discussion

4.1. AMF Colonization

In the literature only some studies report AMF colonization of *C. sativus* roots. In the present study, their presence in potted cultivation was detected in *C. sativus* roots subjected to both AMF treatments (Ri + Fm and Ri). Saffron root fragments showed extensive AM fungal colonization, characterized by a moderate to high intensity of colonization and arbuscule formation. Saffron root colonization in the present pot cultivation trial was markedly superior to the results obtained in the open field test. This could be due to the antagonistic action of the naturally occurring fungi in the soil that compete with the AMF and by the different cultivation substrate used. However, to the best of our knowledge, this is the first report clearly indicating and measuring successful symbiosis between *C. sativus* roots and AMF under pot cultivation conditions.

Our open field data are equal or lower than findings obtained in other open field trials as reported by Aimo et al. [40] and Lone et al. [43]. Applying the percent colonization method for root AMF evaluation, these authors reached a maximum of 30% and 60% mycorrhizal colonization in saffron roots in Italy and Kashmir cultivation fields when using a mix of AMF belonging to the genus Glomus, respectively. In a similar study conducted in Iranian fields, the percentage of root colonization of saffron was of 39% [42], while in a field in Kashmir, ranged between 15 and 90% on the basis of the season [43]. As reported in Supplementary Table S1, P Olsen values measured at the experimental sites are high (>69.2 mg kg^{-1}), indicating the potential for a detrimental effect of P on AMF colonization in our experiment. As the cost of the symbiosis to the plant outweighs the benefit of access to P via the fungal pathway, plants reduce fungal access to carbohydrate [54]. Similar data were reported also in other species such as *Zea mays* L. in which the AMF root colonization was reduced with a soil P content of 90 mg kg^{-1} [55]. In other geophyte plants, such as *Allium tricoccum* Aiton., a low level of AMF symbiosis was observed in the absence of leaves and photosynthetic activity. However, once leaves elongate in early spring, root colonization increases rapidly. This is similar to the pattern of *Maianthemum racemosum* L., where AMF colonization peaked during vegetative growth [56]. Here, AMF sampling was performed during maximum leaf elongation, and therefore, the detection of low colonization is likely more related to soil characteristics than to other physiological or biochemical parameters.

Taken together, all these findings indicated that under open field conditions in alpine environments, AMF colonization was substantially lower than under pot conditions as already indicated in literature. This is in agreement with the meta-analysis of Berruti et al. [27], in which successful outcomes of AMF inoculation were more often found in controlled (greenhouse and growth chamber) conditions. In this condition, environmental extremes and variation are minimized or absent [38]. Moreover, one of the most important confounding factors in pot or field experiments is the effect of root temperature on the AMF growth [57]. The higher temperatures typical of greenhouse conditions favor greater growth and superior colonization by AMF [58].

4.2. AMF Modulate Crop Performance and Spice Quality

Flower yield is a difficult parameter to forecast in saffron since it is influenced by a combination of agronomic, biological, and environmental factors [11]. Generally, a saffron field may produce from 200 to 3000 mg m^{-2} of spice, depending on the cultivation factors [11] and obviously, by the planting density, which may vary considerably. By planting at a 55 corms m^{-2} density in southern Italy (Sicily), Gresta et al. [3] obtained more than 1200 mg m^{-2}. In the area of Navelli (central Italy) [59], with a similar corm density, the average yield ranged between 1000 and 1600 mg m^{-2}. In Iranian fields with a density of 150 and 100 corms m^{-2}, Mollafilabi et al. [60] and Koocheki et al. [61] obtained 740 and 370 mg m^{-2} of saffron, respectively. Recently, the path coefficient analysis conducted by Bayat et al. [62] highlighted that fresh stigma weight, flower number, dry stigma and flower weight, leaf size, and number and size of replacement corms have the highest positive correlation with saffron yield.

Arbuscular mycorrhizal fungi are known to be beneficial to several important plants, including some medicinal plants [30]. Unfortunately, very scarce reports of the effective role of AMF in saffron yield are available. Only, Aimo et al. [40] indicate an increase in flower production m^{-2} (equal to 68%, compared to control) using a mix of AMF species belonging to the genus Glomus. Our results are generally more supportive of the benefits of AMF inoculation with an increase of flower production m^{-2} of circa 140%. Taken together, these findings suggest a beneficial effect of AMF inoculation with a mixture of *R. intraradices* and *F. mosseae* on saffron yield performance.

Few spices are able to provide the combination of color, taste, and aroma to the foods and possess several nutraceutical properties for human health as saffron. Most of the beneficial effects of saffron, recognized since ancient times, are due mainly to its total phenolic content (TPC) and antioxidant activity (FRAP and ABTS assays). *R. intraradices* alone was found to induce an increase in secondary metabolite contents, such as terpenes and phenolics, in *Salvia officinalis* L. [63] and *Echinacea purpurea* L. [64]. Overall, the saffron produced in the studied alpine areas had very high TPC (ranges between 1340.7 and 2355.5 mg GAE 100 g^{-1} DW), which was more than saffron cultivated in different areas of Lebanon (160 mgGAE 100 g^{-1} DW) [65], and is much greater when compared with other common food additives and spices, such as *Eugenia caryophylate* (Thunb.), *Lavandula* spp., *Curcuma domestica* Val, and *Curcuma longa* L. (0.26, 0.22, 23, and 36 mg GAE 100 g^{-1} DW, respectively) (Table 4) [66,67]. Results of ABTS and FRAP assays also demonstrated elevated antioxidant activity (Table 4). ABTS assay values were comparable to what was found in Greek saffron by Ordoudi et al. [68]. FRAP assay values (between 408.9 and 1937.1 mmol Fe^{2+} kg^{-1}) were generally higher in comparison to the Iranian samples (circa 570 mmol Fe^{2+} kg^{-1}) analyzed by Karimi et al. [69]. The saffron produced in the west Italian Alps also had different bioactive compounds (Table 4) known for their health-promoting activity, that is, cinnamic acids, flavonols, benzoic acids, catechins, and carotenoids [50]. Other studies report that water-soluble carotenoids such as crocins have antioxidant effects superior to α-tocopherol [67]. It was recently observed in a clinical study that high crocin I and crocin II contents (4000 and 1000 mg, respectively) inhibit β-amyloid and tau aggregation [70]. Apart from crocins, Asdaq and Inamdar [71] suggest that flavonols are responsible for the synergistic antihyperlipidemic and antioxidant potential of saffron. Amin et al. [72] indicated that a concentration of 1 mg of safranal attenuated the behavioral symptoms of neuropathic pain. Our data indicate that the saffron produced presented high crocin II content (27.7−38.8 mg 100 g^{-1} DW), almost in line with the saffron produced in Sardinia (Italy, DOP Zafferano di Sardegna) [73], while also presenting a higher content of gallic acid compared to what was found in Iranian and Greek saffron (2 mg and 1.2 100 g^{-1} DW) by Karimi et al. [69] and Proestos et al. [74], respectively. Thus, the saffron obtained could be of particular interest for its elevated antioxidant properties.

5. Conclusions

Saffron quality may vary greatly by site on the basis of several factors, among which are climatic conditions and cultivation techniques. We hereby provide data indicating the production of high quality saffron in marginal alpine areas, thus confirming that this crop is a strategic resource and good alternative for mountainous areas building multifunctional economies. Besides the phytochemical profile highlighted, the crop had many bioactive compounds. The use of arbuscular mycorrhizal symbionts as biostimulants positively affected saffron cultivation, mainly by increasing crop productivity, and partially by increasing the content of important nutraceutical compounds. Specifically, the inoculum composed by *R. intraradices* and *F. mosseae* was particularly effective in increasing flower production and saffron yield, while *R. intraradices* alone increased the content of some bioactive compounds—picrocrocin, quercitrin, crocin II—as well as antioxidant activity. Since saffron is the world's highest priced spice, the increases in yield and quality obtained using AMF should allow for an increase in profitability.

Furthermore, a new perspective can be envisaged. Since AMF symbiosis was more effective under soilless pot cultivation, this system may be a valuable alternative for saffron production and further work is underway to assess the potential of AMF inocula in saffron soilless cultivation.

Supplementary Materials

Figure S1: Climatic conditions of the Alpine experimental sites, Figure S2: Effects of AMF inoculum composed by *Rhizophagus intraradices* and *Funneliformis mosseae* (Ri + Fm), *R. intraradices* alone (Ri), or control (AMF-) on flower production m-2 during the first (a) and second (b) cultivation cycle, Table S1: Physical and chemical properties of the soils collected in the three saffron experimental fields located in the municipality of Saint Cristophe and Morgex (north west Italy), Table S2: Characteristics of the HPLC methods applied to analyse the bioactive compounds present in the studied saffron samples.

Author Contributions: M.C., A.B., E.L., V.B., and V.S. contributed to the experimental design. M.C., Í.M.M.V., S.D., D.D., A.B., and V.S. acquired and interpreted data. M.C. drafted the manuscript. V.B. and V.S. conceived, coordinated the work, and critically revised the manuscript.

Acknowledgments: The authors acknowledge Dario Sacco for statistical assistance, Alessandro Putzolu for technical assistance, and Azienda Agricola La Branche di Diego Bovard and Azienda Agricola Rosset for providing plant material and hosting cultivation.

References

1. Molina, R.; García-Luis, A.; Coll, V.; Ferrer, C.; Valero, M.; Navarro, Y.; Guardiola, J. Flower formation in the saffron crocus (*Crocus sativus* L.). The role of temperature. *Acta Hortic.* **2004**, *650*, 39–47. [CrossRef]
2. Mollafilabi, A. *Production and Modern Agronomic of Saffron*; Scientific and Industrial Research Organization of Khorasan: Mashhad, Iran, 2004; pp. 22–35.
3. Gresta, F.; Santonoceto, C.; Avola, G. Crop rotation as an effective strategy for saffron (*Crocus sativus* L.) cultivation. *Sci. Hortic.* **2016**, *211*, 34–39. [CrossRef]
4. D'Archivio, A.; Giannitto, A.; Maggi, M.; Ruggieri, F. Geographical classification of Italian saffron (*Crocus sativus* L.) based on chemical constituents determined by high-performance liquid-chromatography and by using linear discriminant analysis. *Food Chem.* **2016**, *212*, 110–116. [CrossRef] [PubMed]
5. Carmona, M.; Zalacain, A.; Pardo, J.; López, E.; Alvarruiz, A.; Alonso, G. Influence of different drying and aging conditions on saffron constituents. *J. Agric. Food Chem.* **2005**, *18*, 3974–3979. [CrossRef] [PubMed]
6. Tong, Y.; Yan, Y.; Zhu, X.; Liu, R.; Gong, F.; Zhang, L.; Wang, P. Simultaneous quantification of crocetin esters and picrocrocin changes in Chinese saffron by high-performance liquid chromatography-diode array detector during 15 years of storage. *Pharmacogn. Mag.* **2015**, *11*, 540–545. [PubMed]
7. ISO 3632-1:2011. *Spices—Saffron (Crocus sativus* L.). *Food Products SC 7, Spices, Culinary Herbs and Condiments*; International Organization for Standardization: Geneva, Switzerland, 2011.
8. Shahi, T.; Assadpour, E.; Jafari, S. Main chemical compounds and pharmacological activities of stigmas and tepals of 'red gold'; saffron. *Trends Food Sci. Technol.* **2016**, *58*, 69–78. [CrossRef]
9. Bagur, M.; Salinas, G.; Jiménez-Monreal, A.; Chaouqi, S.; Llorens, S.; Martinez-Tomé, M.; Alonso, G. Saffron: An Old Medicinal Plant and a Potential Novel Functional Food. *Molecules* **2018**, *23*, 30. [CrossRef]
10. De Juan, J.A.; Córcoles, H.L.; Muñoz, R.M.; Picornell, M.R. Yield and yield components of saffron under different cropping systems. *Ind. Crops Prod.* **2009**, *30*, 212–219. [CrossRef]
11. Gresta, F.; Lombardo, G.; Siracusa, L.; Ruberto, G. Saffron an alternative crop for sustainable agricultural systems. A review. *Agron. Sustain. Dev.* **2008**, *28*, 95–112. [CrossRef]
12. Yarami, N.; Kamgar-Haghighi, A.A.; Sepaskhah, A.R.; Zand-Parsa, S. Determination of the potential evapotranspiration and crop coefficient for saffron using a water-balance lysimeter. *Arch. Agron. Soil Sci.* **2011**, *57*, 727–740. [CrossRef]
13. Maggio, A.; Raimondi, G.; Martino, A.; De Pascale, S. Soilless cultivation of saffron in mediterranean environment. *Acta Hortic.* **2006**, *718*, 515–522. [CrossRef]
14. Molina, R.; Valero, M.; Navarro, Y.; Guardiola, J.; García-Luis, A. Temperature effects on flower formation in saffron (*Crocus sativus* L.). *Sci. Hortic.* **2005**, *103*, 361–379. [CrossRef]
15. Giorgi, A.; Pentimalli, D.; Giupponi, L.; Panseri, S. Quality traits of saffron (*Crocus sativus* L.) produced in the Italian Alps. *Open Agric.* **2017**, *2*, 52–57. [CrossRef]

16. Manzo, A.; Panseri, S.; Bertoni, D.; Giorgi, A. Economic and qualitative traits of Italian Alps saffron. *J. Mt. Sci.* **2015**, *12*, 1542–1550. [CrossRef]
17. De Pascale, S.; Rouphael, Y.; Colla, G. Plant biostimulants: Innovative tool for enhancing plant nutrition in organic farming. *Eur. J. Hortic. Sci.* **2017**, *82*, 277–285. [CrossRef]
18. De Ponti, T.; Rijk, B.; Van Ittersum, M. The crop yield gap between organic and conventional agriculture. *Agric. Syst.* **2012**, *108*, 1–9. [CrossRef]
19. Colla, G.; Nardi, S.; Cardarelli, M.; Ertani, A.; Lucini, L.; Canaguier, R.; Rouphael, Y. Protein hydrolysates as biostimulants in horticulture. *Sci. Hortic.* **2015**, *196*, 28–38. [CrossRef]
20. Rouphael, Y.; Cardarelli, M.; Colla, G. Role of arbuscular mycorrhizal fungi in alleviating the adverse effects of acidity and aluminium toxicity in zucchini squash. *Sci. Hortic.* **2015**, *188*, 97–105. [CrossRef]
21. Mercy, L.; Lucic-Mercy, E.; Nogales, A.; Poghosyan, A.; Schneider, C.; Arnholdt-Schmitt, B. A Functional Approach towards Understanding the Role of the Mitochondrial Respiratory Chain in an Endomycorrhizal Symbiosis. *Front. Plant Sci.* **2017**, *8*, 417. [CrossRef]
22. Błaszkowski, J.; Kovács, G.; Balázs, T.; Orłowska, E.; Sadravi, M.; Wubet, T.; Buscot, F. *Glomus africanum* and *G. iranicum*, two new species of arbuscular mycorrhizal fungi (Glomeromycota). *Mycologia* **2010**, *102*, 1450–1462. [CrossRef]
23. Johnson, N.; Gehring, C. Mycorrhizas: Symbiotic mediators of rhizosphere and ecosystem processes. *Rhizosphere* **2007**, 73–100. [CrossRef]
24. Berruti, A.; Borriello, R.; Lumini, E.; Scariot, V.; Bianciotto, V.; Balestrini, R. Application of laser microdissection to identify the mycorrhizal fungi that establish arbuscules inside root cells. *Front. Plant Sci.* **2013**, *4*, 135. [CrossRef] [PubMed]
25. Borriello, R.; Berruti, A.; Lumini, E.; Della Beffa, M.T.; Scariot, V.; Bianciotto, V. Edaphic factors trigger diverse AM fungal communities associated to exotic camellias in closely located Lake Maggiore (Italy) sites. *Mycorrhiza* **2015**, *25*, 253–265. [CrossRef] [PubMed]
26. Liu, J.; Wu, L.; Wei, S.; Xiao, X.; Su, C.; Jiang, P.; Song, Z.; Wang, T.; Yu, Z. Effects of arbuscular mycorrhizal fungi on the growth, nutrient uptake and glycyrrhizin production of licorice (*Glycyrrhiza uralensis* Fisch). *Plant Growth Regul.* **2007**, *52*, 29–39. [CrossRef]
27. Berruti, A.; Lumini, E.; Balestrini, R.; Bianciotto, V. Arbuscular mycorrhizal fungi as natural biofertilizers: Let's benefit from past successes. *Front. Microbiol.* **2016**, *6*, 1559. [CrossRef] [PubMed]
28. Bianciotto, V.; Victorino, I.; Scariot, V.; Berruti, A. Arbuscular mycorrhizal fungi as natural biofertilizers: Current role and potential for the horticulture industry. *Acta Hortic.* **2018**, *1191*, 207–215. [CrossRef]
29. López-Ráez, J.; Verhage, A.; Fenández, I.; García, J.; Azcón-Aguilar, C.; Flors, V.; Pozo, M. Hormonal and transcriptional profiles highlight common and differential host responses to arbuscular mycorrhizal fungi and the regulation of the oxylipin pathway. *J. Exp. Bot.* **2010**, *61*, 2589–2601. [CrossRef]
30. De Sousa Oliveira, M.; da Silva Campos, M.; de Albuquerque, U.P.; da Silva, F.S.B. Arbuscular mycorrhizal fungi (AMF) affects biomolecules content in *Myracrodruon urundeuva* seedlings. *Ind. Crops Prod.* **2013**, *50*, 244–247. [CrossRef]
31. Bona, E.; Lingua, G.; Manassero, P.; Cantamessa, S.; Marsano, F.; Todeschini, V.; Copetta, A.; D'Agostino, G.; Massa, N.; Avidano, L.; et al. AM fungi and PGP pseudomonads increase flowering, fruit production, and vitamin content in strawberry grown at low nitrogen and phosphorus levels. *Mycorrhiza* **2015**, *25*, 181–193. [CrossRef]
32. Lazzara, S.; Militello, M.; Carrubba, A.; Napoli, E.; Saia, S. Arbuscualr mycorrhizal fungi altered the hypericin, pseudohypericin, and hyperforin conten in flowers of Hypericum perforatum grown under contrasting P availability in a highly organic substrate. *Mycorrhiza* **2017**, *27*, 345–354. [CrossRef]
33. Jurkiewicz, A.; Ryszka, P.; Anielska, T.; Waligórski, P.; Bialońska, D.; Góralska, K.; Tsimilli-Michael, M.; Turnau, K. Optimization of culture conditions of *Arnica montana* L.: Effects of mycorrhizal fungi and competing plants. *Mycorrhiza* **2010**, *20*, 293–306. [CrossRef] [PubMed]
34. Zeng, Y.; Guo, L.; Chen, B.; Hao, Z.; Wang, J.; Huang, L.; Yang, G.; Cui, X.; Yang, L.; Wu, Z.; et al. Arbuscular mycorrhizal symbiosis and active ingredients of medicinal plants: Current research status and prospectives. *Mycorrhiza* **2013**, *23*, 253–265. [CrossRef] [PubMed]
35. Giovannetti, M.; Avio, L.; Sbrana, C. Improvement of nutraceutical value of food by plant symbionts. In *Natural Products, Phytochemistry, Botany and Metabolism of Alkaloids, Phenolics and Terpenes*; Ramawat, K.G., Merillon, J.M., Eds.; Springer: Berlin, Germany, 2013; pp. 2641–2662.

36. Rydlová, J.; Jelínkova, M.; Dušek, K.; Dušková, E.; Vosátka, M.; Püschel, D. Arbuscular mycorrhiza differentially affects synthesis of essential oils in coriander and dill. *Mycorrhiza* **2016**, *26*, 123–131. [CrossRef] [PubMed]
37. Hart, M.M.; Antunes, P.M.; Chaudhary, V.B.; Abbott, L.K. Fungal inoculants in the field: Is the reward greater than the risk? *Funct. Ecol.* **2017**, *32*, 126–135. [CrossRef]
38. Ryan, M.H.; Graham, J.H. Little evidence that farmers should consider abundance or diversity of arbuscular mycorrhizal fungi when managing crops. *New Phytol.* **2018**, *220*, 1092–1107. [CrossRef] [PubMed]
39. Berg, G. Plant-microbe interactions promoting plant growth and health: Perspectives for controlled use of microorganisms in agriculture. *Appl. Microbiol. Biotechnol.* **2009**, *84*, 11–18. [CrossRef] [PubMed]
40. Aimo, S.; Gosetti, F.; D'Agostino, G.; Gamalero, E.; Gianotti, V.; Bottaro, M.; Gennaro, M.; Berta, G. Use of arbuscular mycorrhizal fungi and beneficial soil bacteria to improve yield and quality of saffron (*Crocus sativus* L.). *Acta Hortic.* **2010**, *850*, 159–162. [CrossRef]
41. Venkateshwaran, M.; Volkening, J.; Sussman, M.; Ané, J. Symbiosis and the social network of higher plants. *Curr. Opin. Plant Biol.* **2013**, *16*, 118–127. [CrossRef]
42. Mohebi-Anabat, M.; Riahi, H.; Zanganeh, S.; Sadeghnezhad, E. Effects of arbuscular mycorrhizal inoculation on the growth, photosynthetic pigments and soluble sugar of *Crocus sativus* L. (saffron) in autoclaved soil. *Int. J. Agron. Agric. Res.* **2015**, *6*, 296–304.
43. Lone, R.; Shuab, R.; Koul, K. AMF association and their effect on metabolite mobilization, mineral nutrition and nitrogen assimilating enzymes in saffron (*Crocus sativus*) plant. *J. Plant Nutr.* **2016**, *39*, 1852–1862. [CrossRef]
44. Shajari, M.A.; Moghaddam, P.R.; Ghorbani, R.; Koocheki, A. Increasing saffron (*Crocus sativus* L.) corm size through the mycorrhizal inoculation, humic acid application and irrigation managements. *J. Plant Nutr.* **2018**, *41*, 1047–1064. [CrossRef]
45. Moghaddam, P.R.; Koocheki, A.; Molafilabi, A.; Seyyedi, M. Effect of biological and chemical fertilizers on replacement corm and flower yield of saffron (*Crocus sativus* L.). *Iran. J. Crop Sci.* **2013**, *15*, 234–246.
46. Cusano, E.; Consonni, R.; Petrakis, E.A.; Astraka, K.; Cagliani, L.R.; Polissiou, M.G. Integrated analytical methodology to investigate bioactive compounds in *Crocus sativus* L. flowers. *Phytochem. Anal.* **2018**, *29*, 476–486. [CrossRef]
47. Vahedi, M.; Kabiri, M.; Salami, S.A.; Rezadoost, H.; Mirzaie, M.; Kanani, M.R. Quantitative HPLC-based metabolomics of some Iranian saffron (*Crocus sativus* L.) accessions. *Ind. Crops Prod.* **2018**, *118*, 26–29. [CrossRef]
48. Trouvelot, A.; Kough, J.; Gianinazzi-Pearson, V. Mesure du taux de mycorhization VA d'un système radiculaire. Recherche de méthodes d'estimation ayant une signification fonctionnelle. In *Mycorrhizae: Physiology and Genetics*; Gianinazzi-Pearson, V., Gianinazzi, S., Eds.; INRA Press: Paris, France, 1986; pp. 217–221.
49. Caser, M.; D'Angiolillo, F.; Chitarra, W.; Lovisolo, C.; Ruffoni, B.; Pistelli, L.; Pistelli, L.; Scariot, V. Ecophysiological and phytochemical responses of *Salvia sinaloensis Fern.* to drought stress. *Plant Growth Regul.* **2018**, *84*, 383–394. [CrossRef]
50. Donno, D.; Beccaro, G.; Mellano, M.; Cerutti, A.; Bounous, G. Goji berry fruit (*Lycium* spp.): Antioxidant compound fingerprint and bioactivity evaluation. *J. Funct. Foods* **2015**, *18*, 1070–1085. [CrossRef]
51. Caser, M.; D'Angiolillo, F.; Chitarra, W.; Lovisolo, C.; Ruffoni, B.; Pistelli, L.; Pistelli, L.; Scariot, V. Water deficit regimes trigger changes in valuable physiological and phytochemical parameters in *Helichrysum petiolare* Hilliard & B.L. Burtt. *Ind. Crops Prod.* **2016**, *83*, 680–669.
52. Urbani, E.; Blasi, F.; Simonetti, M.; Chiesi, C.; Cossignani, L. Investigation on secondary metabolite content and antioxidant activity of commercial saffron powder. *Eur. Food Res. Technol.* **2016**, *242*, 987–993. [CrossRef]
53. Donno, D.; Mellano, M.G.; Prgomet, Z.; Beccaro, G.L. Advances in *Ribes* x *nidigrolaria* Rud. Bauer & A. Bauer fruits as potential source of natural molecules: A preliminary study on physico-chemical traits of an underutilized berry. *Sci. Hortic.* **2018**, *237*, 20–27.
54. Gosling, P.; Mead, A.; Proctor, M.; Hammond, J.P.; Bending, G.D. Contrasting arbuscular mycorrhizal communities colonizing different host plants show a similar response to a soil phosphorus concentration gradient. *New Phytol.* **2013**, *198*, 546–556. [CrossRef]
55. Watt-Williams, S.J.; Smith, F.A.; Jakobsen, I. Soil phosphorus availability is a driver of the responses of maize (*Zea mays*) to elevated CO_2 concentration and arbuscular mycorrhizal colonization. *Symbiosis* **2018**, 1–10. [CrossRef]

56. Brundrett, M.; Kendrick, B. The roots and mycorrhizas of herbaceous woodland plants. II. Structural aspects of morphology. *New Phytol.* **1990**, *114*, 469–479. [CrossRef]
57. Gavito, M.E.; Olsson, P.A.; Rouhier, H.; Medina-Peñafiel, A.; Jakobsen, I.; Bago, A.; Azcón-Aguilar, C. Temperature constraints on the growth and functioning of root organ cultures with arbuscular mycorrhizal fungi. *New Phytol.* **2005**, *168*, 179–188. [CrossRef] [PubMed]
58. Gavito, M.E.; Azcón-Aguilar, C. Temperature stress in arbuscular mycorrhizal fungi: A test for adaptation to soil temperature in three isolates of *Funneliformis mosseae* from different climates. *Agric. Food Sci.* **2012**, *21*, 2–11. [CrossRef]
59. Tammaro, F. Saffron (*Crocus sativus* L.) in Italy. In *Saffron: Crocus sativus L.*; Negbi, M., Ed.; Harwood Academic Publishers: Amsterdam, NL, 1999; pp. 53–62.
60. Mollafilabi, A.; Koocheki, A.; Moeinerad, H.; Kooshki, L. Effect of plant density and weight of corm on yield and yield components of saffron (*Crocus sativus* L.) under soil, hydroponic and plastic tunnel cultivation. *Acta Hortic.* **2013**, *997*, 51–58. [CrossRef]
61. Koocheki, A.; Seyyedi, S. How irrigation rounds and mother corm size control saffron yield, quality, daughter corms behavior and phosphorus uptake. *Sci. Hortic.* **2016**, *213*, 132–143. [CrossRef]
62. Bayat, M.; Amirnia, R.; Rahimi, M. Phenotypic and genotypic relationships between traits in saffron (*Crocus sativus* L.) as revealed by path analysis. *J. Appl. Res. Med. Aromat. Plants* **2017**, *5*, 33–40. [CrossRef]
63. Geneva, M.; Stancheva, I.; Boychinova, M.; Mincheva, N.; Yonova, P. Effects of foliar fertilization and arbuscular mycorrhizal colonization on *Salvia officinalis* L. growth, antioxidant capacity, and essential oil composition. *J. Sci. Food Agric.* **2010**, *90*, 696–702.
64. Araim, G.; Saleem, A.; Amason, J.; Charest, C. Root colonization by an arbuscular mycorrhizal (AM) fungus increases growth and secondary metabolism of purple coneflower, *Echinacea purpurea* (L.) Moench. *J. Agric. Food Chem.* **2009**, *57*, 2255–2258. [CrossRef]
65. Makhlouf, H.; Saksouk, M.; Habib, J.; Chahine, R. Determination of antioxidant activity of saffron taken from the flower of *Crocus sativus* grown in Lebanon. *Afr. J. Biotechnol.* **2011**, *10*, 8093–8100.
66. Gülçin, I.; Güngör, Ş.; Beydemir, Ş.; Elmastas, M.; İrfan, K. Comparison of antioxidant activity of clove (*Eugenia caryophylata* Thunb) buds and lavender (*Lavandula stoechas* L.). *Food Chem.* **2004**, *87*, 393–400. [CrossRef]
67. Chen, Y.; Zhang, H.; Tian, X.; Zhao, C.; Liu, Y.; Jia, L.; Yin, H.X.; Chen, C. Antioxidant potential of crocins and ethanol extracts of *Gardenia jasminoides* ELLIS and *Crocus sativus* L. A relationship investigation between antioxidant activity and crocin contents. *Food Chem.* **2008**, *109*, 484–492. [CrossRef]
68. Ordoudi, S.; Befani, C.; Nenadis, N.; Koliakos, G.; Tsimidou, M. Further Examination of Antiradical Properties of *Crocus sativus* Stigmas Extract Rich in Crocins. *J. Agric. Food Chem.* **2009**, *57*, 3080–3086. [CrossRef] [PubMed]
69. Karimi, E.; Oskoueian, E.; Hendra, R.; Jaafar, H. Evaluation of *Crocus sativus* L. stigma phenolic and flavonoid compounds and its antioxidant activity. *Molecules* **2010**, *15*, 6244–6256. [CrossRef] [PubMed]
70. Inoue, E.; Shimizu, Y.; Masui, R.; Hayakawa, T.; Tsubonoya, T.; Hori, S.; Sudoh, K. Effects of saffron and its constituents, crocin-1, crocin-2, and crocetin on α-synuclein fibrils. *J. Nat. Med.* **2018**, *72*, 274–279. [CrossRef] [PubMed]
71. Asdaq, S.; Inamdar, M. Potential of garlic and its active constituent, S-allyl cysteine, as antihypertensive and cardioprotective in presence of captopril. *Phytomedicine* **2010**, *17*, 1016–1026. [CrossRef] [PubMed]
72. Amin, S.; Parker, M.; Armbrust, E. Interactions between diatoms and bacteria. *Microbiol. Mol. Boil. Rev.* **2012**, *76*, 667–684. [CrossRef] [PubMed]
73. Del Campo, P.; Carmona, M.; Maggi, L.; Kanakis, C.; Anastasaki, E.; Tarantilis, P.; Polissiou, M.; Alonso, G. Picrocrocin Content and Quality Categories in Different (345) Worldwide Samples of Saffron (*Crocus sativus* L.). *J. Agric. Food Chem.* **2010**, *58*, 1305–1312. [CrossRef]
74. Proestos, C.; Chorianopoulos, N.; Nychas, G.; Komaitis, M. RP-HPLC Analysis of the Phenolic Compounds of Plant Extracts. Investigation of Their Antioxidant Capacity and Antimicrobial Activity. *J. Agric. Food Chem.* **2005**, *53*, 1190–1195. [CrossRef]

8

Paramylon Treatment Improves Quality Profile and Drought Resistance in *Solanum lycopersicum* L., cv. Micro-Tom

Laura Barsanti [1], Primo Coltelli [2] and Paolo Gualtieri [1,*]

[1] Istituto di Biofisica, Consiglio Nazionale delle Ricerche (CNR), Via Moruzzi 1, 56124 Pisa, Italy
[2] Istituto Scienza e Tecnologie dell'Informazione, CNR, Via Moruzzi 1, 56124 Pisa, Italy
* Correspondence: paolo.gualtieri@pi.ibf.cnr.it

Abstract: Tomatoes, the most cultivated vegetables worldwide, require large amounts of water and are adversely affected by water stress. *Solanum lycopersicum* L., cv. Micro-Tom was used to assess the effects of β-(1,3)-glucan (paramylon) purified from the microalga *Euglena gracilis* on drought resistance and fruit quality profile. Plants were grown in an aeroponic system under three cultivation conditions: optimal water regimen, water scarcity regimen, and water scarcity regimen coupled with a root treatment with paramylon. Eco–physiological, physicochemical and quality parameters were monitored and compared throughout the lifecycle of the plants. Drought stress caused only a transient effect on the eco–physiological parameters of paramylon-treated plants, whereas physicochemical and biochemical parameters underwent significant variations. In particular, the fruits of paramylon-treated plants reached the first ripening stage two weeks before untreated plants grown under the optimal water regime, while the fruits of stressed untreated plants did not ripe beyond category II. Moreover, antioxidant compounds (carotenoids, phenolic acid, and vitamins) of fruits from treated plants underwent a two-fold increase with respect to untreated plants, as well as soluble carbohydrates (glucose, fructose, and sucrose). These results show that paramylon increases plant resistance to drought and highly improves the quality profile of the fruits with respect to untreated plants grown under drought stress.

Keywords: Biostimulants; *Euglena gracilis*; algal polysaccharide; β-glucan; water stress; tomato; aeroponics

1. Introduction

Paramylon is the storage product of the unicellular alga *Euglena gracilis.* This polysaccharide is a β-(1,3)-glucan endogenously synthetized as 1–2 μm granules consisting of 100% glucose [1]. The granules are composed of concentric segments, which possibly indicates the successive deposition of unbranched linear β-(1,3)-glucan chains on a central nucleus [2–4]. Wild type (WT) photosynthetic cells can accumulate paramylon up to 60% of cell dry weight (DW) [2], while the WZSL mutant of *E. gracilis* (spontaneous, non-chloroplastic, osmotrophic mutant; W describes the white color of the cells; Z means *E. gracilis* Klebs, Z strain; S means spontaneous mutant, and L means light grown parent culture) [5] can accumulate large amounts of it (up to 95% DW) when grown in the dark with an adequate carbon source [2].

B-glucans are PAMPs (pathogen-associated molecular patterns) recognized by specific membrane receptors (pattern recognition receptors, PRRs) which trigger the activation of the innate immune system [6,7]. Both the molecular structure and degree of polymerization affect the strength and efficacy of β-glucans recognition by PRRs, as well as their successive reactions [2,8–10].

Linear β-(1,3)-glucans bind preferentially to Dectin-1, a C-type lectin receptor expressed on most cells of the innate immune system [11,12]. A minimum of 10 units of glucose is necessary to trigger an immune response [8,13,14].

Evaluations of the effective potential of linear β-(1,3)-glucans have been often made by testing preparations from plant/algae/fungal sources that are always contaminated by pigments, proteins, and membranes resulting in non-specific immunoresponses. β-(1,3)-glucan purified from paramylon synthetized by the WZSL mutant lacks any kind of contaminations from cellular components, which are always present in the paramylon extracted from WT cells. β-1-3-glucan purified from the WZSL mutant is further processed to produce linear nanofibers suited for binding to Dectin-1 receptors of target cell membranes. The effect of these nanofibers has been already investigated in our laboratory on tomato plants, animals, and humans [4,15–17]. In tomatoes, paramylon nanofibers modulate conductance to carbon dioxide (CO_2) diffusion from air to the carboxylation sites by regulating hormone levels and water-use efficiency, leading to an increase of plant defense capacity against drought [16].

Tomatoes are the most cultivated vegetable worldwide, being one of the most nutritionally and economically important crops. They require large amounts of water and are adversely affected by drought, which limits photosynthesis and, consequently, plant growth and yield worldwide [18]. Hence, we investigated the role of β-1,3-glucan nanofibers as elicitors of tomato plants response to drought to understand their physiological and photosynthetic responses to this stress.

In this study, tomato cv. Micro-Tom was chosen because of its small size (10–20 cm in height) and short life cycle of about 3 months. Plants were grown in an aeroponic system under three cultivation conditions: optimal water regimen, water scarcity regimen (drought), and water scarcity regimen (drought) coupled with a root treatment with paramylon to monitor and compare eco–physiological (leaf water potential, CO_2 assimilation rate, stomatal conductance, internal CO_2 concentration, photosystem II (PSII) photochemical efficiency, actual photon yield of PSII, and photochemical quenching of PSII), physicochemical (dry biomass, ashes, dry matter, moisture, microelements, weight, and size), and quality parameters (antioxidant compositions and activities, as well as soluble carbohydrates) throughout the lifecycle of the plants.

2. Materials and Methods

2.1. Aeroponic Culture System

The aeroponic cultivation system used was the Nutriculture Twin Amazon 16 (Nutriculture DGS, UK), which consists of a 100 L reservoir tank (160 × 75 × 46 cm), a root chamber housing the delivery system and a molded plastic lid holding sixteen 75 mm mesh pots (Figure 1). The Maxi Jet 1000 pump (14 W, flow rate 1000 Lh^{-1}, max head height 142 cm, NEWA Tecno Industria SRL, Italy) supplies a powerful spray creating a miniature rainstorm inside the chamber through eight 360 degree sprinklers that shower the whole root area, leaving no blind spots. Since the aeroponic container can be assimilated to a closed system, the condensation of growth medium eventually balances its evaporation, reducing the amount of evapotranspiration.

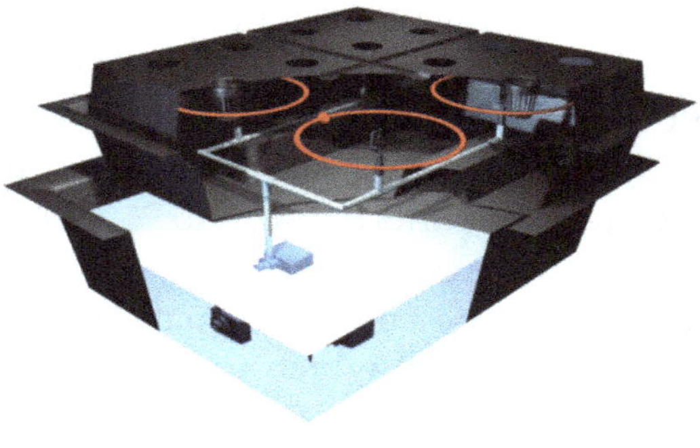

Figure 1. The aeroponic cultivation system.

2.2. *Paramylon Nanofibers Preparation*

Paramylon granules were extracted and purified from two-days-old cultures of *Euglena gracilis* WZSL mutant according to Barsanti et al. [2]. Nanofibers were obtained by alkaline degradation of the granules [2]. Assuming a 20% loss during the procedure, the final concentration of β-1,3-glucan nanofibers was about 0.8% w/v.

2.3. *Plant Material, Growth Conditions and Paramylon Treatment*

Seeds of *Solanum lycopersicum* L., cv. Micro-Tom [19] were surface sterilized with a bleach solution (commercial bleach 30% v/v, Triton X-100 0.02% v/v) for 15 min, washed 3 times with sterile water, and placed overnight in the dark at 4 °C. Seeds were then sown on filter paper covering the bottom of 150 mm diameter Petri dishes (about 20 seeds per plate); the paper was wetted with sterile demineralized water, the dishes were wrapped with Parafilm, and they were placed in the dark for about 2 days to aid germination. Upon germination, the cover was removed to allow the development of the roots (2–3 days). Seedlings were then transferred to rock wool (Grodan® Pro Plug) plugs, one seed per plug, to grow under controlled climate conditions (16/8 h light/dark; 400 $\mu molm^{-2} s^{-1}$ PAR, Photosynthetic Active Radiation; 22 °C) for approximately 2 weeks in a plastic tray containing half strength nutrient solution [$Ca(NO_3)_2$ $4H_2O$, 0.944 gL^{-1}; KNO_3, 0.808 gL^{-1}; $MgSO_4$ $2H_2O$, 0.492 gL^{-1}; $NH_4H_2PO_4$, 0.152 gL^{-1}; microelements, 0.050 gL^{-1}; pH 5.5] according to the protocol by Motohashi et al. [20] in a non-circulating hydroponic method. Upon appearance of the first true leaf, seedlings selected for uniform development were transferred to 3 aeroponic cultivation systems, 16 plants per system, in a glasshouse with a mean temperature of 24 °C, a 14/10 h light/dark photoperiod, and 800 $\mu molm^{-2}$ s^{-1} PAR irradiance. Plants were transferred into the greenhouse on May 14, 2018, 39 days after germination (dag). Each aeroponic tank was filled with 100 L of half strength nutrient solution, average pH (±SE) 5.50 ± 0.02.

Three cultivation conditions were tested: optimal water regimen (WW_P^-), water scarcity regimen (drought, WS_P^-), and water scarcity regimen (drought) coupled with root treatment with paramylon (WS_P^+).

Paramylon was added to the nutrient solution of one of the 3 tanks to a final concentration of 500 mg L^{-1}. The paramylon concentration was chosen according to the results of previous experiments [16]. The atomization spray time and interval time were 3-s on/5 min off in the control system (optimal water regimen), and 3-s on/120 min off in the other two systems (water scarcity regimen with and without paramylon). As a consequence, the irrigation supply was 8.64 $Lplant^{-1}$ d^{-1} in WW_P^- and 0.36 $Lplant^{-1}d^{-1}$ in both WS_P^+ and WS_P^-. The parameters of the aeroponics system used in the experiment are shown in Table 1.

The optimal water regimen was chosen according to Johnstone et al. [21], who used an aeroponic system for nutritional studies on *Lycopersicon esculentum* Mill. cv Cannery Row and established a misting regimen to plant roots of about 10 Ld^{-1} $plant^{-1}$. The water scarcity regimen (0.36 Ld^{-1} $plant^{-1}$) was assessed previously by a watering threshold experiment and by checking the representative features of the plants for yellowing and wilting symptoms in 100 tomato plants 30 dag in early spring. We want to stress that the irrigation supply did not correspond to the water needed by the plants.

The levels of nutrient solution and pH were monitored daily; a fresh solution was added in order to maintain a volume of 100 L in each tank. Controlled conditions were maintained throughout the experiment (14/10 h light/dark; 800 $\mu molm^{-2}$ s^{-1} PAR; 24 °C; pH (±SE) 5.5 ± 0.022).

Table 1. Parameters of the aeroponic system used in the experiment.

Parameters	Units	WW_P$^-$	WS_P$^-$ and WS_P$^+$
Root available space	m^{-3}	0.012	0.012
Resident water	L	100	100
Emitter capacity	mLs^{-1}	20	20
Emitters per system		8	8
Plants per system		16	16
Irrigation duration	sh^{-1}	36	1.5
Irrigation supply per emitter	mLh^{-1}	720	30
Irrigation supply per system	Lh^{-1}	5.76	0.24
Irrigation supply per plant	Ld^{-1}	8.64	0.36

2.4. *Scanning Electron Microscopy (SEM) Preparations*

Root samples were fixed in 100% methanol for 20 min and then transferred in 100% dry ethanol for 30 min with a further change into fresh 100% ethanol overnight. After dehydration, samples were dried in a critical-point dryer apparatus, coated with gold and viewed using a Philips-SEM 505 microscope (Eindhoven, The Netherlands).

2.5. *Water Potential, Gas Exchanges and Chlorophyll a Fluorescence*

Preliminary measurements of water potential, gas exchanges and chlorophyll *a* fluorescence were done to assess whether there were variations of these parameters in the different leaves of each plant. Since no variation was detected, measurement was performed on a single leaf per plant, using the same leaf for all the measurement of the experiment.

The first measure of all the eco–physiological parameters was performed upon the transfer of the plants to the aeroponic system, i.e., 39 dag. The predawn leaf water potential (Ψ_w) was measured on one fully expanded mature leaf per plant (n = 16), using a Scholander-type pressure chamber (model 600, PMS Instrument, Albany, OR, USA) and N_2 for the application of pressure, following the precautions suggested by Turner and Long [22].

Leaf gas exchanges and chlorophyll *a* fluorescence measurements were determined between 10:00 AM and 1:00 PM (solar time) on one fully-expanded mature leaf plant (n = 16). Instantaneous measurements of steady state photosynthetic carbon dioxide (CO_2) assimilation rate (A), stomatal conductance (g_s), and internal CO_2 concentration (C_i) were performed using an LI-6400 portable photosynthesis system (Li-Cor, Lincoln, NE, USA), according to Scartazza et al. [16]. The modulated chlorophyll *a* fluorescence and the status of the electron transport of photosystem II (PSII) were measured with a PAM-2000 fluorometer (Walz, Effeltrich, Germany) on the same leaves used for gas exchange after 40 min of dark-adaptation. Fluorescence in light-adapted leaves was induced according to Scartazza et al. [16] and Schreiber et al. [23]. The maximum efficiency of PSII photochemistry was calculated as $F_v/F_m = (F_m-F_0)/F_m$; the actual photon yield of PSII photochemistry (Φ_{PSII}) was calculated as $(F'_m-F')/F'_m$; the photochemical quenching qP was calculated as $(F'_m-F_t)/(F'_m-F'_0)$. The used variables in dark-adapted state were as follows: F_v was the variable fluorescence, F_m was the maximal fluorescence, and F_0 was the minimum fluorescence. The used variables in light-adapted state were: F' was the fluorescence at the actual state of PSII reaction centers, F'_m was the maximal fluorescence, F'_0 was the minimal fluorescence, and F_t was the transient fluorescence.

2.6. *Physicochemical Parameters and Mineral Content*

At the end of the experiment, all plants were sampled and separated in fruits, leaves, stems, and roots. Fruits were harvested at the ripening category intermediate between VI and VII [24]. All the fruits produced by each plant were gathered and samples (16 fruits) for the analysis were taken from this pool. Several quality attributes were determined on the plants: fresh and dry weights, dry matter content, size (using Vernier calipers), number of fruits, and total ash. The roots, stems, leaves, and fruits of all the plants were dried at 40 °C for 96 h in a ventilated oven. One gram of dry matter was ashed at 550 °C for 6 h. Total micronutrient content (Ca, Cu, Fe, K, Mg, Mn, Na and Zn) was determined in a sub-sample after digestion with HNO_3 and $HClO_4$ [25]. The digests were analyzed by means of a Varian AA240 FS (Varian, USA) hydride generation atomic absorption spectrophotometer equipped with flow vapor generation accessory VGA 77 (Agilent Technologies, USA). Eight independent replicates were used for chemical analysis, and data are presented as mean ± standard deviation ($n = 8$).

2.7. Antioxidant Compounds, Total Antioxidant Capacity and Carbohydrates

Retinol content was obtained using standard conversion formula (1 μg retinol = 1 retinol equivalent (RE) method used; 1 μg β-carotene = 0.167 μg RE), according to Aremu and Nweze [26]. Lycopene and β-carotene contents were measured spectrophotometrically according to Georgé et al. [27]. Dried tomatoes (500 mg) were homogenized in a mortar with 100 mL of hexane/acetone/ethanol (50/25/25, v/v/v) in the dark, ultrasonically disrupted, and centrifuged at 12,000× *g* for 20 min at 4 °C. The supernatants were filtered through 0.2 μm Minisart SRT 15 filters and transferred into a separating funnel. The organic phase was washed three times with 20 mL of distilled water (in order to remove acetone and ethanol). The aqueous phase was discarded, and the remaining water in the organic phase was removed by adding anhydrous sodium sulphate. The final volume was made up to 50 mL with hexane. The reaction mixture absorbance was measured at 436 and 450 for β-carotene determination and 503 nm for lycopene determination.

Reduced (ascorbic acid, AsA) and oxidized (dehydroascorbate, DHA) ascorbate contents were measured spectrophotometrically according to Kampfenkel et al. [28]. Fresh fruit samples (250 mg) were homogenized in a mortar with 0.5 mL of 2% (w/v) phosphoric acid and centrifuged at 12,000× *g* for 15 min at 4 °C. The AsA assay mixture contained 50 μL of sample extract, 150 μl of K/P 200 mM (pH 7.4), 50 μl of trichloroacetic acid (TCA) 6% (w/v) and 50 μL of water. The total ascorbate (AsA + DHA) assay mixture contained 50 μl of sample extract, 150 μl of K/P 200 mM (pH 7.4), 50 μl of TCA 6% (w/v) and 50 μL of dithiothreitol (DTT) 10 mM. The reaction mixture was left at room temperature for 15 min; 50 μL of N-ethylmaleimide 0.5% (w/v) were added after the reduction of DHA to AsA. The color was developed in both assays by adding the reagents in the following sequence: 250 μL of TCA 10% (w/v), 200 μL of ortho-phosphoric acid 42% (v/v), 200 μL of 2,2-dipyridyl 4.0% (w/v) in ethanol 70% (v/v) and 100 μL of $FeCl_3$ 3% (v/v) to a final volume of 1 mL. Controls were also run, and the solution was allowed to stand at 40 °C for an Fe^{2+}-bathophenanthroline complex to develop. The DHA levels were estimated on the basis of the difference between total ascorbate and AsA values. A standard calibration curve covering 0–10 nM of AsA or DHA range was used.

Tocopherols were determined by HPLC according to Döring et al. [29]. Fresh fruit samples (250 mg) were homogenized in a mortar with 0.4 mL of 100% HPLC-grade methanol and incubated overnight at 4 °C in the dark. The supernatant was filtered through 0.2 μm Minisart SRT 15 filters and immediately analyzed at room temperature with a reverse-phase Dionex column (Acclaim 120, C18, 5 μm particle size, 4.6 mm internal diameter × 150 mm length). Tocopherols were eluted at a flow rate of 1 mL min^{-1} using 100% solvent A (acetonitrile/methanol, 75/25, v/v) for the first 14 min, followed by a 3 min linear gradient to 100% solvent B (methanol/ethylacetate, 68/32, v/v) and 15 min with 100% solvent B. Tocopherols were detected at 280 nm. Authentic standards (Sigma-Aldrich, Italy) were used to quantify the tocopherols content of each sample.

The content of carbohydrates in fruits was determined spectrophotometrically according to Aguiar et al. [30] and quantified using a K-SUFRG commercial kit (Megazyme, Wicklow, Ireland), following the manufacturer's protocol.

The antioxidant properties of the fruits were assessed spectrofluorimetrically by the oxygen radical absorption capacity (ORAC) and hydroxyl radical antioxidant capacity (HORAC) assays [31,32].

Fresh fruit samples (10 mg) were added to 0.75 mL of 100% ethanol/methanol/water/formic acid (35:35:28:2, v/v/v/v) and centrifuged at 12,000× *g* for 10 min at 4 °C. The supernatant was collected, and 10 μL were mixed with 170 μL of 48 nM fluorescein (FL). The reagents were transferred into the main reagent wells (OptiPlate 96 F plates, Perkin Elmer, Waltham, MA, USA) and incubated at 37 °C for 20 min before recording the initial fluorescence (excitation/emission = 485/527 nm). After incubation, 20 μL of the 2,2′-azobis(2-methylpropionamidine) dihydrochloride reagent (51.5 mM final concentration) were added, and fluorescence readings were taken every minute for 60 min. A phosphate buffer (75 mM, pH 7.4) was used as a blank, and a Trolox solution (0.78–25 μM) was used as a standard. The final ORAC values were calculated by using a regression equation between the Trolox concentration and the net area under the FL decay curve and expressed as μmol Trolox equivalents (TE) per gram of fresh weight (FW). In the HORAC assay, 10 μL of supernatant were mixed with 170 μL of 48 nM fluorescein (605 mM final concentration) and incubated at 37 °C for 10 min, before recording the initial fluorescence (excitation/emission = 485/520 nm). After incubation, 10 μL of H_2O_2 (27.5 mM final concentration) and 10 μL of Co(II) (230 μM final concentration) solutions were added, and fluorescence readings were taken every minute for 60 min. A phosphate buffer was used as a blank, and a gallic acid solution (100–600 μM) was used as a standard. The final HORAC values were calculated using a regression equation between the standard antioxidant concentration and the net area under the curve. One HORAC unit was assigned to the net protection area provided by 1 μM gallic acid, and the activity of the sample was expressed as μmol gallic acid equivalent (GAE) per gram of fresh weight (FW).

The content of total phenolic compounds was determined spectrophotometrically according to Waterhouse [33]. Fresh fruit samples (100 mg) were homogenized in a mortar with 5 mL of methanol acidified with 1 % HCl (v/v) for 20 h in the dark at 4 °C. Extracts were centrifuged for 15 min at 12,000× *g* at 4 °C, and the supernatants were filtered through 0.2 μm Minisart SRT 15 filters and stored in test tubes at −20 °C. Fifty μL of a 4-times diluted extract was mixed with 2.45 mL of distilled water and 250 μL of Folin–Ciocalteu's phenol reagent. After incubation at room temperature for 6 min, 750 μL sodium carbonate 7.5% (w/v) and 500 μL of deionized water were mixed. After 120 min incubation at room temperature, the reaction mixture absorbance was measured at 760 nm. A calibration curve was prepared using a standard solution of gallic acid (range 0–1 mg mL^{-1}). Eight independent replicates were used for chemical analysis, and data are presented as mean ± standard deviation ($n = 8$).

2.8. Statistical Analysis

The statistical analysis was performed using JMP 12 (SAS institute, Cary, NC, USA). The normality of the data was preliminary tested by the Shapiro–Wilk W test. If measurements were carried out for more than two time points, data were analyzed using one-way repeated measures ANOVA, and comparison among means was determined by Tukey's HSD (honestly significant difference) multiple comparison test ($p < 0.05$). All the other data were analyzed by Student's *t* test.

3. Results and Discussion

Aeroponics is a soil-less cultivation system considered a specialized version of hydroponics [34]. It is an air-water system in which the roots of the plant extend and grow inside a closed container in the dark, are exposed to air, and directly sprayed with a nutrient-water mix through atomizers (Figure 1). The aerial portions of the plant (leaves, stem and crown) extend above the wet zone separated from the root. Aeroponics systems are very useful for plant root studies under controlled conditions; we chose it to monitor the response of tomato plants to drought and feasibility of paramylon nanofibers, directly applied to the root system, in modulating the response of the whole plant to this stressor.

The representative features of the plants (aerial parts and root system) 60 dag under the three cultivation conditions tested are shown in Figures 2 and 3. Under the optimal water regimen (Figure 2A), plants showed a normal compact growth habit with short internodes, fully expanded leaves, and regular fruit size typical of the determinate growth of Micro-Tom. The root system was fully developed with extremely long roots and plenty lateral roots, indicating a superior growth with respect to the plants under the other cultivation conditions (Figure 3A). Drought stressed plants (Figure 2B) showed wilting symptoms, with yellowing and rolling of the lower leaves, as well as a reduced fruit size. The root system appeared reduced in density and length (Figure 3B). Stressed plants with paramylon root treatment (Figure 2C) did not show any wilting; the growth habit was quite compact, the internode length was greater with respect to plants grown under optimal water regimen, and the fruit sizes were comparable. The root system showed a dramatic reduction of both density and length, coupled with an increase of the lateral rootlets (Figure 3C).

Figure 2. The representative features of the tomato plants (aerial parts) under the three cultivation conditions tested 60 days after germination (dag): (**A**) Optimal water regimen; (**B**) water scarcity regimen; (**C**) water scarcity regimen coupled with paramylon root treatment.

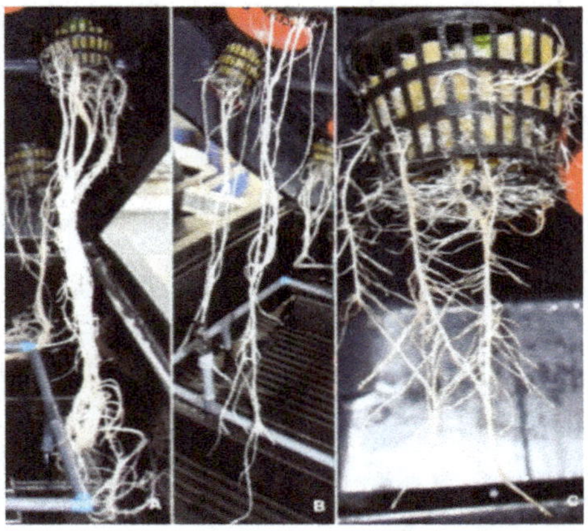

Figure 3. The root system of the tomato plants under the three cultivation conditions tested 60 dag. (**A**) Optimal water regimen, root length about 100 cm; (**B**) water scarcity regimen, root length about 80 cm; (**C**) water scarcity regimen coupled with paramylon root treatment, root length about 10 cm.

Figures 4 and 5 show the variations of the main eco–physiological parameters, monitored to highlight the effect of the paramylon root treatment during the life cycle of the plants.

Figure 4 shows that the leaf water potential (Ψ_w) (Figure 4A), CO_2 assimilation rate (A) (Figure 4B), stomatal conductance (gs) (Figure 4C), and internal CO_2 concentration (Ci) (Figure 4D) had time series with a similar trend in each growth condition tested. In WW_P^- plants, the values of these parameters were almost constant and comparable with the values present in the literature [35] (Figure 4A–D). These data confirm that the chosen water regimen (8.64 Ld^{-1} $plant^{-1}$) was optimal for Micro-Tom growth [21]. In WS_P^- plants, the four parameters underwent a steep decrease after a week of treatment (46 dag), which reached saturation before the end of experiment (Figure 4A–D). The reduction of the stomatal conductance as a reaction to water stress was the cause of this decreasing trend [36,37]. Additionally, in WS_P^+ plants, the four parameters underwent a steep decrease after a week of treatment (46 dag), but the values recovered to those of control plants (WW_P^-) after one or two weeks (Figure 4A–D). This delayed effect of paramylon nanofibers was mainly due to the time necessary to colonize the root system (Figure 6).

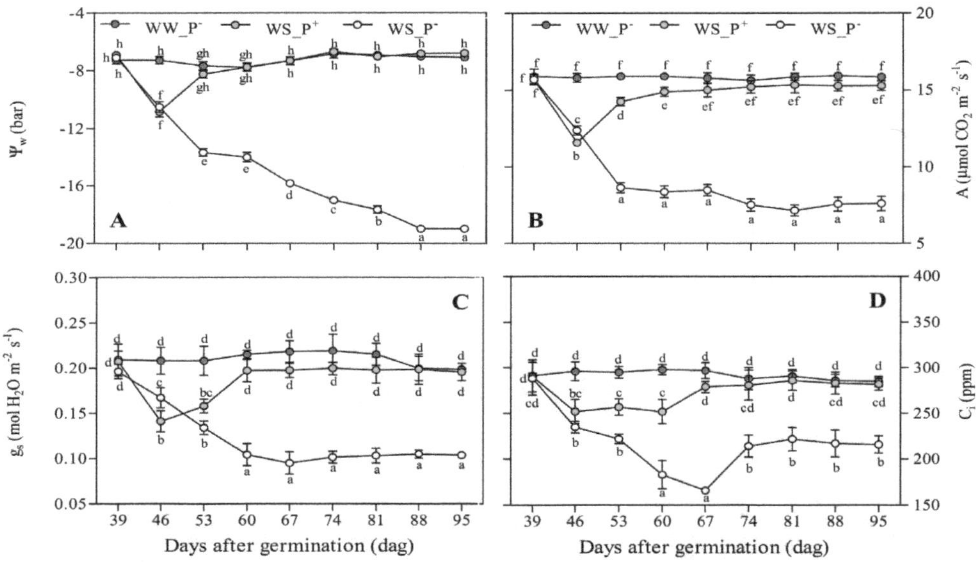

Figure 4. (**A**) Leaf water potential (Ψ_W), (**B**) CO_2 assimilation rate (A), (**C**) stomatal conductance (g_s), and (**D**) internal CO_2 concentration (C_i) in leaves of *Solanum lycopersicum* cv. Micro-Tom grown under well-watered (WW) and water-stressed (WS) conditions coupled (or not) with root treatment with paramylon (P^+ and P^-). Data are shown as mean ± standard deviation (n = 16), and measurements were made starting from 39 dag. In each graph, different letters indicate significant differences among treatments ($p < 0.05$, Tukey's HSD post hoc test).

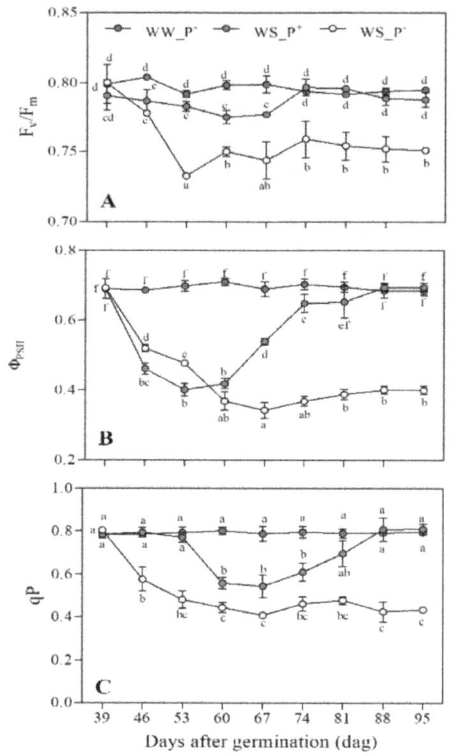

Figure 5. (**A**) Photosystem II (PSII) photochemical efficiency (F_v/F_m), (**B**) actual photon yield of PSII photochemistry (Φ_{PSII}), and (**C**) photochemical quenching state of PSII (qP) in leaves of *Solanum lycopersicum* cv. Micro-Tom grown under well-watered (WW) and water-stressed (WS) conditions coupled (or not) with root treatment with paramylon (P^+ and P^-). Data are shown as mean ± standard deviation (n = 16), and measurements were made starting from 39 dag. In each graph, different letters indicate significant differences among treatments ($p < 0.05$, Tukey's HSD post hoc test).

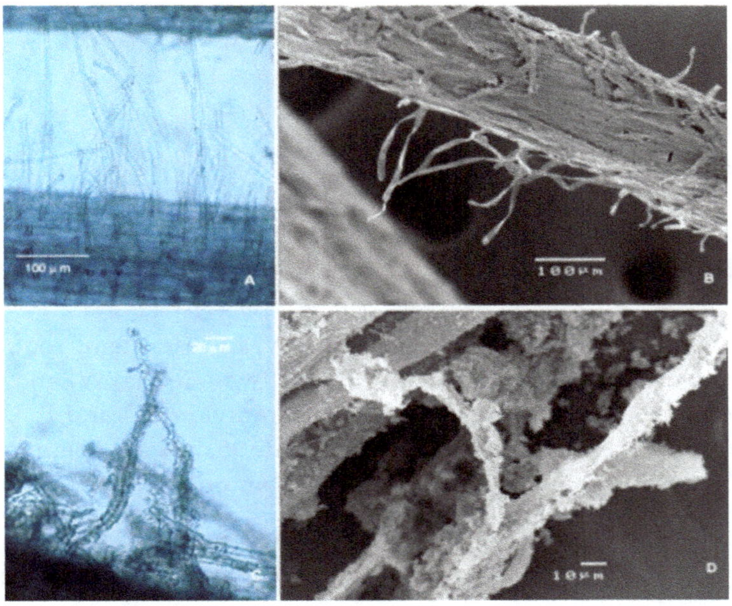

Figure 6. Optical microscopy and scanning electron microscopy images of tomato root hairs 50 dag. (**A** and **B**) Hair surface of non-treated plants. (**C** and **D**) Hair surface of paramylon-treated plants.

Figure 5 shows that the PSII photochemical efficiency (F_v/F_m) (Figure 5A), photon yield of PSII photochemistry (Φ_{PSII}) (Figure 5B), and photochemical quenching of PSII (qP) (Figure 5C) also had time series with a similar trend for each growth condition tested.

In WW_P$^-$ plants, the values of the three parameters were almost constant (Figure 5A–C). In WS_P$^-$ plants, they underwent a steep decrease after a week of treatment (46 dag), which reached saturation before the end of experiment (Figure 5A–C). Additionally, in WS_P$^+$ plants, the three parameters underwent a steep decrease after a week of treatment (46 dag), but their values recovered to those of control plants (WW_P$^-$) after several weeks (Figure 5A–C).

We can say that paramylon nanofibers induced dehydration tolerance and improved intrinsic water use efficiency by influencing stomatal behaviour. Scartazza et al. [16] monitored the effects of water removal on the Ψ_w and Φ_{PSII} of paramylon-treated tomato plants. They suggested that the paramylon caused an increase of CO_2 diffusional constraints but also promoted the ability of tomato plants to reduce water losses and counteract the reduction of Φ_{PSII} caused by the drought. According to these authors [16], β-glucan nanofibers play a potential role in reducing the sensitivity of PSII to potential dehydration damages thanks to a strong stomatal control associated with the transient modified profile of the three major plant hormones content (i.e., abscissic acid, jasmonic acid, salicylic acid) in the xylem sap. Furthermore, they stated that paramylon induced a consistent increase of g_m/g_s ratio (mesophyll conductance/stomatal conductance) and carbon gained per unit water used, which represents a relevant adaptive trait under water-limited conditions [16,36].

All our eco–physiological data confirmed the potentiality of paramylon nanofibers in the buffering water stress effect on both the photosynthetic rate and the PSII photochemical efficiency. The rate of linear electron transport (Φ_{PSII}) returned to the value of the control (WW_P$^-$ plants) with the consequent increase of the proportion of PSII reaction centres that were open (qP). Therefore, PSII photoinhibition and photodamage were counteracted (e.g., a complete recovery of F_v/F_m and Φ_{PSII} values), reducing the sensitivity of PSII to potential dehydration damages.

The values of the root, stem, leaf and fruit dry biomass measured 95 dag on the plants under the three different cultivation conditions are shown in Figure 7. WS_P$^-$ plants showed a significant reduction only in leaf DW (−23% compared with WW_P$^-$) data relative to the fruits are not shown because the fruit did not ripe beyond the II category (mature green) [24]. Drought stress conditions were so severe that all the eco–physiological parameters were deeply altered (Figure 2, Figure 5, and Figure 6) and inhibited the progress of ripening.

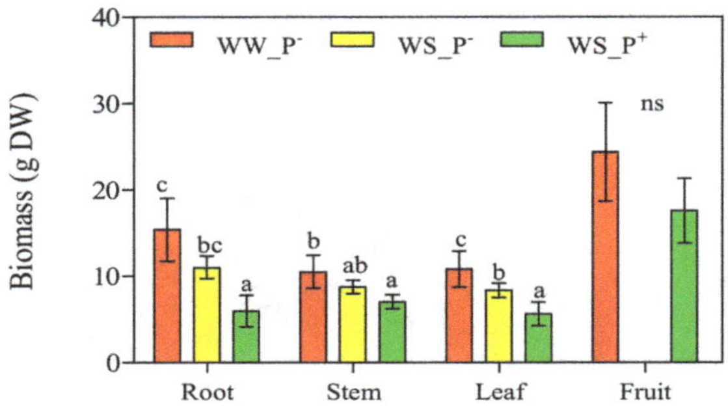

Figure 7. Dry biomass expressed as g dry weight (DW) of leaf, stem, root, and red ripe fruits of *Solanum lycopersicum* cv. Micro-Tom grown under well-watered (WW) and water-stressed (WS) conditions coupled (or not) with root treatment with paramylon (P^+ and P^-). Data are shown as mean ± standard deviation ($n = 8$). Measurements were made 95 days after germination (dag). For each parameter, different letters indicate significant differences among treatments ($p < 0.05$, Tukey's HSD post hoc test).

A significant reduction of root, stem and leaf DW was observed in WS_P$^+$ plants (−61%, −33% and −48% compared with WW_P$^-$). This reduction of growth could be explained as an adaptation strategy to reduce resource spending and allow the plants to divert energy in the fruits. No significant differences ($p > 0.05$) were observed between red ripe fruits DW of WW_P$^-$ and WS_P$^+$ plants. It is clear that WS induced a negative effect on leaf biomass and fruit ripening, as confirmed by the reduction of leaf DW and the alteration of fruit-ripening processes. Our results are not consistent with those by other authors [38–40] but are in accordance with Khan et al. [41]. The differences observed in the WS sensitivity may be due to the severity of drought.

The other physicochemical parameters (ashes, microelements, weight, size and yield) and the quality parameters (antioxidant compositions and activities, soluble carbohydrates) were measured on fruits assigned to a ripening category intermediate between the VI (red ripe) and the VII (red overripe) categories (Figure 8) [24].

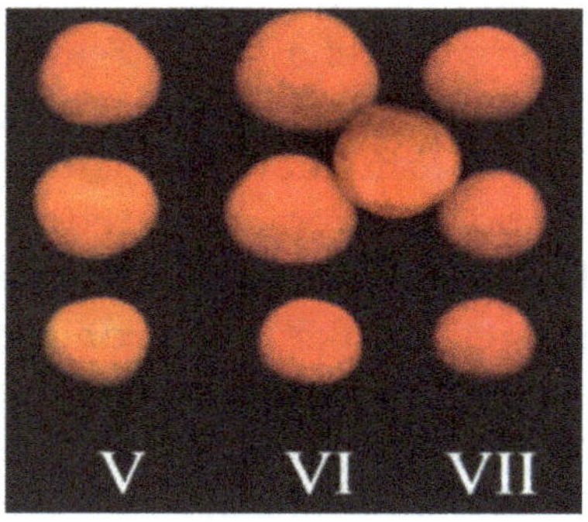

Figure 8. Ripening categories table according to Leide et al. [24]: Fruits used for the measurement of physicochemical and quality parameters belong to a ripening category intermediate between the VI (red ripe) and the VII (red overripe) categories.

No significant differences were observed between the number, weight and dimensions of the fruits of WW_P$^-$ and WS_P$^+$ (data not shown), though paramylon-treated plants showed precocious fruiting and ripening (Figure 9). As already stated, data relative to fruits of the drought stressed plants without paramylon (WW_P$^-$) are not shown because the fruits did not ripe beyond the II category (mature green).

The effects of cultivation conditions on red ripe fruits were monitored by the quantification of biometric parameters (ashes, dry matter, and moisture; Figure 10), microelement content, and bioactive compounds (antioxidant compositions and activities, soluble carbohydrates; Figures 11–13). In fruits harvested from WS_P$^+$ plants, a significant increase of ashes and dry matter was observed compared with WW_P$^-$ plants, while moisture showed an opposite trend (Figure 10). These data confirmed that paramylon treatment could improve fruit quality.

Figure 9. Precocious fruiting and ripening of the fruits of WS_P$^+$ plants (**A**) with respect to WW_P$^-$ plants (**B**) 74 dag.

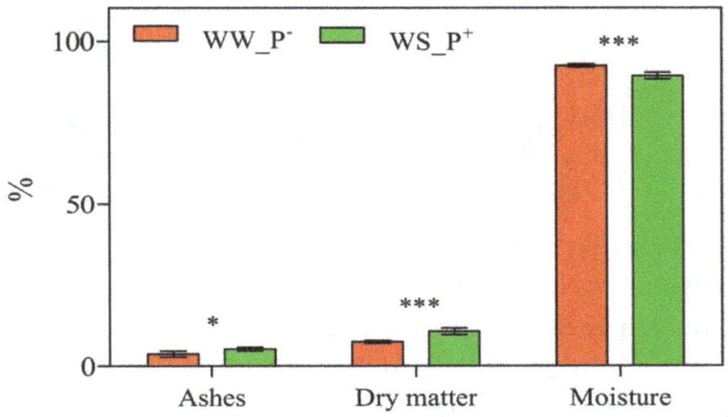

Figure 10. Biometric parameters (dry ash, dry matter and moisture) in red ripe fruits of *Solanum lycopersicum* cv. Micro-Tom grown under well-watered (WW) and water-stressed (WS) conditions coupled (or not) with root treatment with paramylon (P$^+$ and P$^-$). Data are shown as mean ± standard deviation ($n = 8$). Measurements were made 95 dag. For each parameter, the data were analyzed by Student's t test. The significant differences are for: *** = $p < 0.001$ and * = $p > 0.05$.

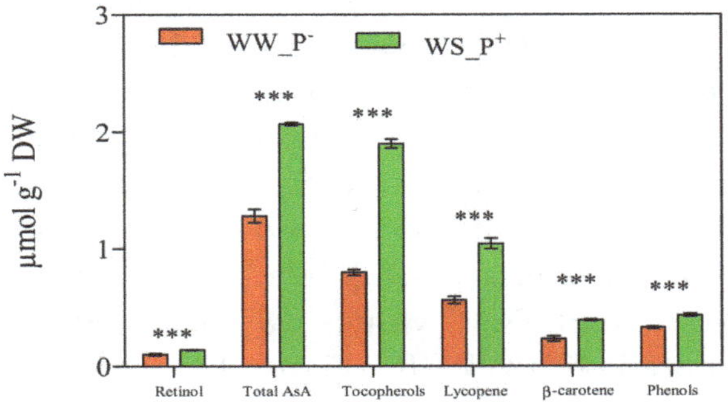

Figure 11. Antioxidant compounds in red ripe fruits of *Solanum lycopersicum* cv. Micro-Tom grown under well-watered (WW) and water-stressed (WS) conditions coupled (or not) with root treatment with paramylon (P$^+$ and P$^-$). Data are shown as mean ± standard deviation ($n = 8$). Measurements were made 95 dag. For each parameter, the data were analyzed by Student's t test. The significant differences are for: *** = $p < 0.001$. Abbreviations: DW, dry weight; Total AsA, total ascorbate (reduced and oxidized forms).

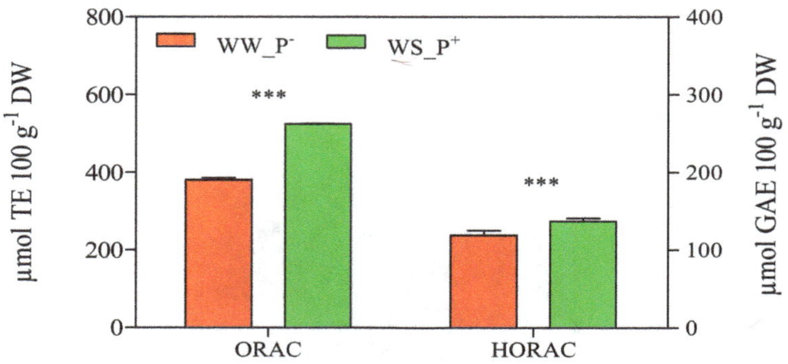

Figure 12. Antioxidant capacity expressed as oxygen radical absorbance capacity (ORAC) and hydroxyl radical antioxidant capacity (HORAC) in red ripe fruits of *Solanum lycopersicum* cv. Micro-Tom grown under well-watered (WW) and water-stressed (WS) conditions coupled (or not) with root treatment with paramylon (P^+ and P^-). Data are shown as mean ± standard deviation ($n = 8$). Measurements were made 95 dag. For each parameter, the data were analyzed by Student's t test. The significant differences are for: *** = $p < 0.001$. Abbreviations: DW, dry weight; GAE, gallic acid equivalent; TE, trolox equivalent.

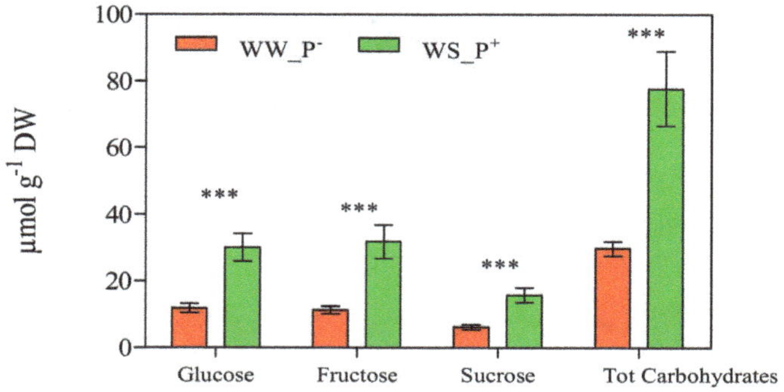

Figure 13. Carbohydrates in red ripe fruits of *Solanum lycopersicum* cv. Micro-Tom grown under well-watered (WW) and water-stressed (WS) conditions coupled (or not) with root treatment with paramylon (P^+ and P^-). Data are shown as mean ± standard deviation ($n = 8$). Measurements were made 95 dag. For each parameter, the data were analyzed by Student's t test. The significant differences are for: *** = $p < 0.001$. Abbreviations: DW, dry weight.

No significant differences were observed between WW_P^- and WS_P^+ fruits regarding microelements content (data not shown).

The cultivation conditions significantly affected the amount of antioxidant compounds of the fruits: all biochemical parameters (retinol, i.e., Vitamin A, tot AsA, i.e., Vitamin C, tocopherols, i.e., Vitamin E, lycopene, β-carotene and phenols) deeply increased in fruits harvested from WS_P^+ plants (about two-fold higher than those recorded in WW_P^- plants; Figure 11). A high AsA content is an important trait in tomato fruits, as it prevents oxidative stress (especially during fruit ripening) and thus enhances shelf life [42]. Similarly, an increase of lycopene (a highly characteristic phytonutrient of tomatoes) could affect the final nutritional quality and commercial value of tomato fruit [43]. In fact, lycopene greatly enhances fruit quality thanks to its intense antioxidant activity that suppresses cell proliferation and interferes with the growth of cancer cells [44,45]. In addition, a marked increase of the antioxidant activity expressed as the ORAC and HORAC values was observed in fruits harvested from WS_P^+ plants (+38% and +15%, compared with WW_P^-; Figure 12). This result confirms the involvement of lycopene, β-carotene, and retinol in the antioxidant response of WS_P^+ fruits. Tocopherols are non-enzymatic lipid-soluble antioxidants that protect the pigments, proteins, and polyunsaturated fatty acids of the photosynthetic apparatus against reactive oxygen species [46]. It has been reported that the tocopherols content in tomato fruits depends on many factors such as the level of irrigation, light, and NaCl [47]. We found that plants grown under WS_P^+ conditions produced fruits with a high tocopherol content, thus indicating an induction of defense mechanisms. Phenolic compounds and

AsA represent the main water-soluble antioxidants in tomatoes [48]. In WS_P$^+$ fruits, a significant increase of total phenols was observed, suggesting that they contribute positively to the antioxidant activity of the tomato water-soluble fraction by reducing the levels of free radicals due to WS (as confirmed by the increase of ORAC and HORAC levels). This response can be considered an adaptive mechanism to water stress that promotes the *de novo* synthesis of these metabolites [40]. At the end of the experiment, the content of carbohydrates showed a similar trend. Glucose, fructose, and sucrose concentrations, as well as total carbohydrates, in the fruits of WS_P$^+$ plants were nearly three-fold higher than in fruits harvested from WW_P$^-$ plants (Figure 13). Carbohydrates are essential for plant growth and survival, as well as maintenance and repair processes, and they are also major sources of cellular energy [49]. These compounds play a key role in regulating overall cellular metabolism, maintaining osmotic equilibrium, and preventing turgor loss in tissue [50,51]. They also can act as scavengers of reactive oxygen species (ROS) and contribute to the protection of membranes and macromolecules [52,53]. Micro-Tom fruits contain the reducing sugars fructose and glucose with trace amounts of sucrose, typical of tomatoes [54]. The carbohydrates composition found here was in agreement with that reported in the literature [55].

4. Conclusions

Drought is by far the most important environmental stressor in agriculture worldwide, and it is expected to contribute to the severe salinization of more than 50% of world arable land by 2050. The research efforts to improve crop productivity under water limiting conditions, focused mainly on natural selection and the breeding activity of tolerant genotypes. Root treatment could be another method to cope with the drought stress. In this paper, we showed the results of the direct application of paramylon on the root system of Micro-Tom tomatoes. Paramylon extracted from the *E. gracilis* WZSL mutant was processed to linear nanofibers that interacted with the Dectin-1 receptors present on the target cell membranes of tomato roots, enhancing the plant defense capacity against drought. Drought tolerance was achieved by influencing stomatal behavior and inducing an effective improvement of water use efficiency, obtained by modulating the conductance to CO_2 diffusion from air to the carboxylation sites through the modulation of hormone levels [16]. We observed that the paramylon treatment allowed the optimal water regimen of about 8.64 L plant^{-1} day^{-1} to be lowered to 0.36 Lplant^{-1} day^{-1} without a detrimental effect on the yield and eco–physiological parameters. The great increase of antioxidant compounds (Vitamin A/C/E, lycopene, β-carotene and phenols) together with the increase of carbohydrates (glucose, fructose and sucrose) in the fruits of paramylon-treated plants improved their nutritional value and sensory quality.

These results confirm the biostimulant activity of paramylon in increasing plant adaptation capacity for abiotic stress.

Author Contributions: Conceptualization, methodology, investigation, writing—L.B. and P.G.; resources, software, and data curation—P.C.

References

1. Barsanti, L.; Passarelli, V.; Evangelista, V.; Frassanito, A.M.; Gualtieri, P. Chemistry, physico-chemistry and applications linked to biological activities of β-glucans. *Nat. Prod. Rep.* **2011**, *28*, 457–466. [CrossRef] [PubMed]
2. Barsanti, L.; Vismara, R.; Passarelli, V.; Gualtieri, P. Paramylon (β-1,3-glucan) content in wild type and WZSL mutant of *Euglena gracilis*: effects of growth conditions. *J. Appl. Phycol.* **2001**, *13*, 59–65. [CrossRef]
3. Shibakami, M.; Tsuboichi, G.; Nakamura, M.; Hayashi, M. Preparation of carboxylic acid-bearing polysaccharide nanofiber made from euglenoid β-1,3-glucans. *Carbohydr. Polym.* **2013**, *15*, 95–101. [CrossRef] [PubMed]

4. Russo, R.; Barsanti, L.; Evangelista, V.; Frassanito, A.M.; Longo, V.; Pucci, L.; Penno, G.; Gualtieri, P. *Euglena gracilis* paramylon activates human lymphocytes by upregulating pro-inflammatory factors. *Food Sci. Nutr.* **2017**, *5*, 205–214. [CrossRef] [PubMed]
5. Rosati, G.; Barsanti, L.; Passarelli, V.; Giambelluca, A.; Gualtieri, P. Ultrastructure of a novel non-photosynthetic *Euglena* mutant. *Micron* **1996**, *27*, 367–373. [CrossRef]
6. Takeuchi, O.; Akira, S. Pattern recognition receptors and inflammation. *Cell* **2010**, *140*, 805–820. [CrossRef] [PubMed]
7. Thompson, I.J.; Oyston, P.C.F.; Williamson, D.E. Potential of the β-glucans to enhance innate resistance to biological agents. *Expert Rev. Anti. Infect. Ther.* **2010**, *8*, 339–352. [CrossRef] [PubMed]
8. Adams, E.L.; Rice, P.J.; Graves, B.; Ensley, H.E.; Yu, H.; Brown, G.D.; Williams, D.L. Differential high-affinity interaction of dectin-1 with natural or synthetic glucans is dependent upon primary structure and is influenced by polymer chain length and side chain branching. *J. Pharmacol. Exp. Ther.* **2008**, *325*, 115–123. [CrossRef]
9. Fu, C.; Iyer, P.; Herkal, A.; Abdullah, J.; Stout, A.; Free, S.J. Identification and characterization of genes required for cell-to-cell fusion in *Neurospora crassa*. *Eukaryot. Cell* **2011**, *10*, 1100–1109. [CrossRef]
10. Bashir, K.M.I.; Choi, J.-S. Clinical and physiological perspectives of β-glucans: the past, present, and future. *Int. J. Mol. Sci.* **2017**, *18*, 1906. [CrossRef]
11. Dennehy, K.M.; Brown, G.D. The role of the beta-glucan receptor dectin-1 in control of fungal infection. *J. Leukoc. Biol.* **2007**, *82*, 253–258. [CrossRef]
12. Monie, T.P. Immune cells and the process of pattern recognition. In *The innate immune system: A compositional and functional perspective*; Monie, T.P., Ed.; Academic Press: London, UK, 2017; pp. 41–83.
13. Palma, A.S.; Feizi, T.; Zhang, Y.; Stoll, M.S.; Lawson, A.M.; Diaz-Rodríguez, E.; Campanero-Rhodes, A.S.; Costa, J.; Brown, G.D.; Chai, W. Ligands for the beta-glucan receptor, dectin-1, assigned using "designer" microarrays of oligosaccharide probes (neoglycolipids) generated from glucan polysaccharides. *J.Biol. Chem.* **2006**, *281*, 5771–5779. [CrossRef] [PubMed]
14. Tsoni, S.V.; Brown, G.D. β-glucans and dectin-1. *Ann. N. Y. Acad. Sci.* **2008**, *1143*, 45–60. [CrossRef] [PubMed]
15. Vismara, R.; Vestri, S.; Frassanito, A.M.; Barsanti, L.; Gualtieri, P. Stress resistance induced by paramylon treatment in *Artemia* sp. *J. Appl. Phycol.* **2004**, *16*, 61–67. [CrossRef]
16. Scartazza, A.; Picciarelli, P.; Mariotti, L.; Curadi, M.; Barsanti, L.; Gualtieri, P. The role of *Euglena gracilis* paramylon in modulating xylem hormone levels, photosynthesis and water-use efficiency in *Solanum lycopersicum* L. *Physiol. Plant.* **2017**, *161*, 486–501. [CrossRef] [PubMed]
17. Kusmic, C.; Barsanti, L.; Di Lascio, N.; Faita, F.; Evangelista, V.; Gualtieri, P. Anti-fibrotic effect of paramylon nanofibers from the WZSL mutant of *Euglena gracilis* on liver damage induced by CCl_4 in mice. *J. Funct. Foods* **2018**, *46*, 538–545. [CrossRef]
18. Morison, J.I.L.; Baker, N.R.; Mullineaux, P.M.; Davies, W.J. Improving water use in crop production. *Philos. Trans. R. Soc. Lond.* **2008**, *363*, 639–658. [CrossRef] [PubMed]
19. Scott, J.W.; Harbaugh, B.K. *Micro-Tom: A Miniature Dwarf Tomato*; Agricultural Experiment Station, Institute of Food and Agricultural Sciences, University of Florida: Gainesville, FL, USA, 1989; pp. 1–6.
20. Motohashi, R.; Enoki, H.; Fukazawa, C.; Kiriiwa, S. Hydroponic culture of Micro-Tom tomato. *Bio-protocol* **2015**, *5*, e1613. [CrossRef]
21. Johnstone, P.R.; Nichols, M.A.; Fisher, K.J.; Reid, J. Nutritional studies with processing tomato grown in aeroponic. *Acta Horticulture* **2001**, 143–150. [CrossRef]
22. Turner, N.C.; Long, M.J. Errors arising from rapid water loss in the measurement of leaf water potential by the pressure chamber technique. *Aust. J. Plant Physiol.* **1980**, *7*, 527–537. [CrossRef]
23. Schreiber, U.; Schilwa, U.; Bilger, W. Continuous recording of photochemical and non-photochemical chlorophyll fluorescence quenching with a new type of modulation fluorometer. *Photosynth. Res.* **1986**, *10*, 51–62. [CrossRef] [PubMed]
24. Leide, J.; Hildebrandt, H.; Reussing, K.; Riederer, M.; Vogg, G. The developmental pattern of tomato fruit wax accumulation and its impact on cuticular transpiration barrier properties: effects of a deficiency in a β-ketoacyl-coenzyme A synthase (LeCER6). *Plant Physiol.* **2007**, *144*, 1667–1679. [CrossRef] [PubMed]
25. Zasoski, R.J.; Burau, R.G. A rapid nitric-perchloric acid digestion method for multi-element tissue analysis. *Comm. Soil Sci. Plant Anal.* **1977**, *8*, 425–436. [CrossRef]

26. Aremu, S.O.; Nweze, C.C. Determination of vitamin A content from selected Nigerian fruits using spectrophotometric method. *Bangladesh J. Sci. Ind. Res.* **2017**, *52*, 153–158. [CrossRef]
27. Georgé, S.; Tourniaire, F.; Gautier, H.; Goupy, P.; Rock, E.; Caris-Veyrat, C. Changes in the contents of carotenoids, phenolic compounds and vitamin C during technical processing and lyophilization of red and yellow tomatoes. *Food Chem.* **2011**, *124*, 1603–1611.
28. Kampfenkel, K.; Van Montagu, M.; Inzé, D. Extraction and determination of ascorbate and dehydroascorbate from plant tissue. *Anal. Biochem.* **1995**, *10*, 165–167. [CrossRef] [PubMed]
29. Döring, A.S.; Pellegrini, E.; Campanella, A.; Trivellini, A.; Gennai, C.; Petersen, M.; Nali, C.; Lorenzini, G. How sensitive is *Melissa officinalis* to realistic ozone concentrations? *Plant Physiol. Biochem.* **2014**, *74*, 156–164. [CrossRef]
30. Aguiar, F.; Menezes, V.; Rogez, H. Spontaneous postharvest fermentation of acai (*Euterpe oleracea*) fruit. *Postharvest Biol. Technol.* **2013**, *86*, 294–299. [CrossRef]
31. Ou, B.X.; Hampsch-Woodill, M.; Prior, R.L. Development and validation of an improved Oxygen Radical Absorbance Capacity Assay using fluorescein as the fluorescent probe. *J. Agric. Food Chem.* **2001**, *49*, 4619–4626. [CrossRef]
32. Ou, B.X.; Huang, D.J.; Hampsch-Woodill, M.; Flanagan, J.A.; Deemer, E.K. Analysis of antioxidant activities of common vegetables employing oxygen radical absorbance capacity (ORAC) and ferric reducing antioxidant power (FRAP) assays: a comparative study. *J. Agric. Food Chem.* **2002**, *50*, 3122–3128. [CrossRef]
33. Waterhouse, A.L. Polyphenolics: determination of total phenolics. In *Current Protocols in Food Analytical Chemistry*; Wrolstad, R.E., Ed.; Wiley: New York, NY, USA, 2002; pp. 1–4.
34. Lakhiar, I.A.; Gao, J.; Syed, T.N.; Chandio, F.A.; Buttar, N.A. Modern plant cultivation technologies in agriculture under controlled environment: a review on aeroponics. *J. Plant Interact.* **2018**, *13*, 338–352. [CrossRef]
35. Alves, F.R.R.; Melo, H.C.; Crispin-Filho, A.J.; Costa, A.C.; Nascimiento, K.J.T.; Carvalho, R.F. Physiological and biochemical responses of photomorphogenic tomato mutants (cv. Micro-Tom) under water withholding. *Acta Physiol. Plant* **2016**, *38*, 155–170. [CrossRef]
36. Centritto, M.; Lauteri, M.; Monteverdi, M.C.; Serraj, R. Leaf gas exchange, carbon isotope discrimination, and grain yield in contrasting rice genotypes subjected to water deficits during the reproductive stage. *J. Exp. Bot.* **2009**, *60*, 2325–2339. [CrossRef] [PubMed]
37. Flexas, J.; Niinements, U.; Gallé, A.; Barbour, M.M.; Centritto, M.; Diaz-Espejo, A.; Douthe, C.; Galmés, J.; Ribas-Carbo, M.; Rodriguez, P.L.; et al. Diffusional conductances to CO_2 as a target for increasing photosynthesis and photosynthetic water-use efficiency. *Photosynth. Res.* **2913**, *117*, 45–59. [CrossRef] [PubMed]
38. Patanè, C.; Cosentino, S.L. Effects of soil water deficit on yield and quality of processing tomato under a Mediterranean climate. *Agric. Water Manag.* **2010**, *97*, 131–138. [CrossRef]
39. Casa, R.; Rouphael, Y. Effects of partial root-zone drying irrigation on yield fruit quality, and water-use efficiency in processing tomato. *J. Hortic. Sci. Biotechnol.* **2014**, *89*, 389–396. [CrossRef]
40. Guida, G.; Sellami, M.H.; Mistretta, C.; Oliva, M.; Buonomo, R.; De Mascellis, R.; Patanè, C.; Rouphael, Y.; Albrizio, R.; Giorio, P. Agronomical, physiological and fruit quality responses of two Italian long-storage tomato landraces under rain-fed and full irrigation conditions. *Agric. Water Manag.* **2017**, *180*, 126–135. [CrossRef]
41. Khan, S.H.; Khan, A.; Litaf, U.; Shah, A.S.; Khan, M.A.; Bilal, M.; Ali, M.U. Effect of drought stress on tomato cv. Bombino. *J. Food Process. Technol.* **2015**, *6*, 465. [CrossRef]
42. Slimestad, R.; Verheul, M. Review of flavonoids and other phenolics from fruits of different tomato (*Lycopersicon esculentum* L.) cultivars. *J. Sci. Food Agric.* **2009**, *89*, 1255–1270. [CrossRef]
43. Dumas, Y.; Dadomo, M.; Licca, G.D.; Grolier, P. Effects of environmental factors and agricultural techniques of antioxidant content of tomatoes. *J. Sci. Food Agric.* **2003**, *83*, 369–382. [CrossRef]
44. Levi, F.; Pasche, C.; Lucchini, F.; La Vecchia, C. Dietary intake of selected micronutrients and breast-cancer risk. *Int. J. Cancer* **2001**, *91*, 260–263. [CrossRef]
45. Stahl, W.; Sies, H. Bioactivity and protective effects of natural carotenoids. *Biochim. Biophys. Acta* **2005**, *30*, 101–107. [CrossRef] [PubMed]

46. Quadrana, L.; Almeida, J.; Otaiza, S.N.; Duffy, T.; Corrêa da Silva, J.V.; de Godoy, F.; Asís, R.; Bermúdez, L.; Fernie, A.R.; Carrari, F.; et al. Transcriptional regulation of tocopherol biosynthesis in tomato. *Plant Mol. Biol.* **2013**, *81*, 309–325. [CrossRef] [PubMed]
47. Raiola, A.; Tenore, G.C.; Barone, A.; Frusciante, L.; Rigano, M.M. Vitamin E content and composition in tomato fruits: beneficial roles and bio-fortification. *Int. J. Mol. Sci.* **2015**, *8*, 29250–29264. [CrossRef] [PubMed]
48. Fanasca, S.; Colla, G.; Maiani, G.; Venneria, E.; Rouphael, Y.; Azzini, E.; Saccardo, F. Changes in antioxidant content of tomato fruits in response to cultivar and nutrient solution composition. *J. Agric. Food Chem.* **2006**, *54*, 4319–4325. [CrossRef]
49. Kusvuran, S.; Ellialtioglu, S.; Polat, Z. Antioxidative enzyme activity lipid peroxidation, and proline accumulation in the callus tissues of salt and drought tolerant and sensitive pumpkin genotypes under chilling stress. *Hortic. Environ. Biotechnol.* **2013**, *54*, 319–325. [CrossRef]
50. Clifford, S.C.; Arndt, S.K.; Corlett, J.E.; Joshi, S.; Sankhla, N.; Popp, M.; Jones, H.G. The role of solute accumulation, osmotic adjustment and changes in cell wall elasticity in drought tolerance in *Ziziphus mauritania* (Lamk.). *J. Exp. Bot.* **1998**, *49*, 967–977. [CrossRef]
51. Sivakumar, P.; Sharmila, P.; Jain, V.; Pardha-Saradhi, P. Sugars have potential to curtail oxygease acitivity of Rubisco. *Biochem. Biophys. Res. Commun.* **2002**, *298*, 247–250. [CrossRef]
52. Hoekstra, F.A.; Golovina, E.A.; Buitink, J. Mechanisms of plant desiccation tolerance. *Trends Plant Sci.* **2001**, *6*, 431–438. [CrossRef]
53. Mahajan, S.; Tuteja, N. Cold, salinity and drought stresses: an overview. *Arch. Biochem. Biophys.* **2005**, *444*, 139–158. [CrossRef]
54. Malundo, T.M.M.; Shewfelt, R.L.; Scott, J.W. Flavor quality of fresh tomato (*Lycopersicon esculentum* Mill) as affected by sugar and acid levels. *Postharvest Biol. Technol.* **1995**, *6*, 103–110. [CrossRef]
55. Gomez, P.; Ferrer, M.A.; Fernandez-Trujillo, J.P.; Calderon, A.; Artes, F.; Egea-Cortines, M.; Weiss, J. Structural changes, chemical composition and antioxidant activity of cherry tomato fruits (cv Micro-Tom) stored under optimal and chilling conditions. *J. Sci. Food Agric.* **2009**, *89*, 1543–1551. [CrossRef]

9

Impact of Biostimulators on the Quality of Seeds

Marek Gugała [1], Anna Sikorska [2,*], Krystyna Zarzecka [1], Pavol Findura [3,4] and Urszula Malaga-Toboła [4]

1 Faculty of Agrobioengineering and Animal Husbandry, University of Natural Sciences and Humanities in Siedlce, ul. Prusa 14, 08-110 Siedlce, Poland; gugala@uph.edu.pl (M.G.); krystyna.zarzecka@uph.edu.pl (K.Z.)

2 Department of Agriculture, Vocational State School of Ignacy Mościcki in Ciechanów, ul. Narutowicza 9, 06-400 Ciechanów, Poland

3 Department of Biosystems Engineering; Slovak University of Agriculture Nitra, Tr. A. Hlinku 2, 949-76 Nitra, Slovak Republic; pavol.findura@uniag.sk

4 Faculty of Production and Power Engineering; University of Agriculture in Krakow, ul. Balicka 116B, 30-149 Krakow, Poland; Urszula.Malaga-Tobola@ur.krakow.pl

* Correspondence: anna.sikorska@puzim.edu.pl

Abstract: Plant growth regulators may reduce the negative effect of environmental stress factors and can contribute to increasing the quality and quantity of the yield. The aim of the research was to determine the effect of biostimulators on the quality of seeds of three winter rape morphotypes. Three varieties of winter rape were used: Poznaniak (population variety), PX104 (hybrid variety restored with a semi-dwarf growth type) and Konkret (hybrid variety restored with a traditional growth type). The varieties were exposed to three treatments: the biostimulator Tytanit®, the biostimulator Asahi®SL and the biostimulator Silvit®, and the control with no biostimulators. Seeds were analysed for content of crude fat, total fat and crude fibres. The biostimulators reduced total protein content (on average from 0.8 to 1.75 $g \cdot kg^{-1}$ of d.m.) and increased the concentration of crude fat (on average from 0.71 to 1.93 $g \cdot kg^{-1}$ of d.m.) and crude fibre (on average from 0.15 to 0.84 $g \cdot kg^{-1}$ of d.m.) compared to the control. PX104 had the highest content of crude fat and total fat protein, and the lowest in crude fibre. The smallest protein content was found in seeds of the long-stem hybrid Konkret, while crude fat was lowest in the population form (Poznaniak), and crude fibre was lowest in long-stem hybrid (Konkret).

Keywords: anti-nutritional substances; fat; fibre; morphotype; protein

1. Introduction

Rapeseed (*Brassica napus* L. var. oleifera) is one of the most important oil-protein crops grown in the world. One of the many factors with a negative effect on the quantity and quality of rapeseed crops include unfavourable soil conditions and drought-related stress. Strong stress leads to damaged cell structures and disturbances in metabolism and as a result, can lead to photosynthesis and plant and metabolism disruption [1]. Rouphael and Colla [2] reports that plant biostimulators are products obtained from various organic or inorganic substances or microorganisms that improve plant growth, productivity and reduce the negative effects of environmental stress. Many authors [3–8] have shown that regulators of plant growth and development reduce the negative impact of abiotic stress factors. Petrozza et al. [9] showed that when a plant experiences stress, the biostimulator strengthens its stress tolerance mechanism.

Colla and Rouphael [10] and Rouphael et al. [11] emphasize that the use of biostimulators is increasingly becoming one of the basic elements of agricultural technology in many crop species around the world. Calvo et al. [12] forecast that the global market for biostimulants in consumption will increase by 14% per year.

According to El-Boray et al. [13], Przybysz et al. [14], Kocira et al. [15] and Zulfiqar et al. [16], preparations stimulating plant growth can be based on extracts of marine algae, free amino acids, humic compounds, effective microorganisms or phenolic compounds. Their use in plant cultivation has a positive effect on photosynthesis, regulation of water management and increasing the content of organic and inorganic compounds, which in turn, has a positive effect on the size and quality of the crop.

Grabowska et al. [17] and Kolomaznik et al. [18] stated that the effectiveness of biostimulators depends on many factors, including the correct selection of preparations, their dose, concentration and methods of application, as well as plant species and varieties and environmental factors.

The study assumes the hypothesis that the use of biostimulators may have a positive effect on the chemical composition of winter rapeseeds.

Due to few studies being available on the beneficial effects of growth bioregulators on the quality characteristics of winter rapeseed, and the wide interest in agricultural practice, research was undertaken to determine the effect of biostimulators on the chemical composition of three winter rapeseed varieties.

2. Materials and Methods

2.1. Arrangement of the Experiment and Research Location

The field experiment was carried out in 2013–2016 in three different fields at the Agricultural Experimental Station—Zawady (52°03′N; 22°33′E), belonging to the University of Natural Sciences and Humanities in Siedlce. The experiment was established in a random split-plot system in three repetitions (total number of plots 3 × 4 = 12, repeated in 3 successive crop rotations from 2013–2016). The surface of one plot was 21 m^{-2}. The examined factors were:

I—three varieties of winter rape: Poznaniak (population variety), PX104 (hybrid variety restored with a semi-dwarf growth type) and Konkret (hybrid variety restored with a traditional growth type).

II—four types of biostimulators:

1. control—no biostimulators;

2. biostimulator Tytanit® (active substance—titanium), applied in three doses of 0.20 dm^3 ha^{-1} in the autumn (2 October 2013, 6 October 2014, 5 October 2015) at the 4–8 leaf stage (BBCH 14–18) according to the rating of Biologische Bundesantalt, Bundessortenamt and Chemische Industrie (BBCH), in the spring (26 March 2014, 23 March 2015, 21 March 2016) after the onset of growth (BBCH 21–36), and at the budding stage—early flowering (30 April 2014, 29 April 2015,4 May 2016) (BBCH 50–61) [19];

3. biostimulator Asahi®SL (active substances: sodium orto nitrophenol, sodium para nitrophenol, sodium 5-nitroguaiacolate), applied in three doses of 0.60 $dm^3{\cdot}ha^{-1}$ in the autumn (25 September 2013, 29 September 2014, 27 September 2015) at the stage of 3–5 leaves (BBCH 13–15), in the spring (26 March 2014, 23 March 2015, 21 March 2016) after the plants resumed growth (BBCH 28–30), and two weeks following the second application (10 April 2014, 7 April 2015, 4 April 2016);

4. biostimulator Silvit® (active substances: active silicon, potassium oxide, boron, zinc), applied in three doses of 0.20 $dm^3{\cdot}ha^{-1}$, three weeks after emergence (2 October 2013, 6 October 2014, 5 October 2015) (BBCH 12–14), in spring (26 March 2014, 23 March 2015, 21 March 2016) after plants resumed growth (BBCH 28–30), and two weeks after the second application (10 April 2014, 7 April 2015, 4 April 2016).

The studies were carried out on soil classified according to WBR FAO (2014) [20] as the Haplic Luvisols group—sandy, belonging to a very good rye soil complex of the IVb botanical class. In the years of the experiment, the soil reaction (pH) ranged from 5.68 to 5.75. The soil was characterised by a low total nitrogen content (average from 0.80 to 0.90 g·kg^{-1}), phosphorus content (average from 0.33 to 0.55 g·kg^{-1}), potassium content (average from 0.61 to 0.67 g·kg^{-1}) calcium content (average from 0.82 to 0.85 g·kg^{-1}), magnesium content (average from 0.38 to 0.46 g·kg^{-1}) and sulphur content (average from 0.11 to 0.15 g·kg^{-1}). It has a low abundance in assimilable forms of phosphorus (average from 75.0 to 80. g·kg^{-1}) and an average assimilability of potassium (from 200.0 to 205.0 g·kg^{-1}) and magnesium (average from 59.0 to 61.0 g·kg^{-1}).

The phosphorus and potassium fertilization at the dose of 40.0 kg P·ha^{-1} and 110.0 kg K·ha^{-1} with the first dose of 40.0 kg N·ha^{-1} was used before sowing. Fertilization was used in the form of Lubofos for Rape at the dose of 600.0 kg, i.e., 21.0 kg N·ha^{-1}, 26.4 kg P·ha^{-1}, 92.1 kg K·ha^{-1}, 34.8 kg S·ha^{-1}, 1.2 kg B·ha^{-1}. Fertilization rates were supplemented by 55.9 kg·ha^{-1} of ammonium nitrate (19.0 kg N·ha^{-1}), 29.6 kg·ha^{-1} of triple superphosphate (13.6 kg P·ha^{-1}) and 29 kg·ha^{-1} of potassium salt (17.9 kg K·ha^{-1}). The second nitrogen dose of 100.0 kg·ha^{-1} was applied in spring before vegetation using ammonium nitrate at the dose of 255.5 kg·ha^{-1} and ammonium sulphate at the dose of 62.5 kg·ha^{-1}. The third dose of nitrogen 60.0 kg·ha^{-1} was applied at the beginning of budding using ammonium nitrate at the dose of 176.5 kg·ha^{-1}.

The three crops of rapeseed were harvested on 11 July 2014, 17 July 2015 and 14 July 2016, respectively.

2.2. *Chemical Analysis of Seeds*

The tests samples of winter rape seeds were analyzed for:

Crude fat (g·kg^{-1}of d.m.)—with the Soxhlet method, which extracted the fat with petroleum ether in a Soxhlet apparatus and determides its quantity by weight, total protein (g·kg^{-1} of d.m.) [21].

Total protein (g·kg^{-1}of d.m.)—with the Kjeldahl method where protein nitrogen was converted to ammonium sulphate with concentrated sulphuric acid in the presence of a catalyst, the solution was alkalised, distilled and titrated with hydrochloric acid-ammonia bound with boric acid, the conversion factor Nx6.25 was used, crude fibre (g·kg^{-1} of d.m.) [22].

Crude fibres (g·kg^{-1}of d.m.)—with the Wenden method consisting of the quantitative determination of organic substances insoluble during cooking in an acid solution.

2.3. *Statistical Analysis*

Research results were statistically analysed by ANOVA. The results of the study were statistically analysed using the analysis of variance. The significance of the sources of variation was tested by the Fischer-Snedecor "F" test, and the assessment of significance of differences at the significance level $p < 0.05$ between the compared averages used Tukey's multiple intervals. Statistical calculations were made based on our own algorithm written in Excel [23].

2.4. *Weather Conditions*

Climatic data from 2013–2016 was obtained from the Hydrological and Meteorological Station in Siedlce. During the years of conducting the experiment, varied weather conditions prevailed (Table 1). In the second growing season, the largest annual rainfall was recorded (average of 599.2 mm) and the smallest mean annual air temperature (average of 8.8 °C). In this period, the annual amount of rainfall was 171.7 mm higher compared to the long-term period. The last year of tests was the warmest and most dry. The annual rainfall was 43.8 mm lower than the average for the long-term period, and the average air temperature was higher by 1.3 °C compared to the average from 1996–2010. Based on the calculated Sielianinov hydrothermal coefficient, the first and last year of the study were optimal, while the growing season 2014–2015 was rather wet ($K = 1.71$).

Table 1. Characteristics of weather conditions in the years 2013–2016 (Poland).

Months	Rainfalls (mm)				Air Temperature (°C)			
	Multiyear Sum	Monthly Sum			Multiyear Mean	Monthly Mean		
	1996–2010	2013–2014	2014–2015	2015–2016	1996–2010	2013–2014	2014–2015	2015–2016
VIII	59.9	15.0	105.7	11.9	18.5	18.8	18.1	21.0
IX	42.3	94.3	26.3	47.1	13.5	11.7	14.1	14.5
X	24.2	32.8	3.0	37.0	7.9	9.3	8.5	6.5
XI	20.2	34.7	32.5	42.2	4.0	5.1	3.4	4.7
XII	18.6	15.4	90.4	16.5	−0.1	1.2	0.1	3.7
I	19.0	28.6	51.4	10.9	−3.2	−4.5	0.6	−4.5
II	16.0	34.0	0.7	29.0	−2.3	0.7	0.7	2.5
III	18.3	29.6	53.1	33.5	2.4	5.8	4.6	3.5
IV	33.6	45.0	30.0	28.7	8.0	9.8	8.2	9.1
V	58.3	92.7	100.2	54.8	13.5	13.5	12.3	15.1
VI	59.6	55.4	43.3	36.9	17.0	15.4	16.5	18.4
VII	57.5	10.0	62.6	35.2	19.7	20.8	18.7	19.1
VIII–VII	427.5	487.5	599.2	383.7	8.2	9.0	8.8	9.5

Sielianinovs hydrothermic coefficients *			
	2013–2014	2014–2015	2015–2016
VIII	0.31	1.87	0.20
IX	2.63	0.66	1.20
X	1.01	0.22	2.15
III	1.48	4.63	3.49
IV	1.41	1.35	1.07
V	2.33	2.91	1.47
VI	1.23	0.84	0.72
VII	0.16	1.20	0.64
VIII–VII	1.32	1.71	1.37

* Index value [24]: extremely dry $k \leq 0.4$, very dry $0.4 < k \leq 0.7$, dry $0.7 < k \leq 1.0$, rather dry $1.0 < k \leq 1.3$, optimal $1.3 < k \leq 1.6$, rather humid $1.6 < k \leq 2.0$, humid $2.0 < k \leq 2.5$, very humid $2.5 < k \leq 3.0$, extremely humid $k > 3.0$.

3. Results and Discussion

3.1. The Content of Total Protein Depending on the Types of Biostimulators Used

Our own research showed that biostimulators significantly affected the reduction of total protein content in rapeseeds (Table 2). The smallest concentration was recorded on object 4, sprayed with the Silvit biostimulator. This value was lower on average by 1.75 g·kg^{-1} of d.m. compared to the control variant. Different results were obtained by Gugała et al. [25], who did not find a significant effect of the biostimulators Tytanit, Asahi SL or Silvit for the value of this feature. Similarly, Jarecki and Bobrecka-Jamro [26,27], Kozak et al. [28] and Matysiak et al. [29,30] did not prove the effect of bioregulators and foliar fertilizers containing micro- and macro-elements for the value of this feature. While Jankowski et al. [31], after a double foliar application with boron, increased the protein content in seeds by an average of 8.8 g·kg^{-1} of d.m. compared to the control object. In regards to seed protein, the present study's research showed the interaction of the types of biostimulators used in relation to the protein content in the seeds of the rapeseed morphotype varieties studied, which indicated the individual response of the rapeseed varieties to the biopreparations used (Table 2). The lowest protein was in the treatment of Konkret with Silvit and PX104 with Silvit, and in Pozniak with Tytanit. In all morphotypes, the highest protein content was recorded on the object where no natural growth stimulants were used. In the cultivar with the traditional growth type, the lowest protein content was found after the application of the Tytanit biostimulator, while in the other varieities it was after the use of the Silvit biostimulator. Equal protein content was found in the restored hybrids of the semi-dwarf type (PX104) after the application of the Asahi SL and Tytanit preparations. A similar tendency was observed in the restored hybrid with the traditional growth type.

The content of total seed protein was dependant on the genetic factor (Table 2). In our own research, the content of protein in the seeds of the studied winter rape varieties averaged from 361.37 to 373.42 g·kg^{-1} of d.m. The highest concentration was found in the semi-dwarf hybrid PX104, while in the long-stem hybrid (Konkret), it was lower on average by 12.05 g·kg^{-1} of d.m. Different results were obtained by Gugała et al. [25], who received the highest value of this feature in a hybrid with a traditional type of growth and the lowest in a semi-dwarf hybrid. Ratajczak et al. [32] did not show significant differences between heterosis morphotypes with a traditional and semi-dwarf type of growth or in the population Califorium variety.

3.2. The Content of Crude Fat Depending on the Types of Biostimulators Used

The bioregulators used in the experiment significantly influenced the increase of crude fat in winter rapeseeds (Table 2). The greatest value of this feature was noted after the use of the Asahi SL biostimulator, it was significantly smaller on the objects where Tytanit and Silvit were applied. The beneficial effect of the Asahi SL biostimulator on the fat content in seeds was also confirmed by Spychaj-Fabisiak et al. [33] and Gugała et al. [25]. Similarly, Kováčik et al. [34] confirmed that a two-fold application of the Tytanit biostimulator affected the increase of the fat content in rapeseeds compared to the control object. The lack of effect of biostimulators on the fat content in seeds has been demonstrated by Matysiak et al. [29,30]. The authors observed only a slight tendency to increase the value of this feature even by 3.9% in relation to the control object. Similarly, Szczepanek et al. [35] noted a small effect of stimulating plant preparations on this feature. Jankowski et al. [31] after using a boron-containing foliar preparation, found a significant increase in the content of crude fat only after its two applications in the BBCH50 and BBCH55 phases. Jarecki and Bobrecka-Jamro [25,26] did not prove the effect of foliar preparations containing micro- and macro-elements on the value of this feature.

The impact of the types of biostimulators used on the crude fat content in rapeseed depended on the genetic factor (as shown in Table 2). The lowest fat content in all tested cultivars was recorded on the control object. The population cultivar had the highest fat content after using the Asahi SL biostimulator, but after the application of all biopreparations in this cultivar, the differences in protein crude fat content were not statistically significant. The seeds of the restored hybrid with the traditional growth type were characterized by the highest content of crude fat after the application of Asahi SL, and under the influence of the other biostimulators, they were the same as on the control object. A similar tendency was observed in the semi-dwarf hybrid, with differences in the value of this trait on the objects with the Tytanit and Silvit biostimulator were not statistically significant.

The content of crude fat depending on the genetic factor is shown in Table 2. Our own research proved that the highest content of fat was a characteristic of the PX104 (restored hybrid with a semi-dwarf type of growth), it was significantly smaller by 17.66 $g{\cdot}kg^{-1}$ of d.m. in the long-stem hybrid (Konkret), while the smallest on average by 20.57 $g{\cdot}kg^{-1}$ was in the population form (Poznaniak). Different results of studies were obtained by Gugała et al. [25] who showed that the highest value of this feature was characteristic for a restored hybrid with a traditional type of growth, while the smallest the population (Monolit).

3.3. The Content of Crude Fibre Depending on the Types of Biostimulators Used

Natural plant preparations influenced the increase of the crude fibre content in winter oilseed rapeseeds on average from 0.15 to 0.84 $g{\cdot}kg^{-1}$ of d.m. (Table 2). The highest value of this feature was noted on object 3 with the Asahi SL biostimulator. Different results were obtained by Gugała et al. [25]. In this study, the biostimulants did not significantly alter the crude fibre content in seeds of the rapeseed cultivars (Table 2).

The content of crude fibre depending on the genetic factor is shown in Table 2. Our own studies indicate that the highest content of crude fibre was observed in the seeds of the PX104 variety, while the smallest was in the long-stem morphotype (Konkret). Different results were obtained by Gugała et al. [25], who did not find any statistical differences in the value of this feature between the studied morphotypes.

3.4. Chemical Composition Depending on Weather Conditions

The chemical composition of seeds depending on climatic conditions in the study years is shown in Table 2. In our own research, the highest content of total protein, fat and crude fibre was obtained in seeds collected in the second year of research, in which the total precipitation was 41.9 mm higher in May, and the average monthly temperature was smaller by 0.5 °C from the average multi-year. Similar results were obtained by Chmura et al. [36] and Gugała et al. [25].

According to the authors, during the period from the end of flowering to the technical maturity stage of high protein content, temperatures of 16.2 °C were maintained on average, regardless of the sum of rainfall. M ̨aczy ́nska et al. [37] recorded a higher concentration of fat in colder years with a greater sum of precipitation, while it was lower in warm years. In our own studies, differences in the content of total protein and crude fat in the growing season of 2013–2014 and 2015–2016 were statistically insignificant, while the lowest content of crude fibre was found in seeds collected in the first year of research.

Table 2. Chemical composition of winter oilseed rapeseeds depending on factors of experience.

Culticars	Years			Types of Biostimulators Used				Mean
				Objects				
				1.	2.	3.	4.	
	2013–2014	2014–2015	2015–2016	Control Variant	Tytanit®	Asahi®SL	Silvit®	
Total protein [g·kg^{-1} d.m]								
Poznaniak	370.92	372.80	370.78	372.18	370.71	371.83	371.27	371.50a
Konkret	357.84	367.64	358.62	362.79	361.26	361.33	360.09	361.37b
PX104	373.13	374.33	372.81	374.11	373.62	373.49	372.46	373.42c
Mean	367.29a	371.59b	367.40a	369.69a	368.53ab	368.89bc	367.94d	
$LSD_{0.05}$ for: years—0.68; cultivars—0.68; types of biostimulators used—0.52; interaction: years x cultivars—1.18; cultivars x types of biostimulators used—0.82								
Crude fat [g·kg^{-1} d.m]								
Poznaniak	468.98	470.43	469.43	468.88	469.71	470.10	469.77	469.61a
Konkret	472.33	472.93	472.30	472.00	472.41	473.14	472.53	472.52b
PX104	489.78	490.70	490.06	488.78	489.64	492.19	490.10	490.18c
Mean	477.03a	478.02b	477.26a	476.55a	477.26be	478.48d	477.47ce	
$LSD_{0.05}$ for: years—0.35; cultivars—0.35; types of biostimulators used—0.52; interaction: years x cultivars—n.s.; cultivars x types of biostimulators used—0.5.								
Crude fiber [g·kg^{-1} d.m]								
Poznaniak	89.31	90.19	89.73	89.40	89.70	90.32	89.56	89.74a
Konkret	85.28	86.11	85.58	85.41	85.70	86.02	85.49	85.66b
PX104	69.48	70.44	70.09	69.61	70.00	70.58	69.82	70.00c
Mean	81.36a	82.25b	81.80c	81.47a	81.80b	82.31c	81.62d	
$LSD_{0.05}$ for: years—0.2; cultivars—0.2; types of biostimulators used—0.17; interaction: years x cultivars—n.s.; cultivars x types of biostimulators used—n.s.								

n.s.—non-significant differences. Different letters above the bars denote significant differences $p \leq 0.05$.

4. Conclusions

In summary, the applied biostimulators had an effect on reducing the total protein content (on average from 0.8 to 1.75 $g \cdot kg^{-1}$ of d.m.) and increasing the concentration of crude fat (on average from 0.71 to 1.93 $g \cdot kg^{-1}$ of d.m.) and crude fibre (on average from 0.15 to 0.84 $g \cdot kg^{-1}$ of d.m.) compared to the control object. The best quality of seeds was characteristic for the semi-dwarf PX104 variety. The smallest protein content was found in seeds of the long-stem hybrid Konkret, while crude fat was lowest in the population form (Poznaniak), and crude fibre was lowest in long-stem hybrid (Konkret). Diverse climatic conditions prevailing in the years of conducting the experiment influenced the chemical composition of rapeseeds. The highest content of total protein, crude fat and fibre were obtained in the second year of studies.

Author Contributions: Methodology, K.Z. and M.G.; software, M.G.; writing—original draft preparation, A.S.; review and editing—P.F. and U.M.-T.

References

1. Kałużewicz, A.; Krzesiński, W.; Spiżewski, T.; Zaworska, A. Effect of biostimulants on several physiological characteristics and chlorophyll content in broccoli under drought stress and re-watering. *Not. Bot. Horti Agrobo.* **2017**, *45*, 197–202. [CrossRef]
2. Rouphael, Y.; Colla, G. Synergistic biostimulatory action: Designing the next generation of plant biostimulants for sustainable agriculture. *Front. Plant Sci.* **2018**, *9*, 1655. [CrossRef] [PubMed]
3. Wierzbowska, J.; Sienkiewicz, S.; Bowszys, T. Effect of growth regulators on the mineral balance in spring triticale. *J. Elem.* **2010**, *15*, 745–756. [CrossRef]
4. Marjańska-Cichoń, B.; Sapieha-Waszkiewicz, A. The effect of preparations Asahi SL and Tytanit on the growth and yielding of strawberry variety salut. *Prog. Plant Prot.* **2010**, *50*, 383–388. (In Polish)
5. Stadnik, M.J.; de Freitas, M.B. Algal polysaccharides as source of plant resistance inducers. *Trop. Plant Pathol.* **2014**, *39*, 111–118. [CrossRef]
6. Panda, D.; Pramanik, K.B.; Naya, R. Use of sea weed extracts as plant growth regulators for sustainable agriculture. *Int. J. Stress Manag.* **2012**, *3*, 404–411.
7. Sharma, S.H.S.; Fleming, C.; Selby, C.H.; Rao, J.R.; Trevor, M. Plant biostimulants: A review on the processing of macroalgae and use of extracts for crop management to reduce abiotic and biotic stresses. *J. Appl. Phycol.* **2014**, *26*, 465–490. [CrossRef]
8. Yakhin, O.I.; Lubyanov, A.A.; Yakhin, I.; Brown, P.H. Biostimulants in plant science: A global perspective. *Front. Plant Sci.* **2017**, *7*, 2049. [CrossRef]
9. Petrozza, A.; Santaniello, A.; Summerer, S.; di Tommaso, G.; di Tommaso, D.; Paparelli, E.; Piaggesi, A.; Perata, P.; Cellini, F. Physiological responses to Megafol® treatments in tomato plants under drought stress: A phenomic and molecular approach. *Sci. Hortic.* **2014**, *174*, 185–192. [CrossRef]
10. Colla, G.; Rouphael, Y. Biostimulants in horticulture. *Sci. Hortic.* **2015**, *196*, 1–2. [CrossRef]
11. Rouphael, Y.; de Micco, V.; Arena, C.; Raimondi, G.; Colla, G.; de Pascale, S. Effect of *Ecklonia maxima* seaweed extract on yield, mineral composition, gas exchange and leaf anatomy of zucchini squash grown under saline conditions. *J. Appl. Phycol.* **2017**, *29*, 459–470. [CrossRef]
12. Calvo, P.; Nelson, L.; Kloepper, J. Agricultural uses of plant biostimulants. *Plant Soil* **2014**, *383*, 3–4. [CrossRef]
13. El-Boray, M.S.; Mostafa, M.F.M.; Salem, S.E.; El-Sawwah, O.A.O. Improving yield and fruit quality of washington navel orange using foliar applications of some natural biostimulants. *Int. J. Plant Prod.* **2015**, *6*, 1317–1332. [CrossRef]
14. Przybysz, A.; Gawronska, H.; Gajc-Wolska, J. Biological mode of action of a nitrophenolates-based biostimulant: Case study. *Front. Plant Sci.* **2014**, *5*, 713. [CrossRef]
15. Kocira, A.; Świeca, M.; Kocira, S.; Złotek, U.; Jakubczyk, A. Enhancement of yield, nutritional and nutraceutical properties of two common bean cultivars following the application of seaweed extract (*Ecklonia maxima*). *Saudi J. Biol. Sci.* **2018**, *25*, 563–571. [CrossRef]
16. Zulfiqar, F.; Casadesús, A.; Brockman, H.; Munné-Bosch, S. An overview of plant-based natural biostimulants for sustainable horticulture with a particular focus on moringa leaf extracts. *Plant Sci.* 2019. [CrossRef]

17. Grabowska, A.; Kunicki, E.; Sekara, A.; Kalisz, A.; Wojciechowska, R. The effect of cultivar and biostimulant treatment on the carrot yield and its quality. *Veg. Crops Res. Bull.* **2012**, *77*, 37–48. [CrossRef]
18. Kolomaznik, K.; Pecha, J.; Friebrova, V.; Janacova, D.; Vasek, V. Diffusion of biostimulators into plant tissues. *Heat Mass Transf.* **2012**, *48*, 1505–1512. [CrossRef]
19. Meier, U.; Bleiholder, H.; Buhr, L.; Feller, C.; Hack, H.; Heß, M. The BBCH system to coding the phenological growth stages of plants-history and publications. *J. Kult.* **2009**, *61*, 41–52.
20. World Reference Base for Soil Resources. International soil classification system for naming soils and creating legends for soil. In *World Soil Resources Reports 106*; Field Experiment; Food and Agriculture Organization: Rome, Italy, 2014; Available online: http://www.fao.org (accessed on 1 November 2019).
21. PN-ISO 6492:2005, Feed—Determination of fat content. Available online: http://sklep.pkn.pl/pn-iso-6492-2005p.html (accessed on 1 November 2019).
22. PN-EN ISO 5983-2:2009, Feed—Determination of nitrogen content and calculation of total protein content. Available online: https://infostore.saiglobal.com/en-au/Standards/PN-EN-ISO-5983-2-2009-924549_SAIG_PKN_PKN_2182095/ (accessed on 1 November 2019).
23. Trętowski, J.; Wójcik, R. *Methodology of Agricultural Experiments*; Wyższa Szkoła Rolniczo-Pedagogiczna: Siedlce, Poland, 1991; pp. 331–334. (In Polish)
24. Skowera, B. Changes of hydrothermal conditions in the Polish area (1971–2010). *Fragm. Agron.* **2014**, *31*, 74–87. (In Polish)
25. Gugała, M.; Sikorska, A.; Zarzecka, K.; Kapela, K.; Mystkowska, M. The effect of biostimulators on the content of crude oil and total protein in winter oilseed rape (*Brassica napus* L.) seeds. *Acta Agric. Scand. Sect. B Soil Plant Sci.* **2019**, *69*, 121–125. [CrossRef]
26. Jarecki, W.; Bobrecka-Jamro, D. Reaction of winter rape to the extra feeding of its leaves. *Ann. UMCS Agric.* **2008**, *63*, 86–96. [CrossRef]
27. Jarecki, W.; Bobrecka-Jamro, D. The reaction of the Huzar cultivar of spring oilseed rape to the density of sowing and to leaf feeding with urea. *Rośliny Oleiste* **2011**, *32*, 117–126. (In Polish)
28. Kozak, M.; Malarz, W.; Bzowy-Wójtowicz, A.; Białkowska, A. Influence of biostimulators on yield of winter rapeseed. In Proceedings of the Prosperující Olejniny, Prague, Czech Republic, 9–10 December 2010; pp. 67–69.
29. Matysiak, K.; Adamczewski, K.; Kaczmarek, S. Response of some crops cultivated in Great Poland to application of Asahi SL. *Prog. Plant Prot.* **2011**, *51*, 1849–1857. (In Polish)
30. Matysiak, K.; Kaczmarek, S.; Kierzek, R. Effect of algae *Ecklonia maxima* (Kelpak SL) on winter oilseed rape. *Rośliny Oleiste* **2012**, *33*, 81–88. (In Polish) [CrossRef]
31. Jankowski, K.J.; Sokólski, M.; Dubis, B.; Krzebietke, S.; Żarczyński, P.; Hulanicki, P.; Hulanicki, P.S. Yield and quality of winter oilseed rape (*Brassica napus* L.) seeds in response to foliar application of boron. *Agric. Food Sci.* **2016**, *25*, 164–176. [CrossRef]
32. Ratajczak, K.; Sulewska, H.; Szymańska, G. New winter oilseed rape varieties—Seed quality and morphological traits depending on sowing date and rate. *J. Plant Prod. Sci.* **2017**, *20*, 262–272. [CrossRef]
33. Spychaj-Fabisiak, E.; Murawska, B.; Pacholczyk, Ł. Values of quality traits of oilseed rape seeds depending on the fertilization and plant density. *J. Elem.* **2011**, *16*, 115–124.
34. Kováčik, P.; Šimanský, V.; Wierzbowska, J.; Renčo, M. Impact of foliar application of biostimulator Mg-Titanit on formation of winter oilseed rape phytomass and its titanium content. *J. Elem.* **2016**, *21*, 1235–1251. [CrossRef]
35. Szczepanek, M.; Wilczewski, E.; Grzybowski, K. Response of winter oilseed rape (*Brassica napus* L.) on soil applied humus preparation and foliar potassium fertilizer. *Acta Sci. Pol. Ser. Agric.* **2016**, *15*, 85–94.
36. Chmura, K.; Dzieżyc, H.; Piotrowski, M. Influence of meteorological conditions on the fat and protein content in seed of winter oilseed rape. *Acta Agrophys.* **2016**, *23*, 163–173. (In Polish)
37. Mączyńska, A.; Krzyzińska, B.; Banachowska, J. Effect of fungicides applied at the flowering stage of winter oilseed rape on the yield quantity and quality of seeds. *Prog. Plant Prot.* **2015**, *55*, 231–236. (In Polish)

10

The use of a Plant-Based Biostimulant Improves Plant Performances and Fruit Quality in Tomato Plants Grown at Elevated Temperatures

Silvana Francesca [1], Carmen Arena [2], Bruno Hay Mele [3], Carlo Schettini [4], Patrizia Ambrosino [5], Amalia Barone [1] and Maria Manuela Rigano [1,*]

1 Department of Agricultural Sciences, University of Naples Federico II, Portici, 80055 Naples, Italy; silvana.francesca@unina.it (S.F.); ambarone@unina.it (A.B.)
2 Department of Biology, University of Naples "Federico II", via Cinthia 4, 80126 Napoli, Italy; carena@unina.it
3 Stazione Zoologica Anton Dohrn, Villa Comunale, 80121 Napoli, Italy; bruno.haymele@szn.it
4 ALMA SEGES Soc. Coop., Eboli, 84025 Salerno, Italy; carlo@almaseges.com
5 Agriges S.r.l., San Salvatore Telesino, 82035 Benevento, Italy; patrizia.ambrosino@agriges.com
* Correspondence: mrigano@unina.it

Abstract: Abiotic stresses can cause a substantial decline in fruit quality due to negative impacts on plant growth, physiology and reproduction. The objective of this study was to verify if the use of a biostimulant based on plant and yeast extracts, rich in amino acids and that contains microelements (boron, zinc and manganese) can ensure good crop yield and quality in tomato plants grown at elevated temperatures (up to 42 °C). We investigated physiological responses of four different tomato landraces that were cultivated under plastic tunnel and treated with the biostimulant CycoFlow. The application of the biostimulant stimulated growth (plants up to 48.5% taller) and number of fruits (up to 105.3%). In plants treated with the biostimulant, antioxidants contents were higher compared to non-treated plants, both in leaves and in fruits. In particular, the content of ascorbic acid increased after treatments with CycoFlow. For almost all the traits studied, the effect of the biostimulant depended on the genotype it was applied on. Altogether, the use of the biostimulant on tomato plants led to better plant performances at elevated temperatures, that could be attributed also to a stronger antioxidant defence system, and to a better fruit nutritional quality.

Keywords: antioxidants; biostimulant; tomato; fruit quality; abiotic stress

---

1. Introduction

Tomato (*Solanum lycopersicum* L.) is one of the most consumed vegetables worldwide also owing to the development of products such as soups, juices, purees, and sauces [1]. Tomato is an essential component of the Mediterranean diet and of other traditional diets. However, heat can negatively affect vegetative and reproductive growth phases in tomato resulting in up to 70% harvest losses [2,3]. Indeed, in tomato, when temperatures exceed 35 °C different physiological functions result adversely affected including seed germination, seedling and vegetative growth, flowering and fruit set and ripening [3]. High temperature stress leads also to inhibition of chlorophyll biosynthesis and of photosystem II activity [4]. Indeed, photosynthesis is one of the processes most affected by elevated temperatures [5].

Considering the importance of this crop, the development of new management practices to enhance tolerance to abiotic stresses, including heat stress, could contribute to global food production. The use of biostimulants is proposed as an innovative solution to address the novel challenge to improve the sustainability of agricultural systems and reduce the use of chemical fertilizers [6,7]. The most accepted and complete definition of a biostimulant is the one from Du Jardin that defines a plant biostimulant as "any substance or microorganism that applied to plants, regardless of its nutrients content, is able to enhance nutrition efficiency and also abiotic stress tolerance and quality traits" [8]. Du Jardin allocated the biostimulants into eight classes: humic substances, complex organic materials, beneficial chemical elements, inorganic salts, seaweed extracts, chitin and chitosan derivatives, anti-transpirant and free amino acids and considered other N-containing substances with microorganism a potential ninth category. The mechanisms activated in plants by the different biostimulants are still not known as they can act directly on plant metabolism and physiology or indirectly on soil conditions [9]. The effects of biostimulants compounds include stimulation of enzyme activities of glycolysis, Krebs cycle, nitrate assimilation, and of hormonal activities [10]. It has been also demonstrated that biostimulants application is able to enhance tolerance to different abiotic stresses, such as drought [11,12], salinity [7,13,14], and thermal stresses [15]. For example, it has been demonstrated that applications of algal extracts are able to promote tolerance to drought, salinity, and heat, while extracts rich in amino acids can help increasing tolerance to thermal stresses [16,17]. Lettuce plants (*Lactuca sativa*) treated with a mixture derived from enzymatic hydrolysis of proteins and subjected to cold showed higher fresh weights and better stomatal conductance compared to non-treated plants [18]. In another work, perennial ryegrass (*Lolium perenne* L.) treated with hydrolyzed amino acids had improved photosynthetic efficiency compared to non-treated plants at high temperatures (36 °C) [15]. In general, the application of amino acids was found to exert positive effects on plant growth due to their use for the biosynthesis of a large number of non-protein nitrogenous compounds (pigments, vitamins, coenzymes, purine, and pyrimidine bases). Therefore, amino acids applications could directly influence the physiological activity in plant growth and yield also under abiotic stress [19]. Protein hydrolysates can also improve soil respiration, microbial biomass and activity and impact on plant nutrition by forming complexes and chelates between amino acids and soil nutrients [20].

To improve the tolerance to high temperatures the use of biostimulants has been previously investigated, even if it is presently unclear to what extent these compounds are able to improve the physiological performances of tomato plants under elevated temperatures [7]. We hypothesize that the use of an amino acid-based biostimulant could stimulate natural processes to enhance plant performances also at elevated temperatures. Indeed, the use of protein hydrolysates could directly stimulate carbon and nitrogen metabolism and indirectly enhance nutrient availability, nutrient uptake and nutrient use-efficiency in plants [21]. To verify this hypothesis, we used a novel plant-based biostimulant named CycoFlow and we performed physiological and biochemical analyses on four different tomato landraces grown at elevated temperatures and treated or not with this biostimulant. We reasoned that treatments with CycoFlow could facilitate stress adaptation because of its putative cytokinin-like action and its high concentration of glycine betaine known to mitigate the effect of heat stress [7,22]. Considering climate changes and the expected rise of temperatures in the next few years, to understand the contribution of biostimulants to ensure good plant performances at high temperatures may become increasingly important.

2. Materials and Methods

2.1. Plant Growth, Experimental Design, and Treatments

One-month-old tomato seedlings (landraces E17, E36, E107, PDVIT, described in Table 1) were transplanted in May 2018 under walk-in plastic tunnel (22 × 8 m^2) in Battipaglia in the Campania Region in Southern Italy (40°57′68″N 14°95′97″E). The tunnel was covered in polyethylene sheet and was open on both sides. Microclimatic conditions and temperatures were not regulated but were recorded during the growing season. All four genotypes have an indeterminate growth habit. The genotype E17 is characterized by large fruits (200–500 g), the genotype E107 is characterized by medium-sized fruits (70–100 g) and the E36 and the PDVIT genotypes are characterized by small cherry fruits (Table 1). Only the mature fruits of the E107 genotype are yellow while the fruits of the other genotypes are red. Tomato plants were grown following the standard cultural practices of the area. The experimental design consisted of a completely randomized design with three replicates *per* treatment and ten plant *per* each biological replication. There were two different groups: one control, which did not receive any biostimulant, and one that was treated with the biostimulant. The biostimulant CycoFlow (Agriges, Benevento, Italy) was produced by mixing sugar cane molasses with yeast extract obtained by autolysis of previously grown *Saccharomyces cerevisiae* yeasts. It is rich in high free amino acids, peptides, nucleotides, B-vitamins, trace elements, and other growth factors. Its chemical composition contains total nitrogen of 4.5% and organic carbon of 19.5%. The aminogram of the Biostimulant Cyco Flow is reported in Supplementary Table S1. The product contains also Boron (0.2%), Manganese (1%) and Zinc (1.2%). The biostimulant has a pH of 5.0, a density of 1200 kg/m^3 and an EC value of 15.0 dS/m. The Biostimulant, in liquid formulation, was initially applied directly to the soil (400 mL *per* plant) at the moment of transplanting, and thereafter every 15 days, until the end of the cultivation cycle for a total of four total applications. CycoFlow was applied by fertigation at a final concentration of 3 g/L. The control and the treatment groups received the same amount of water. No fertilizer has been applied. During the whole growing period climatic data (Figure S1) were recorded using the weather station VantagePro2 from Davis Instrument Corp. At the end of the cultivation cycle, plants were harvested and separated into leaves, stems, roots and fully ripe fruits. Plant height, numbers of leaves *per* plant, fresh weight of biomass, total number of fruits, weight of fruit and final yield were recorded. Dry biomass (in grams) was determined by drying plant tissues to constant weight in a forced-air-oven at 80 °C for 72 hours. Measurements were done on three randomly selected plants *per* each biological replication *per* genotypes for each treatment.

Table 1. Details of the tomato genotypes used in this study.

Genotype	Origin	Common Accession	Fruit Size	Fruit Color
E17	Italy	Pantano Romanesco	Big (200–250 g)	red
E36	Italy	Riccia San Vito	Small (25–30 g)	red
E107	Spain	E-L-19	Medium (70–100 g)	yellow
PDVIT	Italy	Caramella	Small (10–15 g)	red

2.2. Pollen Viability

Pollen viability was analyzed using five flowers *per* plant sampled from three different plants *per* replicate. In the laboratory, pollen grains were spread on microscope slides. One droplet of DAB solution (SIGMA) was added on each pollen sample; slides were gently warmed with a gas lighter and mounted with a cover slip [23]. Scoring was made using an LEITZ Laborlux12 microscope.

2.3. Ascorbic Acid Quantification

Reduced ascorbic acid (AsA) and total ascorbic acid (AsA + dehydroascorbate – DHA) measurements were carried out by using a colorimetric method [24] with modifications reported by Rigano et al. [25,26]. Briefly, 500 mg of frozen powder from tomato fruits or leaves were extracted with 300 μL of ice cold 6% trichloroacetic acid (TCA) and the mixture was then incubated for 15 min on ice and centrifuged at 14,000 rpm for 20 min. For reduced AsA evaluation, to 20 μL of supernatant were added 20 μL of 0.4 M phosphate buffer (pH 7.4), 10 μL of double distilled (dd) H_2O and 80 μL of color reagent solution. This solution was prepared by mixing solution A (31% (*w*/*v*) H_3PO_4, 4.6% (*w*/*v*) TCA and 0.6% (*w*/*v*) $FeCl_3$) with solution B (4% (*w*/*v*) 2,2′-Dipyridyl). For total AsA, to 20 μL of sample, 20 μL of 5 mM dithiotreitol in 0.4 M phosphate buffer (pH 7.4) were added and the mixture was incubated for 20 min at 37 °C. Ten microliters of N-ethyl maleimide (NEM; 0.5% (*w*/*v*) in water) were added and left for 1 min at room temperature. Eighty microliters of color reagent were added as previously described for reduced AsA. Both the final mixtures were incubated at 37 °C for 40 min and measured at 525 nm by using a Nano Photometer TM (Implen, Munich, Germany). Three separated biological replicates for each sample and three technical assays for each biological repetition were measured. The concentration was expressed in mg/100 g of fresh weight (FW).

2.4. Total Carotenoids and Chlorophylls Content

The evaluation of total carotenoids and chlorophylls was carried out according to the method reported by Wellburn [27] and by Zouari et al. [28] as modified by Rigano et al. [2]. To obtain the lipophilic extract, 0.25 grams of sample were extracted with 24 mL of acetone/hexane (40/60, *v/v*). The mixture was centrifuged at 15,000 rpm for 5 min at 4 °C. Supernatants were collected and stored at −20 °C until analyses. For carotenoids and chlorophylls a and b levels determination, absorbance of lipophilic extracts was read at 470, 663, and 645 nm, respectively. For lycopene and β-carotene levels absorbance was read at 505 and 453 nm, respectively. Results were converted into mg/100 g FW. Three separated biological replicates for each sample and three technical assays for each biological repetition were measured.

2.5. Antioxidant Activity Determination

Hydrophilic antioxidant activity (HAA) was evaluated in the water-soluble fraction, obtained by adding to 2 g of frozen powder 25 mL of 80% methanol, using the ferric reducing/antioxidant power (FRAP) method [29] with slight modifications. The FRAP assay was carried out by adding in a vial 2.5 mL of acetate buffer at pH 3.6, 0.25 mL of TPTZ solution (10 mM) in 40 mM HCl, 0.25 mL of $FeCl_3 \cdot 6H_2O$ solution (12 mM), and 150 μL of methanolic extract. The mixture was incubated for 30 min in the dark, and then readings of the colored products (ferrous tripyridyltriazine complex) were taken at 593 nm using a spectrophotometer. Results were expressed as micromoles of Trolox equivalents (TE) per 100 g FW. Lipophilic antioxidant activity (LAA) determination was carried out according to the 2,2′-azinobis-(3-ethylbenzothiazoline-6-sulfonic acid) (ABTS) method, using the lipophilic extract obtained as described in the previous paragraph [30]. The ABTS assay was based on the reduction of the ABTS•+ radical action by the antioxidants present in the sample. A solution constituted by 7.4 mM ABTS•+ (5 mL) mixed with 140 mM $K_2S_2O_8$ (88 μL) was prepared and stabilized for 12 h. This mixture was then diluted by mixing ABTS•+ solution with ethanol (1:88) to obtain an absorbance of 0.70 ± 0.10 unit at 734 nm using a spectrophotometer. Methanolic extracts (100 μL) were allowed to react with 1 mL of diluted ABTS•+ solution for 2.5 min, and then the absorbance was taken at 734 nm using a spectrophotometer. All biological replicates of samples were analyzed in triplicate. Results were expressed as micromoles of TE per 100 g FW.

2.6. Fluorescence Emission Measurements

Fluorescence emission measurements were performed on five replicates *per* each treatment, coming from five different plants. A portable FluorPen FP100max fluorometer, equipped with a light sensor (Photon System Instruments, Brno, Czech) was used for measurements, following the procedure reported in Figlioli et al. [31]. The ground fluorescence signal, F_o, was induced on 40′ dark adapted leaves, by a blue LED internal light of about 1–2 μmol m^{-2} s^{-1}. The maximal fluorescence level in the dark, F_m, was induced by a 1s saturating light pulse of 3000 μmol m^{-2} s^{-1}. The maximum quantum efficiency of PSII photochemistry, F_v/F_m, was calculated as $(F_m - F_o)/F_m$, according to Kitajima and Butler [32].

2.7. Leaf Functional Traits Determination

Fully expanded leaves, without apparent damages, were collected to determine the functional leaf traits following Arena et al. [33]. Leaf area (LA) was measured by the program Image J 1.45 (Image Analysis Software) and expressed in *per* square centimeter, specific leaf area (SLA) was measured as the ratio of leaf area to leaf dry mass and expressed as square centimeter *per* gram dry weight (DW). For dry mass determination, leaves were dried at 70 °C for 48 h. Leaf dry matter content (LDMC) was measured as the oven-dry mass of a leaf divided by its water-saturated fresh mass and expressed as gram *per* gram of water saturated leaf mass (WSLM). Relative water content in leaves (RWC) was calculated by dividing the amount of water in the fresh leaf tissue by the water in the leaf tissue after rehydration multiplied by 100 [34].

2.8. Statistical Analysis

Data were subjected to analysis of variance using a two-way ANOVA. To separate means within each parameter, the Tukey-HSD's test was performed. Differences at $p < 0.05$ were considered to be significant. ANOVA was performed by using SPSS (Statistical Package for Social Sciences) Package 6, version 23.0. To explore the overall data, we used the R environment for statistical computing and graphics R Core Team (2018). We first selected variables of interest for each genotype, treatment and plant part ($4 \times 2 \times 2$) then calculated the arithmetic mean (n = 3), and finally used the scale function to center the data around the mean and scale it using the standard deviation. The transformed data were visualized using a heatmap (heatmap function). To aid interpretation of the data, we also performed an SVD-based Principal Component Analysis over the multivariate matrix (function prcomp in base R) after normalization.

3. Results

3.1. Phenotypic and Physiological Analyses

In this study four different tomato genotypes were transplanted under a plastic walk-in tunnel with a delay of one month compared to the usual transplanting period (tomato plants in the South of Italy are usually transplanted in April), thus imposing a high-temperature condition during flowering and fruit setting. Indeed, the maximum temperature of 32 °C during the day, which represents a critical threshold in the sensitive stages of reproductive development, was frequently exceeded in this trial [3](Figure S1).

The four different tomato landraces were treated with a plant-based biostimulant named CycoFlow. According to ANOVA analyses, the treatment with the biostimulant increased the height of genotypes E107 and PDVIT by 48.5% and 30.1%, respectively (Supplementary Table S2). Generally, the number of leaves was lower in the biostimulant treated group compared to the control, independently of the genotype it was applied to (no significant interaction G X T). For the fresh biomass parameter, in PDVIT the treatment with CycoFlow increased the above ground fresh biomass by 68.4% (Figure 1a). Genotypes E17 and E36 showed, instead, lower values in treated plants compared to non-treated ones (−53.8% and −21.1%, respectively).

A slightly higher pollen viability was also observed in the genotypes treated with the biostimulant compared to the respective controls (Figure 1b). In particular, in the genotype E107 the treatment with the biostimulant increased pollen viability by 125%. Generally, the treatment with the biostimulant increased the number of fruits, independently of the genotype (no significant interaction G X T).

In particular, the treatment with the biostimulant increased the number of fruits in the genotype PDVIT by 105.3% (Figure 1c). The medium fruit weight was significantly affected only by the factor genotype (Supplementary Table S2). Generally, the final yield (kg *per* plant) showed a tendency to be higher in all the samples from the treated genotypes, even though these differences were not significant (Figure 1d). Interestingly, the yield was significantly affected only by the factor treatment (Supplementary Table S2).

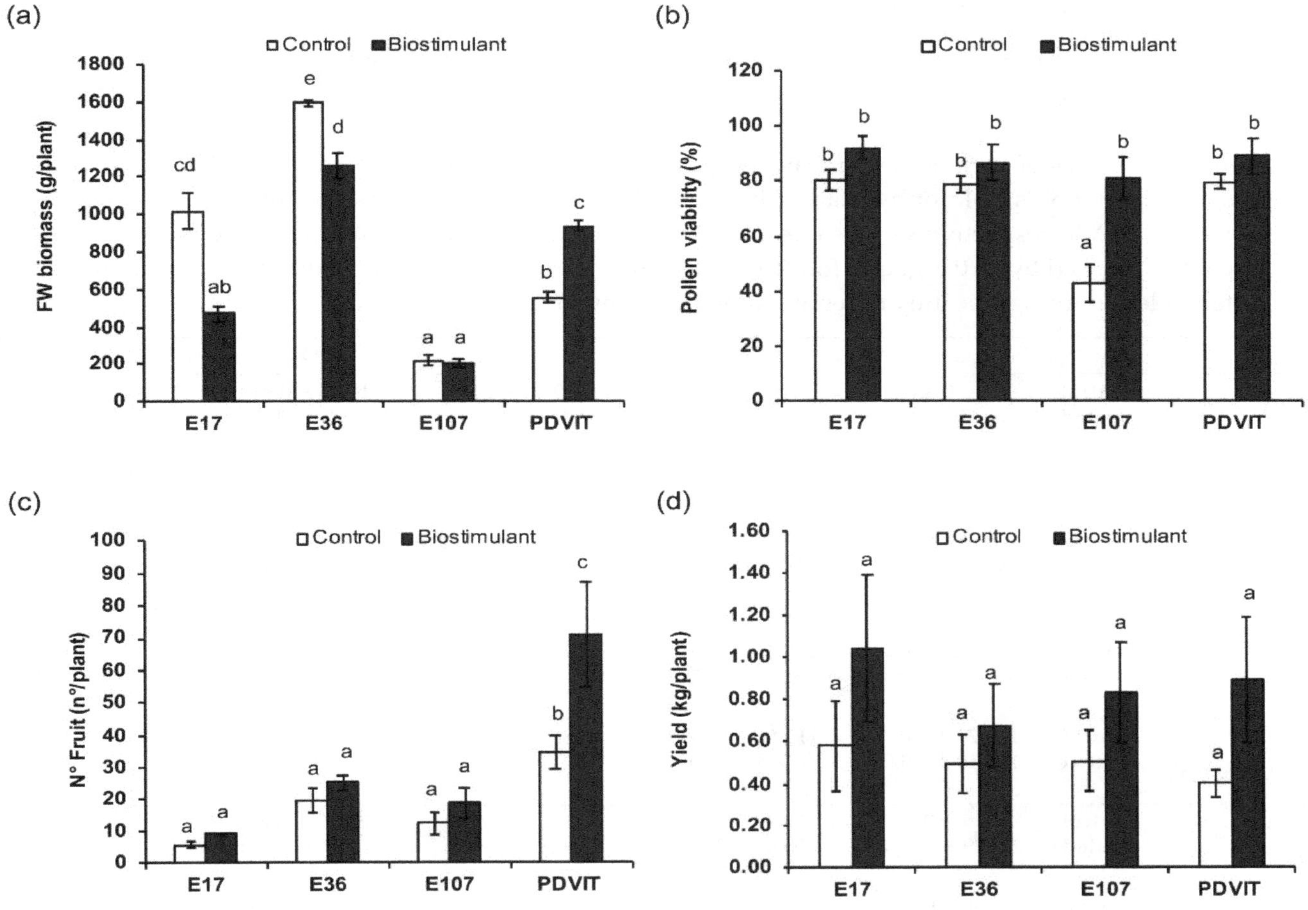

Figure 1. Effect of CycoFlow on: (**a**) Fresh weight (FW) biomass, (**b**) pollen viability, (**c**) fruit number, and (**d**) final yield in four tomato genotypes. Values are mean ± SE. Different letters indicate significant differences based on Tukey-HSD test ($p \leq 0.05$).

The treatment with the biostimulant CycoFlow also increased the maximal PSII photochemical efficiency (F_v/F_m) in the E107 and PDVIT genotypes (Figure 2). The monitoring of leaf functional traits evidenced that biostimulant application did not affect these traits significantly (Supplementary Table S3).

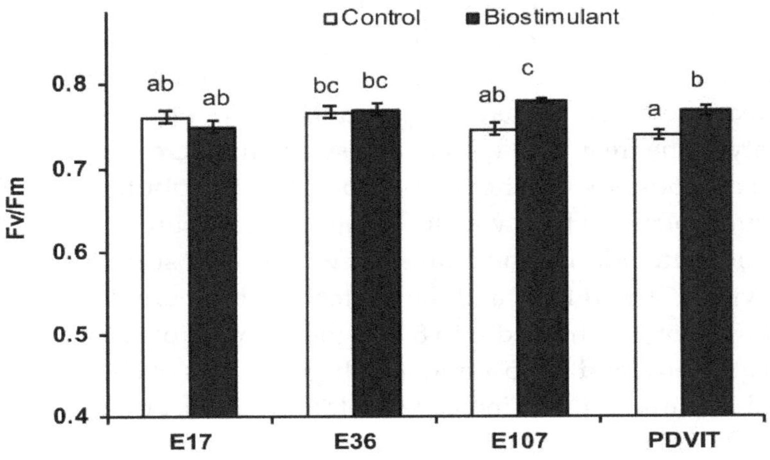

Figure 2. Maximal photochemical efficiency (F_v/F_m) in leaves of four tomato genotypes. Data are mean ± SE (n = 5). Different letters indicate significant differences based on Tukey-HSD test ($p \leq 0.05$).

3.2. Leaf and Fruit Antioxidant Content

The main interaction effects of the biostimulant Cyco Flow on the content of antioxidants in leaves from treated and non-treated plants is reported in Table 2.

Table 2. Analyses of variance and mean comparison for reduced and total ascorbic acid (AsA), total phenols, carotenoids, chlorophylls a and b and total lipophilic and hydrophilic antioxidant activities (LAA and HAA, respectively) in leaves of different tomato cultivars treated with the biostimulant CycoFlow applied by fertirrigation four times. Means ± SD within rows and columns followed by the different letter are significantly different based on Tukey-HSD test ($p \leq 0.05$).

		E17	E36	E107	PDVIT	SIGNIFICANCE	
Reduced AsA (mg/100 g FW)	**control**	6 ± 0.43 a	7.95 ± 1.33 a	10.83 ± 1 ab	18.20 ± 0.91 bc	G	**
	treated	20.05 ± 3.30 c	20.12 ± 1.42 c	17.41 ± 1.91 bc	19.34 ± 1.33 c	T	***
						GXT	**
Total AsA (mg/100 g FW)	**control**	16.79 ± 0.73 ab	14.45 ± 0.51 a	24.52 ± 2.03 bc	21.28 ± 0.86 bc	G	***
	treated	21.15 ± 0.90 bc	24.40 ± 2.55 cd	20.27 ± 0.83 cd	26.85 ± 0.69 d	T	***
						GXT	***
Phenols (mg/100 g FW)	**control**	43.38 ± 0.98 e	26.91 ± 1.19 a	35.14 ± 0.48 c	35.30 ± 0.56 c	G	***
	treated	25.33 ± 1.20 a	25.58 ± 0.27 a	31.57 ± 0.52 b	39.57 ± 0.54 d	T	***
						GXT	***
Carotenoids (mg/100 g FW)	**control**	23.91 ± 1.06 ab	26.06 ± 0.53 abc	23.80 ± 0.75 a	28.73 ± 0.23 de	G	***
	treated	23.78 ± 0.48 a	30.17 ± 0.24 e	28.10 ± 0.47 cde	26.42 ± 0.46 bcd	T	***
						GXT	***
Chl a (mg/100 g FW)	**control**	108.78 ± 3.05 a	113.30 ± 4.3 6 ab	128.22 ± 5.34 bc	140.30 ± 4.25 c	G	***
	treated	110.13 ± 1.37 a	137.08 ± 2.07 c	138.61 ± 3.32 c	130.20 ± 2.80 bc	T	**
						GXT	***
Chl b (mg/100 g FW)	**control**	38.65 ± 3.96 a	37.45 ± 2.12 a	45.84 ± 3.67 ab	55.75 ± 3.74 b	G	***
	treated	37.29 ± 2.73 a	55.41 ± 2.11 b	59.47 ± 2.69 b	45.85 ± 5.72 ab	T	**
						GXT	***
LAA (mg/100 g FW)	**control**	18.88 ± 0.14 a	18.75 ± 0.07 a	18.86 ± 0.04 a	18.62 ± 0.05 b	G	***
	treated	18.98 ± 0.04 a	19.07 ± 0.21 a	19.90 ± 0.08 b	19.75 ± 0.10 a	T	***
						GXT	***
HAA (mg/100 g FW)	**control**	828.58 ± 140.08 a	493.19 ± 220.27 bc	599.85 ± 118.33 ab	434.30 ± 88.34 cd	G	***
	treated	255.57 ± 91.31 d	390.49 ± 25.34 bc	510.33 ± 53.59 ab	438.26 ± 125.38 bc	T	***
						GXT	***

G = genotype; T = treatment; * = $p \leq 0.05$; ** = $p \leq 0.01$; *** = $p \leq 0.001$.

For the hydrophilic antioxidants, the treatment with the biostimulant increased the content of reduced AsA in the genotypes E17 and E36 and of total AsA in the leaves of the genotypes E36 and PDVIT. In particular, in the genotype E107 a 60.8% higher content of total AsA was registered in leaves treated with the biostimulant. As for the content of phenolic compounds, two genotypes (E17 and E107) showed lower contents of total phenols in the leaf after treatment with the biostimulant. In particular, in the E17 genotype a 41.6% decrease in the treated compared to the non- treated samples was demonstrated. Only in the PDVIT genotype the treatment with the biostimulant increased phenols content. It has been reported that phenolics compounds are the most important contributors to HAA [35]. Accordingly, in the leaves of the treated plants, HAA was lower in E17 compared to the respective non-treated control. For the lipophilic antioxidants, the treatment with the biostimulant increased the content of carotenoids in the genotypes E36 and E107 and the content of chlorophylls a and b only in the genotype E36. Particularly, the E36 genotype showed a 15.8% higher content of carotenoids in the treated leaves compared to the non-treated one, and 17.35% and 48% higher levels of chlorophyll a and b, respectively. The treatment with the biostimulant also increased total lipophilic antioxidant activities in E107 and surprisingly also in PDVIT, suggesting that other compounds outside of carotenoids contributed to this parameter.

In Table 3 is reported the content of hydrophilic antioxidants determined in red ripe fruit from genotypes treated or non-treated with the biostimulant CycoFlow. In general, the content of hydrophilic antioxidants in the fruits was higher in almost all the genotypes treated with biostimulants compared to the non-treated ones. The treatment with the biostimulant increased the content of reduced AsA independently of the genotype it was applied on (not significant interaction G X T). The content of reduced AsA was 28.7%–58.7% higher in fruits from treated genotypes compared to non-treated genotypes. Moreover, a content 112.8% higher of total AsA was registered in fruits from PDVIT treated with the biostimulant compared to the respective non-treated control. Contrary to what seen in the leaf, the content of total phenols in berries of treated E17 and E36 genotypes was higher compared to the non-treated control. In particular, in the E17 genotype 72.8% higher values were registered. Moreover, a significantly higher antioxidant activity HAA was demonstrated in fruits from E36 plants treated with CycoFlow, according to ANOVA analyses. Assessing the content of lipophilic antioxidants, the treatment with the biostimulant had no effects on the content of carotenoids and chlorophylls but only on the total lipophilic antioxidant activity, as reported in Supplementary Table S4. In particular, LAA was higher in fruits from the treated genotypes E17, E36, and E107.

Table 3. Analyses of variance and mean comparison for reduced and total ascorbic acid (AsA), total phenols, hydrophilic antioxidant activities (HAA) in fruits of different tomato cultivars treated with the biostimulant CycoFlow applied by fertirrigation four times. Means ± SD within rows and columns followed by the different letter are significantly different based on Tukey-HSD test ($p \leq 0.05$).

		E17	E36	E107	PDVIT	SIGNIFICANCE	
Reduced AsA (mg/100 g FW)	**control**	33.31 ± 2.99 a	39.56 ± 2.30 ab	47.14 ± 1.66 bc	50.36 ± 1.84 bc	G	***
	treated	47.36 ± 1.60 bc	59.87 ± 4.34 cd	74.79 ± 3.25 e	64.83 ± 2.34 de	T	***
						G X T	ns
Total AsA (mg/100 g FW)	**control**	61.97 ± 0.57 ab	78.03 ± 3.29 bc	85.40 ± 3.75 c	52.87 ± 4.24 a	G	**
	treated	79.34 ± 4.44 bc	87.15 ± 2.35 c	93.36 ± 6.19 cd	112.53 ± 4.08 d	T	***
						G X T	***
Phenols (mg/100 g FW)	**control**	9.62 ± 0.46 a	13.70 ± 0.68 b	16.17 ± 0.58 c	22.35 ± 0.37 e	G	***
	treated	16.62 ± 0.46 c	18.92 ± 0.76 d	16.88 ± 0.44 c	22.55 ± 0.19 e	T	***
						G X T	***
HAA (mg/100 g FW)	**control**	129.28 ± 33.95 a	189.22 ± 49.66 b	179.38 ± 20.62 bc	309.06 ± 39.51 d	G	***
	treated	151.57 ± 8.71 c	304.38 ± 30.92 c	212.47 ± 7.08 c	333.03 ± 46.91 d	T	***
						G X T	***

G = genotype; T = treatment; * = $p \leq 0.05$; ** = $p \leq 0.01$; *** = $p \leq 0.001$; ns = not significant.

3.3. Heat Map Analysis

A heat map providing the morphological, biochemical, and physiological changes in leaves and fruits of four different tomato genotypes in response to the addition of one biostimulant is displayed in Figure 3. With regard to leaves, the heat-map identified two main clusters which divided the analyzed samples differently (Figure 3, panel a). The first cluster separated the control genotypes E107 and E17 from the other genotypes and respective treated samples, the second cluster associated the treated genotypes E107, E17, and PDVIT in a sub-group and control PDVIT and E36 genotypes in another sub-group (Figure 3). Our data indicate that biostimulant application was the main clustering factor for E107, E17, and PDVIT genotypes, on the basis of differences in some leaf traits, F_v/F_m, phenols, yield and HAA, suggesting that the biostimulant utilization produces significant effect on many metabolites. The heat map built on tomato fruits clearly separated the treated PDVIT genotype from all others, in particular for number of fruits and reduced AsA (Figure 3, panel b), indicating this genotype as the most responsive to biostimulant application for fruit characteristics. A remarkable separation was also evident for control E107 and E36 compared to treated genotypes, grouped in two sub-clusters on the basis of pigments (chlorophylls and carotenoids) and LAA. A PCA analyses was also performed (Supplementary Figure S2). The PCA output further showed an evident separation between the treated and the non- treated genotypes.

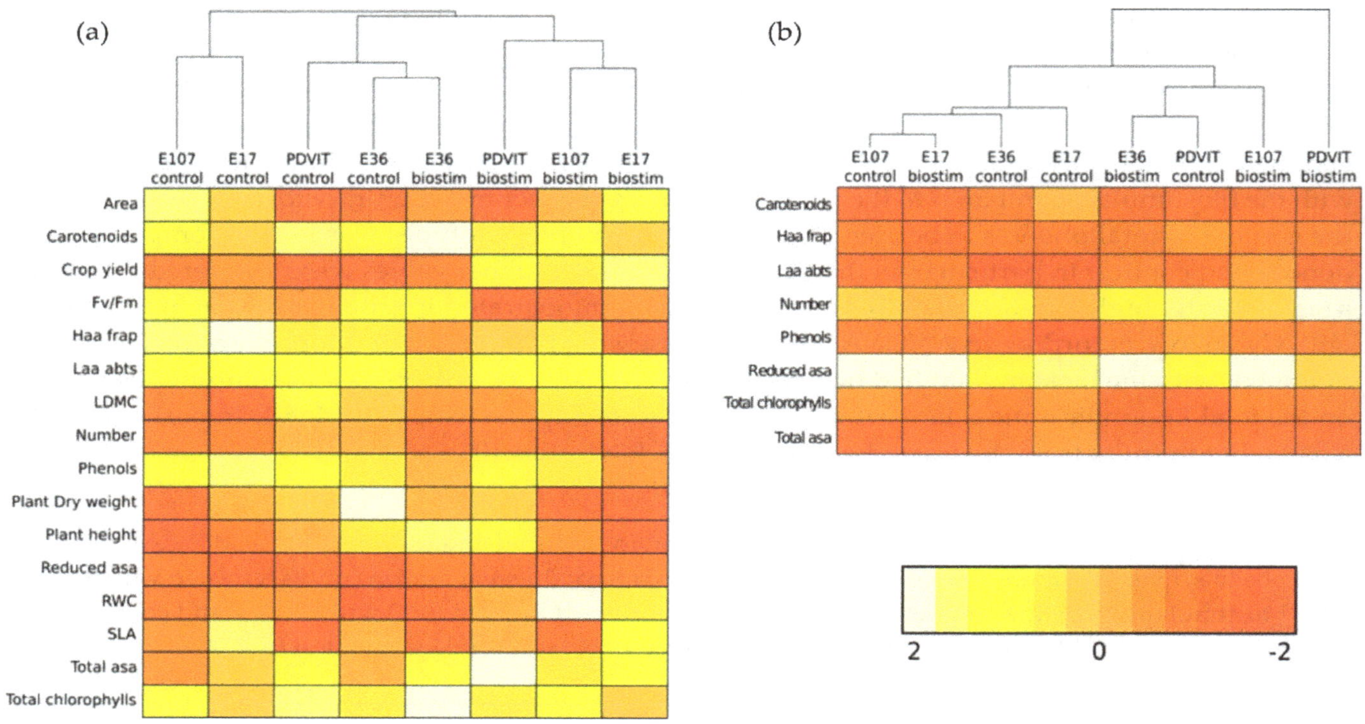

Figure 3. Cluster heat map analysis summarizing the behavior of the different tomato genotypes E36, E17, E107, PDVIT treated or non-treated with the biostimulant CycoFlow in leaf (panel **a**) and in fruit (panel **b**). The heat map was generated using the R environment for statistical computing and graphics (https://www.R-project.org/online) program package with Euclidean distance as the similarity measure and hierarchical clustering with complete linkage.

4. Discussion

In this paper four different tomato landraces were grown at elevated temperatures under a plastic walk-in tunnel and were treated or not with a plant-based biostimulant named CycoFlow. The higher height demonstrated in the majority of the tomato plants treated with CycoFlow compared to non-treated plants is in agreement with previous studies on different plant species and biostimulants [36–40]. Probably, the presence of signaling molecules in the biostimulant, such as free amino acids, promoted endogenous phytohormonal biosynthesis thus stimulating growth and also fruit setting [41]. Indeed, several authors demonstrated that the application of plant-based biostimulants exhibited cytokinin-like activity promoting cell division [42]. Moreover, cytokinins mitigate stresses induced by free radicals by direct scavenging and also by preventing ROS formation inhibiting xanthine oxidation [39]. Also, the treatment with CycoFlow overall increased the number of fruits, as previously demonstrated also in tomatoes treated with other biostimulants [10,36,39–41]. For example, Rouphael et al. [41] demonstrated that application of a protein hydrolysate in tomato increased in one cultivar the fruit mean weight and in another cultivar the number of fruits. In this study, in the genotype E107, the higher number of fruits observed was also linked to a higher pollen vitality observed after CycoFlow treatment. This result could be due to a combination of multiple effects. While the cytokinin-like activity could have favored cell division, the high level of proline present in the biostimulant, an amino acid whose natural content in the flower organs is ten times higher than that in the leaves, may have played an important role [31]. Indeed, it is known that also the amino acid proline promotes the translocation of nutrients towards developing flowers (sink) [43]. The positive effects of biostimulants based on amino acid on growth and yield is also due to the fact that the amino acids present in plant-based biostimulants stimulate plant defenses, participate in the synthesis of organic compounds (such as amines, purines, pyrimidines, vitamins) and affect the uptake of macro and micronutrients [37]. The CycoFlow effects observed in this study on yield and yield components are even more remarkable considering the elevated temperatures (up to 43 °C) reached under the plastic walk-in tunnel in Battipaglia. Indeed, this temperature normally impairs fertilization and reduces pollen viability [10]. It can be hypothesized that the presence of glycine betaine in the CycoFlow may have enhanced the tolerance of tomato plants to elevated temperatures. Indeed, it has been previously demonstrated that during tomato germination glycine betaine applied exogenously improved tolerance to high temperatures and enhanced the expression of heat shock genes [44]. At elevated temperatures, the glycine betaine compound may have also a crucial role in the repair of photodamaged PSII, in maintaining the activity of Rubisco and in alleviating the inhibition of gas exchanges [22]. Accordingly, a higher maximal photochemical efficiency was observed in the genotypes E107 and PDVIT treated with the biostimulant. These results are consistent with other papers, which demonstrated that applications of plant- and animal-based biostimulants are able to enhance photosynthetic rates and ensure a higher carbon assimilation efficiency [45,46]. For example, under drought stress conditions, Arabidopsis plants treated with an *Ascophyllum nodosum*-extract maintained a better photosynthetic performance compared to non-treated plants during the dehydration period, showing a higher capacity to dissipate thermally the excess of energy in the PSII reaction centers [47]. These results were linked to the fact that pre-treatments with the Ascophyllum-extracts induced partial stomatal closures and also modifications of the expression levels of genes involved in ABA-responsive and antioxidant system pathways [47]. Accordingly, our data indicate that biostimulant treatment induced the activation of the antioxidant defense system, as demonstrated by the higher content of reduced and total AsA in treated leaves. Although the precise reasons for these increases are not explained, it is known that biostimulants components, including glycine betaine, can promote the activity of specific enzymes involved in antioxidant homeostasis [22,41,48]. The ability to maintain an optimal chlorophyll content during heat stress is another key heat tolerance trait in tomato [49]. Interestingly, herein we observed higher contents of carotenoids and chlorophylls in two genotypes (E36 and E107) treated with the biostimulant compared to the non-treated samples. The higher chlorophylls content detected in these genotypes could be linked to limited chlorophyll degradation and leaf senescence [9]. In particular, this could be the case for the genotype E107 that demonstrated a higher maximal photochemical efficiency after treatment with the biostimulant.

The biostimulant-mediated effects on photosynthesis and secondary metabolism could also enhance fruit quality [10]. Indeed, one interesting finding of this study is the positive effect of the biostimulant CycoFlow on the quality of the tomato fruits. In general, the content of hydrophilic antioxidants in the fruits, including AsA, was higher in almost all the genotypes treated with biostimulants compared to the non-treated ones. Higher content of reduced AsA was observed in all the genotypes and of total AsA in the genotypes E17 and PDVIT. This result confirms data previously obtained in other studies that demonstrated an increase in AsA content in tomato, in kiwi fruits and in peppers after the application of plant-based biostimulants [36,41]. Contrary to what seen in the leaf, the content of total phenols in berries of treated E17 and E36 genotypes was higher compared to the non-treated control. Moreover, a significantly higher antioxidant activity HAA was demonstrated in fruits from E36 plants treated with CycoFlow. These results are in agreement with results previously obtained in other crops (soybean seeds, common bean, tomato, corn), even if the reported effects depended on the type of biostimulants, their concentrations and the number of applications [37]. Assessing the content of lipophilic antioxidants, the treatment with the biostimulant had no effects on the content of carotenoids and chlorophylls but only on the total lipophilic antioxidant activity. Similar results were obtained by Chehade et al. [36] in tomato. On the contrary, Rouphael et al. [41] demonstrated that in tomato foliar applications of a legume-derived protein hydrolysate had an effect also on lycopene content. Also, Colla et al. [10] demonstrated that foliar applications of protein hydrolysate, plant and seaweed extract affected lycopene content in greenhouse tomato. In the future, foliar application of CycoFlow will be also tested in order to verify if the results obtained in this study are also linked to the used application regimen.

Altogether, the genotypic factors remain decisive in the response obtained in the different tomato lines to the biostimulant. Indeed, for almost all the traits considered the effect of the biostimulant depended on the cultivar it was applied to, as seen by the interaction between the effect of the biostimulant and cultivars in most of the studied parameters. These variations can be explained by the differences in the genetic background between the different cultivars that were used in this study [33]. Indeed, the four genotypes here tested differed in terms of fruit shape and size and also in terms of fruit color (e.g., fruit of E107 is yellow). The geographical origin is also different with the E107 genotype coming from Spain and the other coming from Italy. These further highlight the fact that one biostimulant should be tested on a certain number of cultivars in order to assess its mechanisms of action.

5. Conclusions

In this paper we investigated the effects of the application of one plant-based biostimulant named CycoFlow on the nutritional quality and yield of tomatoes grown in walk-in tunnel under elevated temperatures. The application of the CycoFlow biostimulant had a clear effect on plant growth and final crop quality. Indeed, CycoFlow application had a significant effect on the content of hydrophilic antioxidants in both tomato leaves and fruits. In particular, the content of AsA increased after treatments with CycoFlow. Herein, the biostimulant application improved plant performances and fruit quality mostly in the genotypes E107 and PDVIT. In particular, in the genotype PDVIT application with CycoFlow determined a higher plant height, a higher number of fruits, a higher pollen vitality, a higher photochemical efficiency, a higher accumulation of AsA and a higher antioxidant activity. Additional studies are now planned in order to investigate if different applications regimen, such as foliar application, can also influence the observed effects.

Supplementary Materials
Table S1: Amino acid composition expressed in g/100 g of the biostimulant CycoFlow, Table S2: Analyses of variance and mean comparison for height, number of leaves, fresh weight (FW) and dry weight (DW) biomass, number of fruits, medium fruit weight, yields, and pollen viability (%) per plants of different tomato cultivars treated with the biostimulant CycoFlow applied by fertirrigation four times. Means ± SD within rows and columns followed by the different letter are significantly different based on Tukey-HSD test ($p \leq 0.05$). Table S3: Analyses of variance and mean comparison for maximal PSII photochemical efficiency (F_v/F_m), leaf area (LA), specific leaf area (SLA), leaf dry matter content (LDMC) and relative water content (RWC) *per* plants of different tomato cultivars treated with the biostimulant CycoFlow applied by fertirrigation four times. Means ± SD within rows and columns followed by the different letter are significantly different based on based on Tukey-HSD test ($p \leq 0.05$). Table S4: Analyses of variance and mean comparison for total lipophilic antioxidant activities (LAA), carotenoids, chlorophylls a and b (Chl A and Chl B, respectively) content in fruit of different tomato cultivars treated with the biostimulant CycoFlow applied by fertirrigation four times. Means ± SD within rows and columns followed by the different letter are significantly different based on Tukey-HSD test ($p \leq 0.05$). Figure S1: Maximum temperatures recorded in the experimental field located in Battipaglia during the day from May to August 2018. Figure S2: Principal component analysis (PCA) of phenotypic and physiological traits in tomato plants treated or not with the biostimulant CycoFlow. The treated genotypes are indicated by the letter T after the name.

Author Contributions: Conceptualization: S.F., A.B. and M.M.R.; Data Curation: S.F., M.M.R., C.A., B.H.M.; Funding acquisition: A.B.; Writing—original draft: S.F., M.M.R., C.A., A.B.; Writing—review and editing: S.F., M.M.R., C.A., A.B., P.A., C.S. All authors have read and agreed to the published version of the manuscript.

References

1. Del Giudice, R.; Petruk, G.; Raiola, A.; Barone, A.; Monti, D.M.; Rigano, M.M. Carotenoids in Fresh and Processed Tomato (*Solanum lycopersicum*) Fruits Protect Cells from Oxidative Stress Injury. *J. Sci. Food Agric.* **2016**, *97*, 1616–1623. [CrossRef] [PubMed]
2. Rigano, M.M.; Arena, C.; Di Matteo, A.; Sellitto, S.; Frusciante, L.; Barone, A. Eco-physiological response to water stress of drought-tolerant and drought-sensitive tomato genotypes. *Plant Biosyst.* **2016**, *150*, 682–691. [CrossRef]
3. Ruggieri, V.; Calafiore, R.; Schettini, C.; Rigano, M.M.; Olivieri, F.; Frusciante, L.; Barone, A. Exploiting genetic and genomic resources to enhance heat-tolerance in tomatoes. *Agronomy* **2019**, *9*, 22. [CrossRef]

4. Wen, J.; Jiang, F.; Weng, Y.; Sun, M.; Shi, X.; Zhou, Y.; Yu, L.; Wu, Z. Identification of heat-tolerance QTLs and high-temperature stress-responsive genes through conventional QTL mapping, QTL-seq and RNA-seq in tomato. *BMC Plant Biol.* **2019**, *19*, 398. [CrossRef] [PubMed]
5. Szymanska, R.; Slesak, I.; Orzechowska, A.; Kruk, J. Physiological and biochemical responses to high light and temperature stress in plants. *Environ. Exp. Bot.* **2017**, *139*, 165–177. [CrossRef]
6. Povero, G.; Mejia, J.F.; Di Tommaso, D.; Pioggesi, A.; Warrior, P. A systematic approach to discover and characterize natural plant biostimulant. *Front. Plant Sci.* **2016**, *7*, 435. [CrossRef]
7. Di Stasio, E.; Van Oosten, M.J.; Silletti, S.; Raimondi, G.; Dell'Aversana, E.; Carillo, P.; Maggio, A. *Ascophyllum nodosum*- based algal extracts act as enhancers of growth, fruit quality, and adaptation to stress in salinized tomato plants. *J. Appl. Physiol.* **2018**, *30*, 2675–2686. [CrossRef]
8. Du Jardin, P. Plant biostimulant: Definition, concept, main categories and regulation. *Sci. Hortic.* **2015**, *196*, 3–14. [CrossRef]
9. Di Mola, I.; Ottaiano, L.; Cozzolino, E.; Senatore, M.; Giordano, M.; El-Nakhel, C.; Sacco, A.; Youssef, R.; Colla, G.; Mori, M. Plant-Based Biostimulants Influence the Agronomical, Physiological, and Qualitative Responses of Baby Rocket Leaves under Diverse Nitrogen Conditions. *Plants* **2019**, *8*, 522. [CrossRef]
10. Colla, G.; Cardarelli, M.; Bonini, P.; Rouphael, Y. Foliar Application of Protein Hydrolysate, Plant and Seaweed Extracts Increase Yield but Differentially Modulate Fruit Quality of Greenhouse Tomato. *Hortic. Sci.* **2017**, *52*, 1214–1220. [CrossRef]
11. Feitosa de Vasconcelos, A.C.; Zhang, X.; Ervin, E.H.; De Castro Kiehl, J. Enzymatic antioxidant responses to biostimulants in maize and soybean subjected to drought. *Sci. Agric.* **2009**, *66*, 395–402. [CrossRef]
12. Petrozza, A.; Santaniello, A.; Summerer, S.; Di Tommaso, G.; Di Tommaso, D.; Paparelli, E.; Piaggesi, A.; Perata, P.; Cellini, F. Physiological responses to Megafol®treatments in tomato plants under drought stress: A phenomic and molecular approch. *Sci. Hortic.* **2014**, *174*, 185–192. [CrossRef]
13. Ertani, A.; Schiavon, M.; Muscolo, A.; Nardi, S. Alfalfa plant-dericed biostimulant stimulate short-term growth of salt stressed *Zea mays* L. plants. *Plant Soil* **2013**, *364*, 145–158. [CrossRef]
14. Lucini, L.; Rouphael, Y.; Cardarelli, M.; Canaguier, R.; Kumar, P.; Colla, G. The effect of a plant-derived biostimulant on metabolic profiling and crop performance of lettuce grown under saline conditions. *Sci. Hortic.* **2015**, *182*, 124–133. [CrossRef]
15. Botta, A. Enhancing plant tolerance to temperature stress with amino acids: An approach to their mode of action. *Acta Hortic.* **2013**, *1009*, 29–35. [CrossRef]
16. Verkleij, F.N. Seaweed extracts in agriculture and horticulture: A review. *Biol. Agric. Hortic.* **1992**, *8*, 309–324. [CrossRef]
17. Battacharyya, D.; Babgohari, M.Z.; Rathor, P.; Prithiviraj, B. Seaweed extracts as biostimulants in horticulture. *Sci. Hortic.* **2015**, *30*, 39–48. [CrossRef]
18. Van Oosten, M.J.; Pepe, O.; De Pascale, S.; Silletti, S.; Maggio, A. The role of biostimulants and bioeffectors as alleviators of abiotic stress in crop plants. *Chem. Biol. Technol. Agric.* **2017**, *4*, 5. [CrossRef]
19. Hammad, S.A.R.; Ali, A.M.O. Physiological and biochemical studies on drought tolerance of wheat plants by application of amino acids and yeast extract. *Ann. Agric. Sci.* **2014**, *59*, 133–145. [CrossRef]
20. De Pascale, S.; Rouphel, Y.; Colla, G. Plant biostimulants: Innovative tool for enhancing plant nutrition in organic farming. *Eur. J. Hortic. Sci.* **2017**, *82*, 277–285. [CrossRef]
21. Paul, K.; Sorrentino, M.; Lucini, L.; Rouphael, Y.; Cardarelli, M.; Bonini, P.; Moreno, M.B.M.; Reynaud, H.; Canaguier, R.; Trtilek, M.; et al. A combined phenotypic and metabolic approach for elucidating the biostimulant action of a plant-derived protein hydrolysate on tomato grown under limited water availability. *Front. Plant Sci.* **2019**, *10*, 493. [CrossRef] [PubMed]
22. Sorwong, A.; Sakhonwasee, S. Foliar Application of Glycine Betaine Mitigates the Effect of Heat Stress in Three Marigold (Tageteserecta) Cultivars. *Hortic. J.* **2015**, *84*, 161–171. [CrossRef]
23. Dafni, A. *Pollination Ecology: A Practical Approach*; Oxford University Press: Oxford, UK, 1992; p. xiv +250.
24. Stevens, R.; Buret, M.; Garchery, C.; Carretero, Y.; Causse, M. Technique for rapid small-scale analysis of vitamin C levels in fruit and application to a tomato mutant collection. *J. Agric. Food Chem.* **2006**, *54*, 6159–6165. [CrossRef] [PubMed]
25. Rigano, M.M.; Raiola, A.; Tenore, G.C.; Monti, D.M.; Del Giudice, R.; Frusciante, L.; Barone, A. Quantitative trait loci pyramiding can improve the nutritional potential of tomato (*Solanum lycopersicum*) fruits. *J. Agric. Food Chem.* **2014**, *62*, 11519–11527. [CrossRef] [PubMed]

26. Rigano, M.M.; Lionetti, V.; Raiola, A.; Bellincampi, D.; Barone, A. Pectic enzymes as potential enhancers of ascorbic acid production through the D-galacturonate pathway in Solanaceae. *Plant Sci.* **2018**, *266*, 55–63. [CrossRef] [PubMed]
27. Wellburn, A.R. The spectral determination of chlorophylls a and b, as well as total carotenoids, using various solvents with spectrophotometers of different resolution. *J. Plant Physiol.* **1994**, *144*, 307–313. [CrossRef]
28. Zouari, I.; Salvioli, A.; Chialva, M.; Novero, M.; Miozzi, L.; Tenore, G.C.; Bagnaresi, P.; Bonfante, P. From root to fruit: RNA-Seq analysis shows that arbuscular mycorrhizal symbiosis may affect tomato fruit metabolism. *BMC Genom.* **2014**, *15*, 221. [CrossRef]
29. Benzie, I.F.F.; Strain, J.J. The ferric reducing ability of plasma (FRAP) as a measure of "antioxidant power": The FRAP assay. *Anal Biochem.* **1996**, *239*, 70–76. [CrossRef]
30. Miller, J.N.; Rice-Evans, C.A. Factors influencing the antioxidant activity determined by the ABTS•+ radical cation assay. *Free Radic. Res.* **1997**, *26*, 195–199. [CrossRef]
31. Figlioli, F.; Sorrentino, M.C.; Memoli, V.; Arena, C.; Maisto, G.; Giordano, S.; Capozzi, F.; Spagnuolo, V. Overall plant responses to Cd and Pb metal stress in maize: Growth pattern, ultrastructure, and photosynthetic activity. *Environ. Sci. Pollut. Res. ESPR* **2019**, *26*, 1781–1790. [CrossRef]
32. Kitajima, M.; Butler, W.L. Quenching of chlorophyll fluorescence and primary photochemistry in chloroplasts by dibromothymoquinone. *Biochim. Biophys. Acta-Bioenerg.* **1975**, *376*, 105–115. [CrossRef]
33. Arena, C.; De Micco, V.; De Maio, A.; Mistretta, C.; Aronne, G.; Vitale, L. Winter and summer leaves of *Cistus incanus*: Differences in leaf morpho-functional traits, photosynthetic energy partitioning and poly-ADP-ribose polymerase (PARP) activity. *Botany* **2013**, *91*, 805–813. [CrossRef]
34. Hossain, M.I.; Khatun, A.; Talukder, M.S.A.; Dewan, M.M.R.; Uddin, M.S. Effect of drought o physiology and yield contributing characters of sunflower. *Bangladesh J. Agric Res.* **2010**, *35*, 113–124. [CrossRef]
35. Paradikovic, N.; Vinkovic, T.; VinkovicVrcek, I.; Zuntar, I.; Bojic, M.; Medic-Saric, M. Effect of natural biostimulants on yield and nutritional quality: An example of sweet yellow pepper (*Capsicum annuum* L.) plants. *J. Sci. Food Agric.* **2011**, *91*, 2146–2152. [CrossRef]
36. Chehade, L.A.; Al Chami, Z.; De Pascali, S.A.; Cavoski, I.; Fanizzi, F.P. Biostimulants from food processing by-poducts: Agronomic, quality and metabolic impacts on organic tomato (*Solanum lycopersicum* L.). *J. Sci. Food Agric.* **2017**, *98*, 1426–1436. [CrossRef]
37. Kocira, S.; Szparaga, A.; Kocira, A.; Czerwinska, E.; Wojtowicz, A.; Bronowicka-Mielniczuk, U.; Koszel, M.; Findura, P. Modeling biometric traits, yield and nutritional and antioxidant properties of seeds of three soybean cultivars through the application of biostimulant containing seaweed and amino acids. *Front. Plant Sci.* **2018**, *9*, 388. [CrossRef]
38. Cristiano, G.; Pallozzi, E.; Conversa, G.; Tufarelli, V.; De Lucia, B. Effect of an Animal-Derived Biostimulant on the Growth and Physiological Parameters of Potted Snapdragon (*Antirrhinum majus* L.). *Front. Plant Sci.* **2018**, *9*, 861. [CrossRef]
39. Polo, J.; Mata, P. Evaluation of a biostimulant (pepton) based in enzymatic hydrolyzed animal protein in comparison to seaweed extracts on root development, vegetative growth, flowering, and yield of gold cherry tomatoes grown under low stress ambient field conditions. *Front. Plant Sci.* **2018**, *8*, 2261. [CrossRef]
40. Hamasaki, T.; Kitagawa, T.; Yasuhara, T. Efficacy of yeast cell wall extract, a Byproduct of beer brewing, in Tomato (*Solanum lycopersicum*) culture. *IPCBEE* **2014**, *76*, 21–25. [CrossRef]
41. Rouphael, Y.; Colla, G.; Giordano, M.; El-Nakhel, C.; Kyriacou, M.C.; De Pascale, S. Foliar applications of a legume-derived protein hydrolysate elicit dose-dependent increases of growth, leaf mineral composition, yield and fruit quality in two greenhouse tomato cultivars. *Sci. Hortic.* **2017**, *226*, 353–360. [CrossRef]
42. Matsuo, S.; Kikuchi, K.; Fukuda, M.; Honda, I.; Imanishi, S. Roles and regulation of cytokinins in tomato fruit development. *J. Exp. Bot.* **2012**, *63*, 5569–5579. [CrossRef] [PubMed]
43. Sato, S.; Kamiyama, M.; Iwara, T.; Makita, N.; Furukawa, H.; Ikeda, H. Moderate increase of mean daily temperature adversely affects fruit set of *Lycopersicum esculentum* by distrupting specific physiological processes in male reproductive development. *Ann. Bot.* **2006**, *97*, 731–738. [CrossRef] [PubMed]
44. Li, S.; Li, F.; Wang, J.; Zhang, W.; Meng, Q.; Chen, T.H.H.; Murata, N.; Yang, X. Glycinebetaine enhances the tolerance of tomato plants to high temperature during germination of seeds and growth of seedlings. *Plant Cell Environ.* **2011**, *34*, 1931–1943. [CrossRef] [PubMed]

45. Cristiano, G.; Pallozzi, E.; Conversa, G.; Tufarelli, V.; De Lucia, B. Effects of an animal-derived biostimulant on the growth and physiological parameters of potted snapdragon (*Antirrhinum majus* L.). *Front. Plant Sci.* **2017**, *9*, 861. [CrossRef]
46. Xu, J.; Wolters-Arts, M.; Mariani, C.; Huber, H.; Rieu, I. Heat stress affects vegetative and reproductive performance and trait correlations in tomato (*Solanum lycopersicum*). *Euphytica* **2017**, *213*, 156. [CrossRef]
47. Santaniello, A.; Scartazza, A.; Gresta, F.; Loreti, E.; Biasone, A.; Di Tommaso, D.; Piaggesi, A.; Perata, P. *Ascophyllum nodosum* Seeweed Extract Alleviates Drought Stress in Arabidopsis by Affecting Photosynthetic Performance and Related Gene Expression. *Front. Plant Sci.* **2017**, *8*, 1362. [CrossRef]
48. Ertani, A.; Pizzeghello, D.; Francioso, O.; Sambo, P.; Sanchez-Cortes, S.; Nardi, S. *Capsicum chinensis* L. growth and nutraceutical properties are enhanced by biostimulants in a long-term period: Chemical and metabolomic approaches. *Front. Plant Sci.* **2014**, *5*, 375. [CrossRef]
49. Parrado, J.; Bautista, J.; Romero, E.J.; Garcìa-Martìnez, A.M.; Friaza, V.; Tejada, M. Production of a carob enzymatic extract: Potential use as a biofertilizer. *Bioresour. Technol.* **2008**, *99*, 2312–2318. [CrossRef]

11

Effect of Vegetal- and Seaweed Extract-based Biostimulants on Agronomical and Leaf Quality Traits of Plastic Tunnel-Grown Baby Lettuce under Four Regimes of Nitrogen Fertilization

Ida Di Mola [1,*], Eugenio Cozzolino [2,*], Lucia Ottaiano [1], Maria Giordano [1], Youssef Rouphael [1], Giuseppe Colla [3] and Mauro Mori [1]

1 Department of Agricultural Sciences, University of Naples Federico II, Portici 80055, Italy; lucyottaiano@gmail.com (L.O.); maria.giordano@unina.it (M.G.); youssef.rouphael@unina.it (Y.R.); mori@unina.it (M.M.)

2 Council for Agricultural Research and Economics (CREA)-Research Center for Cereal and Industrial Crops, Caserta 81100, Italy

3 Department of Agriculture and Forest Sciences, University of Tuscia, Viterbo 01100, Italy; giucolla@unitus.it

* Correspondence: ida.dimola@unina.it (I.D.M.); eugenio.cozzolino@crea.gov.it (E.C.)

Abstract: Nitrogen (N) fertilizers play a crucial role in agriculture, representing a powerful tool for farmers for increasing yields throughout the seasons under both optimal and suboptimal conditions. At the same time, their synthetic/chemical nature could have several influences on ecosystems and human health. For this reason, there is an urgent need to find new and more sustainable means of production to increase plant productivity and optimize nitrogen use. An experiment was conducted in a plastic tunnel to assess the response of baby lettuce crop to the foliar application of three plant biostimulants (PBs): Legume-derived protein hydrolysate (LDPH) 'Trainer®', tropical plant extract (TPE) 'Auxym®' and seaweed extract (SwE) from *Ecklonia maxima* 'Kelpak®' under different N rates of 0, 10, 20 and 30 kg N·ha^{-1}. The responses of baby lettuce plants were assessed in terms of yield, growth parameters and physicochemical composition of the leaves. The fresh yield of baby lettuce in both biostimulant-treated and untreated plants was positively affected by increasing N rates from 0 to 20 kg N·ha^{-1}, reaching a plateau thereafter indicating luxury N conditions at 30 kg N·ha^{-1}. However, high N fertilizer application (20 and especially 30 kg N·ha^{-1}) resulted in undesirable decreases in antioxidant activities and total ascorbic acid (TAA). Under non-fertilized regimens, foliar PBs application boost growth and yield of baby lettuce in comparison to non-treated plants. Foliar spray with LDPH and especially SwE elicited significant increases in marketable fresh yield (averaging 14%, 6% and 7% at 10, 20 and 30 kg N·ha^{-1}, respectively) compared to TPE and untreated plants. Improved agronomical performance of baby lettuce under optimal (10 kg N ha^{-1}) and especially suboptimal N regimens (0 kg N ha^{-1}) was associated with increasing photochemical efficiency and a better activity of photosystem II (higher Soil Plant Analysis Development-SPAD index and chlorophyllous pigments biosynthesis). The application of LDPH enhanced antioxidant capacity and TAA in baby lettuce leaf and did not increased nitrate content as recorded in SwE and TPE treatments. Overall, plant biostimulants may be considered as a sustainable tool of production to increase leafy vegetable productivity in low fertility soils.

Keywords: *Lactuca sativa* L.; legume-derived protein hydrolysate; nitrate; tropical plant extract; seaweed extract

1. Introduction

In recent years, the consumption of fresh-cut leafy vegetables has increased and among them, baby leaf lettuce is very widespread. Baby leaf lettuce is widely cultivated in Italy under both open field and greenhouse conditions [1]. Baby leaf vegetables are characterized by a short cycle but it requires a correct agronomic management to avoid high levels of nitrate accumulation and pesticide residues in the final product [2]. Therefore, there is a paramount interest in enhancing its production and quality, and at the same time reducing the nitrate levels of leafy vegetables, in order to overcome the legal limit for the marketing imposed by the European Community (Reg. n° 1258/2011).

With the aim to boost yield and to contain the risk of nitrate accumulation in the leaves, the research community is focusing on the use of sustainable production technologies, including application of beneficial microorganisms (Plant growth promoting Rhizobacteria, mycorrhiza and *Trichoderma*) and plant biostimulants [3–6]. In function of their origin, non-microbial plant biostimulants can be classified into five categories: (i) Seaweed extracts and microalgae, (ii) protein hydrolysate (PH) and amino acid containing products, (iii) plant extracts, (iv) humic substances and (v) silicon, with the first three groups commanding 75% of the market share [7–10]. Protein hydrolysate and amino acids containing products are normally obtained by enzymatic and/or chemical hydrolysis depending on the organic matrix (animal or vegetal) and are characterized by high percentages of amino acids and peptides, followed by carbohydrates and small amounts of micronutrients [7,8]. Moreover, plant extracts are normally produced through the fermentation of tropical plants and contain amino acids and peptides, carbohydrates but also vitamins and micronutrients with small quantities of phytohormones [8], while seaweed extracts particularly the brown macroalgae are obtained through a process called 'cold cell burst' and contain polysaccharides, osmolytes (proline and betaines), macro- and micro-nutrients, brassinosteroids and phytohormones (auxins, cytokinins and gibberellins; [11,12]).

Recent studies carried out on vegetable crops including leafy greens have demonstrated that foliar and/or root applications of plant or seaweed-based biostimulants elicit several physiological and molecular processes, thus resulting in improvements in growth, productivity, nutritional quality and nutrient use efficiency (NUE) and tolerance to abiotic stressors such as drought, soil and water salinity, nutrient deprivation and extreme temperatures [12–24]. The beneficial effects of vegetal- and seaweed-based biostimulants have been attributed to direct and indirect stimulation mechanisms [9]. The direct stimulation action of biostimulants include: (i) Activity enhancement of key enzymes involved in carbon and nitrogen (N) metabolism [13,20,25], (ii) eliciting hormone-activity in particular auxin- and gibberellin-like activities through bioactive peptides [26–28], (iii) physiological, biochemical and anatomical changes such as the production of antioxidant enzymes, pigments, secondary metabolites and smallest cell guard length and width [7,8,12,29]. In addition to direct mechanisms, indirect modes of actions on agronomical performance and nutrient uptake and assimilation have been also reported when vegetal- or seaweed extract-based biostimulants were applied as substrate drench and/or foliar spray [8,29]. The better nutritional status in biostimulant-treated plants in comparison to untreated plants has been mostly associated with root system modulation (increases in root biomass, root length and diameter and lateral root branching [8,12,29]).

Among the different agronomical claims of plant biostimulants, the capacity to improve NUE in particular, N is one of the most important claims supporting their placement in the market for both economic and environmental reasons [9]. However, limited scientific literature are available regarding the effects of plant biostimulants on vegetable crops under sub-optimal N regimens [20,22,30–32]. For instance, Sestili and co-workers [20] demonstrated that the application of a PH at optimal and sub-optimal N regimens enhanced hydroponically grown tomato performance, especially substrate drench. Interestingly, the same authors observed that protein hydrolysate at low N conditions upregulated gene expression for amino acid transporter and glutamine synthetase, leading to a higher assimilation of N with a positive impact on plant growth. Similarly, Carillo et al. [22] reported that foliar application of PH, especially under suboptimal N fertilization regimes (0 or 15 kg N ha^{-1}) boost marketable yield of greenhouse spinach due to an enhancement of nutrient acquisition and to an increase in total amino acids in plants as well as to an improvement of photochemical efficiency, thus boosting yield.

Since there is ample evidence of species-specific response to plant biostimulants, especially that of leaf biostimulant permeability (through leaf cuticule and stomatal aperture) and thus the efficacy of the biostimulant product is species-dependent [9]; there is an urgent need among researchers to assess the effect of vegetal- and seaweed based-biostimulants on baby lettuce performance at different N fertilization regimes.

Taking into account all the previous considerations, an experiment was conducted in a plastic tunnel to assess the response of baby lettuce crop to the foliar application of the legume-derived PH 'Trainer®', tropical plant extract 'Auxym®' and seaweed extract from *Ecklonia maxima* 'Kelpak®' under different N rates of 0, 10, 20 and 30 kg N·ha^{-1}. The responses of baby lettuce plants were assessed in terms of yield, leaf morphometric parameters and leaf quality traits.

2. Materials and Methods

2.1. Experimental Setting, Plant Material and Design

The experiment was carried out in a unheated plastic tunnel covered by polyethylene during the winter 2018 growing season (January 16—March 12) at Gussone Park, experimental site of the Department of Agricultural Sciences (40°48.870′ N; 14°20.821′ E; 70 m above sea level) located in Portici, southern Italy. The trend of daily maximum and minimum air temperature inside the plastic tunnel is reported in Figure 1. The baby leaf lettuce (*Lactuca sativa* L.) cv. 'Zarina' (ISI Sementi SpA, Parma, Italy) was used as the selected crop.

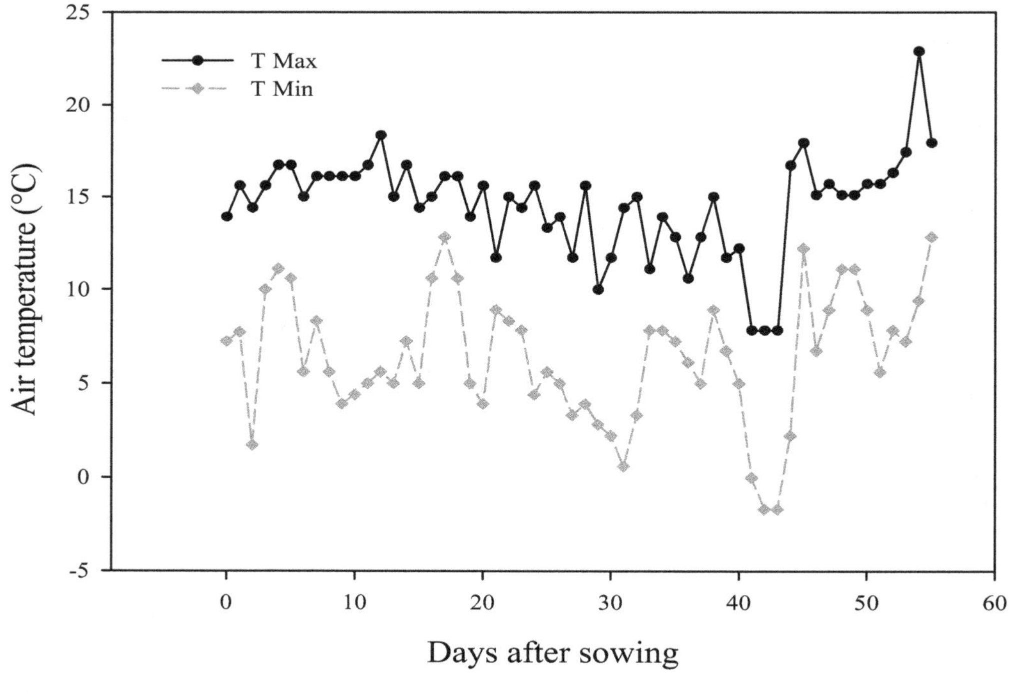

Figure 1. Trend of the maximum and minimum air temperature inside the plastic tunnel during the growing period of baby lettuce.

A factorial combination of N fertilization (N) and biostimulant application (B) was applied based on four increasing N fertilization levels (0, 10, 20 or 30 kg N·ha^{-1}; N0, N10, N20 and N30, respectively) and three plant biostimulants (seaweed extract—SwE, legume-derived protein hydrolysate—LDPH and tropical plant extract—TPE) and a non-treated control. The experimental design was a randomized complete-block design with three replications, yielding 48 experimental units (4N × 4B × 3 replications) established in large lysimeters of reinforced fiber glass with a diameter of 0.70 m and a depth of 0.60 m.

Each experimental unit consisted of one large lysimeters. The lysimeters were filled with a soil having the following chemical and physical characteristics reported in Table 1.

Table 1. Physical and chemical properties of the soil used in this work.

Soil Properties	Units	Mean Values
Texture		
Coarse sand	%	69.1
Fine sand	%	21.9
Silt	%	4.5
Clay	%	4.5
Chemical properties		
pH	-	6.54
Electrical conductivity	$dS{\cdot}m^{-1}$	0.64
Organic matter	$g{\cdot}kg^{-1}$	32.4
Total N (Kjeldahl method)	$g{\cdot}kg^{-1}$	1.2
P_2O_5 (Olsen method)	$mg{\cdot}kg^{-1}$	312.8
K_2O (Tetraphenylborate method)	$mg{\cdot}kg^{-1}$	620.7
NO_3-N	$mg{\cdot}kg^{-1}$	10.0
NH_4-N	$mg{\cdot}kg^{-1}$	9.0

2.2. *Nitrogen Fertilization Levels, Cultural Practices and Biostimulants Application*

The baby leaf lettuce was hand seeded on January 16 at a plant density of 2500 seeds$\cdot m^{-2}$. The N was applied as calcium nitrate (26%) in a single operation 14 days after sowing. The calcium nitrate was used based on standard commercial practices used in Italy.

The three commercial SwE, LDPH and TPE-based biostimulants were made by 'Kelpak®' (Kelp Products (Pty) Ltd., Cape Town, South Africa), 'Trainer®' and 'Auxym®' (Italpollina S.p.A., Rivoli Veronese, Italy), respectively.

The SwE obtained through 'cold cell burst' mainly contained phytohormones (auxins and cytokinins with a very high auxin-to-cytokinin ratio), carbohydrates, amino acids, vitamins (B1, B2, C and E) and macro- and micro-nutrients [18,19]. The LDPH-based biostimulant contained free amino acids and peptides (75%), carbohydrates (22%) and mineral nutrients (3%). The detailed aminogram was reported by Paul et al. [33,34]. The TPE biostimulant obtained by fermentation of tropical plants contained 54% of free amino acids and peptide, 17% carbohydrate, 23% mineral nutrients, 6% vitamins and 0.22% phytohormones as reported in detail by Caruso et al. [23,24]. Baby lettuce leaf plants were sprayed with a biostimulant solution containing 3 $mL{\cdot}L^{-1}$ of SwE and LDPH and 2$\cdot$mL L^{-1} for TPE-based biostimulant, or with water (non-treated control), five times during the growing season at 7-day intervals, starting three weeks after sowing. The relative doses of the three commercial plant biostimulants were used based on manufacturer recommendations. The volume of the solution used during the five foliar applications was 100 mL per square meter.

2.3. *Plant Growth Parameters, Marketable Yield, Leaf Colorimetry and Sampling*

On March 12, the baby leaf lettuce was harvested in all experimental units. The leaf area was measured using an electronic leaf area meter (Li-Cor3000, Li-Cor, Lincoln, NE, USA) in order to calculate the leaf area index (LAI). The marketable fresh yield was also measured and expressed in tons per ha, and a sub-sample was oven dried at 70 °C for 3 days in order to determine the leaf dry matter percentage, and the dry samples were consequently used for the mineral analysis. Furthermore, the specific leaf weight (leaf dry weight per unit area; $mg{\cdot}cm^{-2}$) as well as leaf succulence (leaf water content per unit area; $mg{\cdot}cm^{-2}$) were also recorded.

Leaf colorimetry was measured on the upper side of 10 leaves per experimental unit using Minolta CR-300 Chroma Meter (Minolta Camera Co. Ltd., Osaka, Japan) in order to obtain the color space parameters (L^*, a^* and b^*) and a portable chlorophyll meter SPAD-502 (Konica Minolta, Tokyo, Japan) was also used to measure the SPAD (Soil Plant Analysis Development) index.

Batch samples of fresh leaves from each experimental unit were frozen in liquid nitrogen immediately after harvest, lyophilized Christ, Alpha 1–4 (Osterode, Germany) and stored at −80 °C until further analysis.

2.4. Antioxidant Capacity Analysis

The lipophylic and hydrophilic antioxidant capacities were assessed on extract from freeze-dried baby lettuce leaves (200 mg) added with methanol and distilled water, respectively. The antioxidant activity of the lipophilic and hydrophilic extract fractions was measured spectrophotometrically based on the methods of Re et al. [35] and Fogliano et al. [36], respectively. The absorbance of the solutions for lipophilic and hydrophilic extract fractions were measured at 734 and 505 nm, respectively. Lipophylic and hydrophilic antioxidant activities were expressed as mmol of Trolox and mmol ascorbic acid per 100 g of dry weight (dw) [36].

2.5. Chlorophyllous Pigments and Nitrate Analysis

Chlorophyll and carotenoids content of the baby lettuce leaves were also assayed spectrophotometrically after the extraction of the fresh material (500 mg) using pure acetone as described in detail by Lichtentahler and Buschmann [37], whereas the nitrate content was determined based on the method of Sah [38]. The absorbance of the solutions for chlorophyll a and b, carotenoids and nitrate were measured at 662, 645, 470 and 550 nm. The chlorophyllous pigments were expressed as mg g^{-1} fresh weight (fw), while the nitrate content was expressed as mg kg^{-1} fw.

2.6. Total Ascorbic Acid Analysis

The total ascorbic acid (expressed as mg ascorbic acid on 100 g fw) was also assessed spectrophotometrically based on the protocol by Kampfenkel et al. [39]. The absorbance of the solution for total ascorbic acid was measured at 525 nm.

2.7. Statistical Processing

Morphological and qualitative data were statistically analyzed by a 2-way ANOVA using the SPSS 21 software package for Windows 2007. The means were separated by a Duncan's test (significance level 0.05).

3. Results and Discussion

3.1. Effect of N Fertilization Levels and Biostimulant Application on Yield and Growth

The results regarding morphological parameters and marketable yield of baby lettuce are reported in Figure 2; Figure 3 and Table 2. For marketable fresh yield and leaf area index (LAI) significant interaction between fertilization (F) and biostimulant application (B) was observed, whereas leaf succulence and specific leaf weight (SLW) were only influenced by the two tested factors with no significant F × B interaction (Figures 2 and 3, and Table 2). The fresh yield of baby lettuce in both biostimulant-treated and untreated plants was positively affected by increasing N fertilization levels from 0 to 20 kg N·ha^{-1}, reaching a plateau thereafter indicating a luxury N conditions at 30 kg N·ha^{-1} (Figure 2). The marketable fresh yield of baby lettuce at N0 was clearly higher by 19% in biostimulant-treated plants compared untreated plants, with no significant differences between the three plant biostimulants tested (Figure 2). Interestingly, foliar spray with LDPH and especially SwE elicited significant increases (average 14%, 6% and 7% at 10, 20 and 30 kg N·ha^{-1}, respectively) compared to TPE and untreated plants (Figure 2). Similarly to the effects on marketable fresh yield, the leaf area index (LAI) in SwE and LDPH-treated plants at 10, 20 and 30 kg N·ha^{-1} was significantly higher compared to baby lettuce treated with TPE or untreated plants, whereas under non-fertilized conditions LAI was significantly higher in biostimulant compared to untreated plants, irrespective of the commercial biostimulants used (Figure 3).

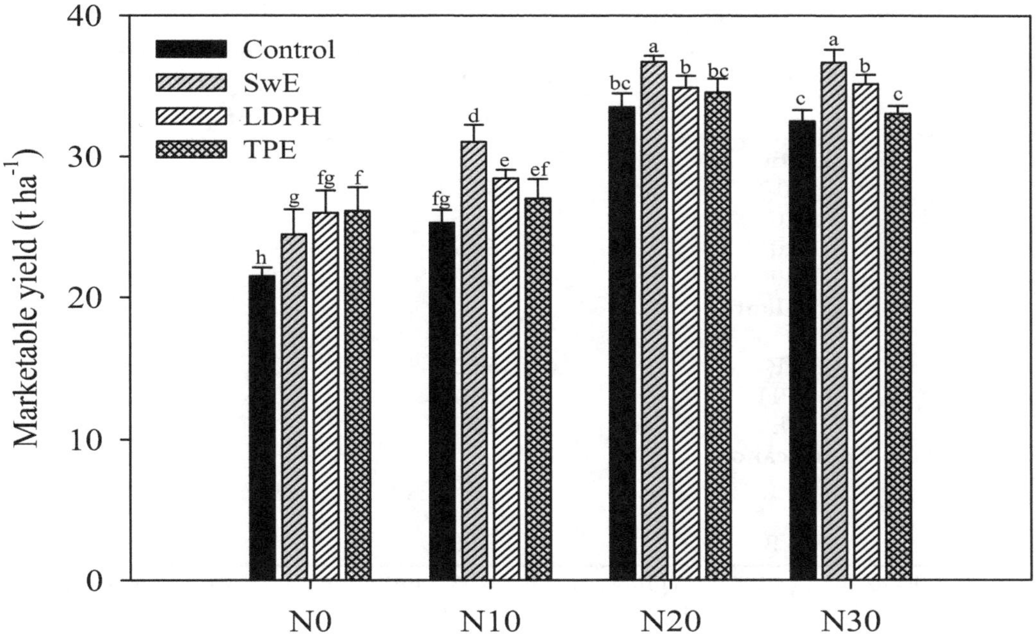

Figure 2. Effects of nitrogen (N) fertilization levels (0, 10, 20 and 30 kg N·ha^{-1}; N0, N10, N20 and N30, respectively) and biostimulant applications (untreated control, SwE: Extract of seaweed *Ecklonia maxima*, LDPH: Legume-derived protein hydrolysate and TPE: Tropical plant extract) on the marketable fresh yield of baby lettuce plants. Different letters indicate significant differences according to the Duncan's test (significance level 0.05).

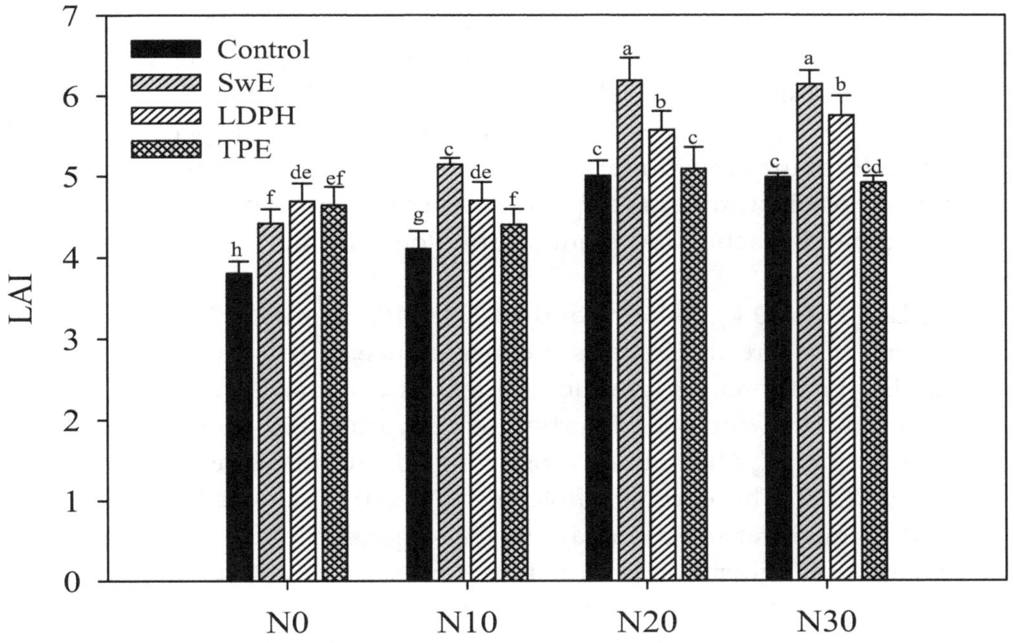

Figure 3. Effects of nitrogen (N) fertilization levels (0, 10, 20 and 30 kg N·ha^{-1}; N0, N10, N20 and N30, respectively) and biostimulant applications (untreated control, SwE: Extract of seaweed *Ecklonia maxima*, LDPH: Legume-derived protein hydrolysate and TPE: Tropical plant extract) on the leaf area index (LAI) of baby lettuce plants. Different letters indicate significant differences according to the Duncan's test (significance level 0.05).

Table 2. Effects of nitrogen (N) fertilization levels (0, 10, 20 and 30 kg N·ha^{-1}; N0, N10, N20 and N30, respectively) and biostimulant applications (untreated control, SwE: Extract of seaweed *Ecklonia maxima*, LDPH: Legume-derived protein hydrolysate and TPE: Tropical plant extract) on leaf succulence and specific leaf weight (SLW) of baby lettuce plants.

Treatments	Succulence	SLW
	(mg H_2O·cm^{-2})	(mg dm·cm^{-2})
Fertilization (F)		
N0	59.9 b	2.99 a
N10	71.5 a	3.26 a
N20	69.9 a	2.66 b
N30	67.7 a	2.60 b
Biostimulant (B)		
Control	60.5 c	2.94 a
SwE	65.2 b	2.56 b
LDPH	70.4 a	2.92 a
TPE	72.9 a	3.10 a
Significance		
F	**	**
B	*	*
F × B	NS	NS

NS, *, and ** indicate non-significant, significant at $p < 0.05$, significant at $p < 0.01$, respectively. Different letters indicate significant differences according to the Duncan's test (significance level 0.05).

When averaged over biostimulant application (F × B = not significant), the leaf succulence increased quadratically by increasing N fertilization levels from 10 to 30 kg N·ha^{-1} with no significant difference among the three N fertilization rates, whereas the SLW declined at 20 and 30 kg N·ha^{-1} (Table 2). Averaged over N fertilization levels, significant differentiation in terms of leaf succulence and SLW was recorded in response to biostimulants application with the higher values of succulence observed with LDPH and TPE followed by SwE as opposed to untreated plants, whereas the lowest values of SLW were recorded in baby lettuce treated with *Ecklonia maxima* extract (Table 2).

The stimulation effect of commercial biostimulants (6%–19%) recorded in the current research is in line with previous studies carried out on greenhouse fresh tomato treated with seaweed extracts of *E. maxima* or *Ascophyllum nodosum*, LDPE and TPE (7%–25%; [12,40]) but far lower than those recorded on greenhouse spinach [19]. The different stimulation effect among tested species indicates a crop-specific differential response to plant biostimulant applications and thus requires additional crop-specific studies to optimize plant biostimulants application, taking into consideration the following factors: environment, management practice and plant morphological traits (e.g., leaf permeability and cuticle thickness [9,28]).

Interestingly, LDPH (at 0 kg N·ha^{-1}) and SwE (at 10, 20 and 30 kg N·ha^{-1}) are likely to boost growth response and crop productivity as a consequence of the presence of bioactive molecules such as amino acids (tryptophan, glutamic and aspartic acids), soluble peptides (in LDPH) and polysaccharides (laminarans, fucoidans and alginates), phenolic compounds, osmolytes (proline, betaine and manitol) and phyohormones (abscisic acid, auxins, brassinosteroids, cytokinins and gibberellins) (in SwE) [8,29]. These former molecules present in seaweed and PH-based biostimulants may have triggered a signal transduction pathway through elicitation of endogenous phytohormone synthesis, thus leading to a higher crop productivity compared to untreated-baby lettuce plants [19,20]. Another possible mechanism of action (indirect mechanism) behind the stimulation of LAI and marketable fresh yield could be the modulation of the root system architecture in terms of root biomass, root volume and length and higher root branching triggered by tryptophan in LDPH and auxins in SwE, which improved nutrient uptake/translocation/assimilation, leading to a higher agronomical performance [12,19,41,42]. Our results are in agreement with those of Carillo and co-workers [22] who reported that foliar application of LDPH at a rate of 4 mL·L^{-1} under suboptimal N fertilization conditions (0 and 15 kg N·ha^{-1}) increased the fresh yield of greenhouse spinach through an increase of the nutritional status (higher macronutrient accumulation), better photosynthetic activity and improving the total acid content.

3.2. *Effect of N Fertilization Levels and Biostimulant Application on Leaf Colorimetry and SPAD Index*

Among the physical properties that may affect the purchasing decisions of vegetable consumers is product appearance, in particular, the color of the vegetable [43]. In the present study, no significant interaction between N fertilization and biostimulants application was recorded for the three leaf colorimetric parameters lightness (L^*), green color intensity (negative values of a^*) and yellow color intensity (positive values of b^*) (Table 3). The colorimetric CIELAB components L^* and b^* were significantly influenced by the two tested factors, whereas a^* was only affected by N fertilization levels (Table 3). Increasing the N fertilization levels from 0 to 30 kg N·ha^{-1} yielded lighter baby lettuce leaf expression by increasing L^* values, but with a decrease in a^* values (Table 3). Moreover, when averaged over N fertilization levels, the foliar application of SwE and TPE-based biostimulants elicited an increase in L^* values compared to the untreated control, whereas LDPH treatment exhibited intermediate values (Table 3).

Table 3. Effects of nitrogen (N) fertilization levels (0, 10, 20 and 30 kg N·ha^{-1}; N0, N10, N20 and N30, respectively) and biostimulant applications (untreated control, SwE: Extract of seaweed *Ecklonia maxima*, LDPH: Legume-derived protein hydrolysate and TPE: Tropical plant extract) on leaf hunter color parameters of baby lettuce plants.

Table	L^*	a^*	b^*
Fertilization (F)			
N0	55.9 d	−20.8 a	39.8 b
N10	56.7 c	−21.2 a	40.1 b
N20	58.7 b	−22.7 b	41.8 a
N30	59.5 a	−22.7 b	42.4 a
Biostimulants (B)			
Control	56.5 c	−21.5	39.8 b
SwE	58.5 a	−22.2	41.6 a
LDPH	57.5 b	−21.8	41.0 a
TPE	58.3 a	−21.9	41.6 a
Significance			
F	**	*	*
B	*	NS	*
F × B	NS	NS	NS

NS, *, and ** indicate non-significant, significant at $p < 0.05$, or significant at $p < 0.01$, respectively. Different letters indicate significant differences according to the Duncan's test (significance level 0.05).

Interestingly, the foliar application of commercial plant biostimulants improved the SPAD index significantly; this is an important physiological parameter having a crucial role on the primary metabolism of plants. With the exception of under N20, where no significant difference in the SPAD index was observed, between biostimulants-treated and untreated plants, the foliar application with SwE (at 10 and 30 kg N·ha^{-1}) and with the three commercial biostimulants (at 0 kg N·ha^{-1}) incurred a significant increase in the SPAD index (Figure 4). Our findings have been also demonstrated in many leafy vegetable species such as jute, lettuce, and spinach [16,19,22]. The highest SPAD values observed after the application of plant biostimulants in particular extracts from brown macroalgae could be attributed to several putative mechanisms like the following: (i) better translocation of soluble sugars via the phloem, (ii) increases in the biogenesis of chloroplast, as well as (iii) limited chlorophyll degradation, and thus, delayed senescence [29,44,45].

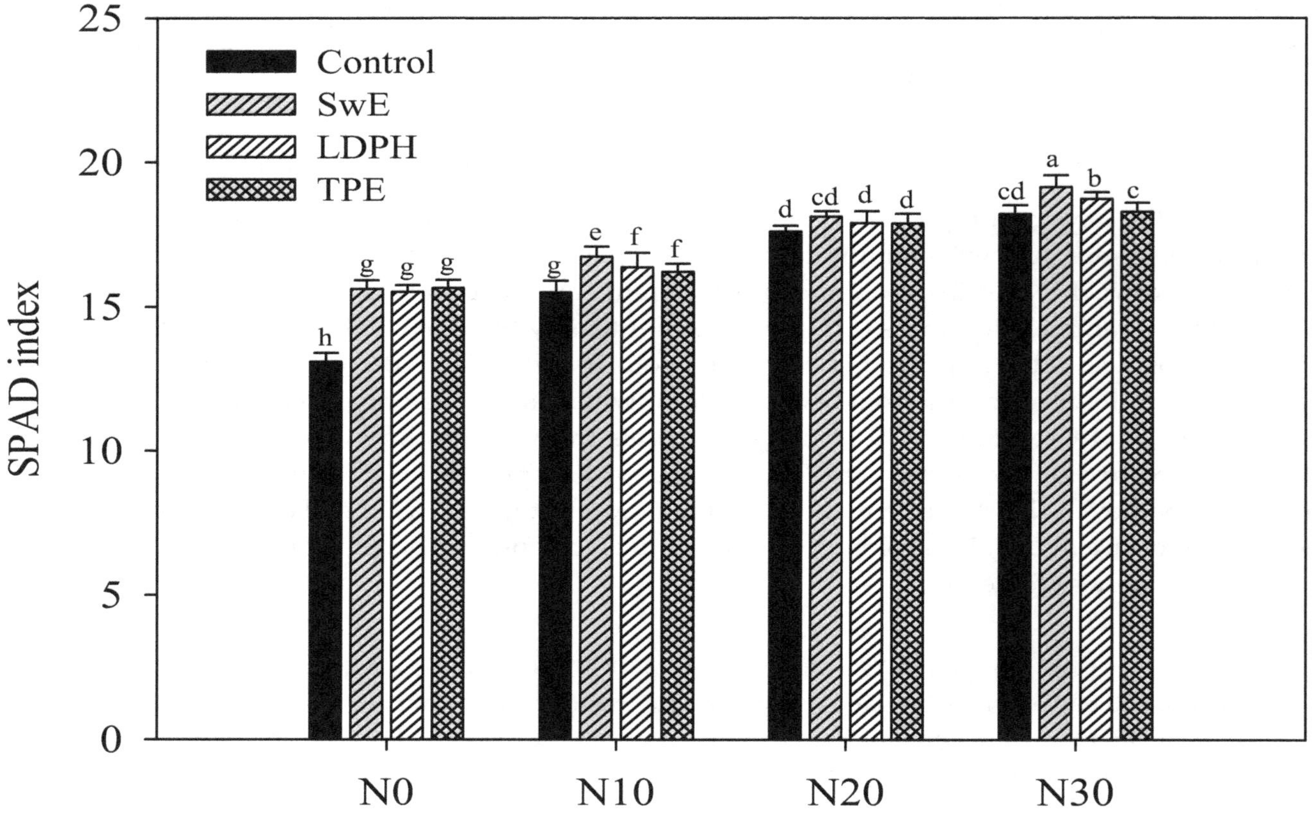

Figure 4. Effects of nitrogen (N) fertilization levels (0, 10, 20 and 30 kg N·ha^{-1}; N0, N10, N20 and N30, respectively) and biostimulant applications (untreated control, SwE: Extract of seaweed *Ecklonia maxima*, LDPH: Legume-derived protein hydrolysate and TPE: Tropical plant extract) on the SPAD index of baby lettuce plants. Different letters indicate significant differences according to the Duncan's test (significance level 0.05).

3.3. Effect of N Fertilization Levels and Biostimulant Application on Nitrate Accumulation and Biochemical Parameters

Nitrate was affected by both N fertilization levels and biostimulant application, without significant F × B interaction (Table 4). The nitrate concentration in baby lettuce leaf was negatively affected by N fertilization levels. Increasing the N fertilization from 0 to 30 kg N·ha^{-1} increased the nitrate accumulation in leaves, especially at 20 and 30 kg N·ha^{-1}, where the content of nitrate was above the upper limits set by the European Union (EU) for safe lettuce marketing (Commission Regulation No. 1258/2011; 3000 to 5000 mg NO_3^-·kg^{-1} of lettuce depending on growing season and cultivation conditions). On the other hand, when averaged over N fertilization levels, the nitrate concentration in LDPH treated plants was significantly lower on average by 21.2% compared to baby lettuce treated with SwE or TPE and it was not significantly different than untreated-baby lettuce plants (Table 4). The capacity of LDPH, which is mainly composed of soluble solids and especially amino acids, to accumulate less nitrate in the leaf tissue, could be attributed to a molecular mechanism such as the up-regulation of genes involved in N metabolism such as nitrate reducatse, and consequently, to an augmenting assimilation of nitrates into amino acids [46,47]. Furthermore, other studies conducted by Sady and Smoleń [31] and Smoleń and Sady [32] on carrots and spinach, respectively, reported that after the foliar application of 'Pentakeep V' containing 5-aminolevulinic acid was able to reduce nitrate accumulation in combination with a 50% N dose, whereas an opposite trend was observed in combination with 100% N. The authors concluded that nitrate accumulation in response to biostimulant application may change in relation to several interacting variables including species, variety and N application rates.

Table 4. Effects of nitrogen (N) fertilization levels (0, 10, 20 and 30 kg N·ha^{-1}; N0, N10, N20 and N30, respectively) and biostimulant applications (untreated control, SwE: Extract of seaweed *Ecklonia maxima*, LDPH: Legume-derived protein hydrolysate and TPE: Tropical plant extract) on nitrate, chlorophyll and carotenoids content of baby lettuce plants.

Treatments	Nitrate	Chlorophyll a	Chlorophyll b	Total chlorophyll	Carotenoids
	(mg·kg^{-1} fw)	(mg·g^{-1} fw)	(mg·g^{-1} fw)	(mg·g^{-1} fw)	(µg·g^{-1} fw)
Fertilization (F)					
N0	703.3 d	0.298 b	0.211	0.508 b	156 b
N10	1476.0 c	0.330 a	0.210	0.540 a	178 a
N20	6206.7 b	0.334 a	0.201	0.535 a	170 a
N30	7288.1 a	0.338 a	0.209	0.546 a	170 a
Biostimulants (B)					
Control	3366.5 b	0.302 b	0.191 c	0.493 b	155 c
SwE	4467.6 a	0.319 b	0.192 c	0.511 b	181 a
LDPH	3504.4 b	0.342 a	0.214 b	0.556 a	173 ab
TPE	4426.5 a	0.337 a	0.232 a	0.569 a	164 bc
Significance					
F	**	*	NS	*	*
B	*	*	**	*	*
F × B	NS	NS	NS	NS	NS

NS, *, and ** indicate non-significant, significant at $p < 0.05$, or significant at $p < 0.01$, respectively. Different letters indicate significant differences according to the Duncan's test (significance level 0.05).

One of the beneficial responses of plant biostimulants application is an increase in chlorophyllous pigments such as chlorophyll a, b and total, as well as carotenoids. This was the case in the current research study, since the foliar application of LDPH and TPE incurred a significant increase in chlorophyll a and b and consequently the total chlorophyll compared to SwE and untreated-baby lettuce plants (Table 4). Furthermore, the content of carotenoids was positively affected by the foliar application of SwE and LDPH compared to the control treatment (Table 4). This beneficial effect of vegetal and seaweed extract-based biostimulants on carotenoids and especially chlorophyll content has been recorded also in corn, jute and eggplant [21,22,48,49]. The increase in chlorophyll a and total content in both LDPH and TPE (characterized by the high percentage of free amino acids (75% and 54%, respectively [23,24]) could be attributed to the higher content of primary amino acids in the vegetal-based treated plants as amino acids (e.g., alanine, aspartate, asparagines and glutamate) which help to boost chlorophyll content, and consequently, increase photosynthetic activity as well as the quantum yield of O_2 evolution [22].

3.4. *Effect of N Fertilization Levels and Biostimulant Application on Antioxidant Capacity and Bioactive Content*

Lipophilic (LAA) and hydrophilic (HAA) antioxidant activities as well as total ascorbic acid (TAA) were significantly affected by both factors with a significant F × B interaction (Table 5). Antioxidant scavenging activity was an important functional quality parameter in assessing the nutritional properties of foods including leafy vegetables, since lipophilic (e.g., β-carotene, lutein, α-tocopherol, etc.) and hydrophilic (e.g., vitamin C, caffeic acid, ferulic acid, quercitin, etc.) antioxidant molecules impart beneficial effects to human health, as these bioactive molecules are known to play a primary role in delaying oxidative damage, thus, preventing a wide range of diseases [50–54]. In the current study, LAA, HAA and TAA of the baby lettuce ranged from 19.9 to 32.3 mmol trolox 100·g^{-1} dw, from 3.0 to 8.2 mmol ascorbic acid 100 g^{-1} dw and from 6.8 to 33.4 mg g^{-1}, respectively (Table 5).

Table 5. Effects of nitrogen (N) fertilization levels (0, 10, 20 and 30 kg N ha^{-1}; N0, N10, N20 and N30, respectively) and biostimulant applications (untreated control, SwE: Extract of seaweed *Ecklonia maxima*, LDPH: Legume-derived protein hydrolysate and TPE: Tropical plant extract) lipophilic (LAA) and hydrophilic (HAA) antioxidant activities and total ascorbic acid (TAA) of baby lettuce plants.

Treatments		LAA	HAA	TAA
		mmol Trolox eq. 100 g^{-1} dw	mmol ascorbic acid eq. 100 g^{-1} dw	mg g^{-1} fw
N0	Control	26.2 b	6.7 c	25.0 b
	SwE	27.6 b	6.6 c	33.4 a
	LDPH	32.3 a	8.2 a	18.0 c
	TPE	25.2 b	7.2 bc	18.6 c
N10	Control	21.0 c	7.6 b	17.3 c
	SwE	20.4 c	7.5 b	16.8 c
	LDPH	20.9 c	7.7 b	18.1 c
	TPE	30.9 a	7.6 b	16.9 c
N20	Control	22.0 c	4.9 d	14.0 de
	SwE	22.2 c	4.8 d	14.3 d
	LDPH	19.9 c	4.4 d	12.6 de
	TPE	22.1 c	3.0 e	14.5 d
N30	Control	21.8 c	3.3 e	13.2 de
	SwE	20.1 c	3.2 e	11.7 e
	LDPH	21.9 c	4.5 d	12.6 de
	TPE	21.5 c	3.4 e	6.8 f
Significance				
Fertilization (F)		**	**	**
Biostimulant (B)		*	*	**
F × B		**	**	**

*, ** significant at $p < 0.05$ and 0.01, respectively. Different letters indicate significant differences according to the Duncan's test (significance level 0.05).

High N fertilizer application (20 and especially 30 kg·N ha^{-1}) resulted in undesirable decreases in HAA and TAA of baby lettuce leaves (Table 5), which is in agreement with the results of Wang et al. [55] who reported that high N fertilization levels can result in undesirable changes in the quality attributes of fruit and leafy vegetables such as soluble solids and ascorbic acid leading to a decrease in commercial, nutritional and functional quality.

The vegetal- and seaweed extract-based biostimulants applied to baby lettuce resulted in higher antioxidant capacity and bioactive content depending on the N fertilization levels compared to untreated control treatment. For instance, at N0 the highest antioxidant activities and TAA compared to the untreated control were recorded in baby lettuce treated with LDPH and SwE-based biostimulant plants, respectively, whereas at N10 and N30 the highest LAA and HAA contents were observed in TPE and LDPH treated plants, respectively (Table 5). Our findings on the effect of plant biostimulants on nutritional and functional quality of the product were in line with previous research on vegetal-based biostimulants (protein hydrolysate and plant extract) conducted by Caruso et al. [23], in which foliar application at weekly interval increases the LAA, HAA and TAA contents of perennial wall rocket compared to the non-treated control. Similarly, Vasantharaja et al. [56] demonstrated that the application of seaweed extract-based biostimulant (*Sargassum swartzii*) boosted the antioxidant activity and the bioactive content (e.g., phenols and vitamin C) of cowpea. A mechanistic explanation of the beneficial effect of plant biostimulants, in particular LDPH, on the biosynthesis of antioxidant molecules could be due to: i) the activity stimulation of key enzymes involved in antioxidant homeostasis in cells, and ii) the higher macro- and micro-nutrient assimilation of biostimulant-treated plants which could contribute to the synthesis of amino acids, phenylalanine and tyrosine [7,40].

4. Conclusions

The idea of working with plant biostimulants to increase yield under both optimal and suboptimal conditions is gaining interest among leafy vegetable growers, as well as private companies and researchers for both economic and environmental reasons. The foliar application of vegetal and seaweed extract-based biostimulants, in particular SwE and LDPH enhanced plant growth, and productivity especially under sub-optimal N regimens, and to a lesser extent; at 20 and 30 kg $N \cdot ha^{-1}$. The foliar application of SwE and LDPH was effective in supporting better physiological and biochemical status in terms of the SPAD index, chlorophyll and carotenoids content leading to a higher agronomical performance. Interestingly, the leaf quality traits of baby lettuce leaf can be improved by biostimulation action, especially with LDPH which delivered leaves with high antioxidant activity and total ascorbic acid as well as low nitrate content. The results of the current experiment highlight the benefit of using vegetal and seaweed extract-based biostimulants in baby lettuce to improve productivity under both optimal and especially suboptimal N regimens, bringing benefits to farmers and to the environment.

Author Contributions: Conceptualization, M.M.; methodology, I.D.M., E.C.; software, L.O.; validation, M.M., I.D.M., E.C. and L.O.; formal analysis, L.O.; investigation, I.D.M.; resources, M.M.; data curation, I.D.M. and L.O.; writing—original draft preparation, I.D.M.; writing—review and editing, Y.R.; visualization, M.G. and G.C.; supervision, G.C. and M.M.; project administration, M.M.; funding acquisition, M.M.

References

1. Colonna, E.; Rouphael, Y.; Barbieri, G.; De Pascale, S. Nutritional quality of leafy vegetables harvested at two light intensities. *Food Chem.* **2016**, *199*, 702–710. [CrossRef] [PubMed]
2. Skovgaard, M.; Renjel Encinas, S.; Jensen, O.C.; Andersen, J.H.; Condarco, G.; Jørs, E. Pesticide residues in commercial lettuce, onion, and potato samples from Bolivia—A threat to public health? *Environ. Health Insights* **2017**, *11*, 1–8. [CrossRef] [PubMed]
3. De Pascale, S.; Rouphael, Y.; Colla, G. Plant biostimulants: Innovative tool for enhancing plant nutrition in organic farming. *Eur. J. Hortic. Sci.* **2017**, *82*, 277–285. [CrossRef]
4. Rouphael, Y.; Franken, P.; Schneider, C.; Schwarz, D.; Giovannetti, M.; Agnolucci, M.; De Pascale, S.; Bonini, P.; Colla, G. Arbuscular mycorrhizal fungi act as biostimulants in horticultural crops. *Sci. Hortic.* **2015**, *196*, 91–108. [CrossRef]
5. Rouphael, Y.; Kyriacou, M.C.; Petropoulos, S.A.; De Pascale, S.; Colla, G. Improving vegetable quality in controlled environments. *Sci. Hortic.* **2018**, *234*, 275–289. [CrossRef]
6. Rouphael, Y.; Spíchal, L.; Panzarová, K.; Casa, R.; Colla, G. High-throughput plant phenotyping for developing novel biostimulants: From lab to field or from field to lab? *Front. Plant Sci.* **2018**, *9*, 1197. [CrossRef] [PubMed]
7. Colla, G.; Nardi, S.; Cardarelli, M.; Ertani, A.; Lucini, L.; Canaguier, R.; Rouphael, Y. Protein hydrolysates as biostimulants in horticulture. *Sci. Hortic.* **2015**, *196*, 28–38. [CrossRef]
8. Colla, G.; Hoagland, L.; Ruzzi, M.; Cardarelli, M.; Bonini, P.; Canaguier, R.; Rouphael, Y. Biostimulant action of protein hydrolysates: Unraveling their effects on plant physiology and microbiome. *Front. Plant Sci.* **2017**, *8*, 2202. [CrossRef]
9. Rouphael, Y.; Colla, G. Synergistic biostimulatory action: Designing the next generation of plant biostimulants for sustainable agriculture. *Front. Plant Sci.* **2018**, *9*. [CrossRef]
10. Chiaiese, P.; Corrado, G.; Colla, G.; Kyriacou, M.C.; Rouphael, Y. Renewable sources of plant biostimulation: Microalgae as a sustainable means to improve crop performance. *Front. Plant Sci.* **2018**, *9*, 1782. [CrossRef]
11. Di Stasio, E.; Rouphael, Y.; Colla, G.; Raimondi, G.; Giordano, M.; Pannico, A.; El-Nakhel, C.; De Pascale, S. The Influence of *Ecklonia maxima* seaweed extract on growth, photosynthetic activity and mineral composition of *Brassica rapa* L. subsp. sylvestris under nutrient stress conditions. *Eur. J. Hortic. Sci.* **2017**, *82*, 286–293. [CrossRef]

12. Rouphael, Y.; De Micco, V.; Arena, C.; Raimondi, G.; Colla, G.; De Pascale, S. Effect of *Ecklonia maxima* seaweed extract on yield, mineral composition, gas exchange and leaf anatomy of zucchini squash grown under saline conditions. *J. Appl. Phycol.* **2017**, *29*, 459–470. [CrossRef]
13. Ertani, A.; Cavani, L.; Pizzeghello, D.; Brandellero, E.; Altissimo, A.; Ciavatta, C.; Nardi, S. Biostimulant activity of two protein hydrolyzates in the growth and nitrogen metabolism of maize seedlings. *J. Plant Nutr. Soil Sci.* **2009**, *172*, 237–244. [CrossRef]
14. Ertani, A.; Schiavon, M.; Muscolo, A.; Nardi, S. Alfalfa plant-derived biostimulant stimulate short-term growth of salt stressed *Zea mays* L. plants. *Plant Soil* **2013**, *64*, 145–158. [CrossRef]
15. Botta, A. Enhancing plant tolerance to temperature stress with amino acids: An approach to their mode of action. *Acta Hortic.* **2013**, *1009*, 29–35. [CrossRef]
16. Lucini, L.; Rouphael, Y.; Cardarelli, M.; Canaguier, R.; Kumar, P.; Colla, G. The effect of a plant-derived biostimulant on metabolic profiling and crop performance of lettuce grown under saline conditions. *Sci. Hortic.* **2015**, *182*, 124–133. [CrossRef]
17. Lucini, L.; Rouphael, Y.; Cardarelli, M.; Bonini, P.; Baffi, C.; Colla, G. A vegetal biopolymer-based biostimulant promoted root growth in melon while triggering brassinosteroids and stress-related compounds. *Front. Plant Sci.* **2018**, *9*, 472. [CrossRef]
18. Rouphael, Y.; Colla, G.; Giordano, M.; El-Nakhel, C.; Kyriacou, M.C.; De Pascale, S. Foliar applications of a legume-derived protein hydrolysate elicit dose dependent increases of growth, leaf mineral composition, yield and fruit quality in two greenhouse tomato cultivars. *Sci. Hortic.* **2017**, *226*, 353–360. [CrossRef]
19. Rouphael, Y.; Giordano, M.; Cardarelli, M.; Cozzolino, E.; Mori, M.; Kyriacou, M.; Bonini, P.; Colla, G. Plant and seaweed-based extracts increase yield but differentially modulate nutritional quality of greenhouse spinach through biostimulant action. *Agronomy* **2018**, *8*, 126. [CrossRef]
20. Sestili, F.; Rouphael, Y.; Cardarelli, M.; Pucci, A.; Bonini, P.; Canaguier, R.; Colla, G. Protein hydrolysate stimulates growth in tomato coupled with N-dependent gene expression involved in N assimilation. *Front. Plant Sci.* **2018**, *9*, 1233. [CrossRef]
21. Carillo, P.; Colla, G.; El-Nakhel, C.; Bonini, P.; D'Amelia, L.; Dell'Aversana, E.; Pannico, A.; Giordano, M.; Sifola, M.I.; Kyriacou, M.C.; et al. Biostimulant application with a tropical plant extract enhances *Corchorus olitorius* adaptation to sub-optimal nutrient regimens by improving physiological parameters. *Agronomy* **2019**, *9*, 249. [CrossRef]
22. Carillo, P.; Colla, G.; Fusco, G.M.; Dell'Aversana, E.; El-Nakhel, C.; Giordano, M.; Pannico, A.; Cozzolino, E.; Mori, M.; Reynaud, H.; et al. Morphological and physiological responses induced by protein hydrolysate-based biostimulant and nitrogen rates in greenhouse spinach. *Agronomy* **2019**, *9*, 450. [CrossRef]
23. Caruso, G.; Stoleru, V.; De Pascale, S.; Cozzolino, E.; Pannico, A.; Giordano, M.; Teliban, G.; Cuciniello, A.; Rouphael, Y. Production, leaf quality and antioxidants of perennial wall rocket as affected by crop cycle and mulching type. *Agronomy* **2019**, *9*, 194. [CrossRef]
24. Caruso, G.; De Pascale, S.; Cozzolino, E.; Cuciniello, A.; Cenvinzo, V.; Bonini, P.; Colla, G.; Rouphael, Y. Yield and nutritional quality of Vesuvian Piennolo tomato PDO as affected by farming system and biostimulant application. *Agronomy* **2019**, *9*, 505. [CrossRef]
25. Schiavon, M.; Ertani, A.; Nardi, S. Effects of an alfalfa protein hydrolysate on the gene expression and activity of enzymes of the tricarboxylic acid (TCA) cycle and nitrogen metabolism in *Zea mays* L. *J. Agric. Food Chem.* **2008**, *56*, 11800–11808. [CrossRef] [PubMed]
26. Matsumiya, Y.; Kubo, M. Soybean peptide: Novel plant growth promoting peptide from soybean. In *Soybean and Nutrition*; El-Shemy, H., Ed.; InTech Europe Publisher: Rijeka, Croatia, 2011; pp. 215–230. [CrossRef]
27. Colla, G.; Rouphael, Y.; Canaguier, R.; Svecova, E.; Cardarelli, M. Biostimulant action of a plant-derived protein hydrolysate produced through enzymatic hydrolysis. *Front. Plant Sci.* **2014**, *5*, 448. [CrossRef] [PubMed]
28. Colla, G.; Rouphael, Y.; Lucini, L.; Canaguier, R.; Stefanoni, W.; Fiorillo, A.; Cardarelli, M. Protein hydrolysate-based biostimulants: Origin, biological activity and application methods. *Acta Hortic.* **2016**, *1148*, 27–34. [CrossRef]
29. Battacharyya, D.; Babgohari, M.Z.; Rathor, P.; Prithiviraj, B. Seaweed extracts as biostimulants in horticulture. *Sci. Hortic.* **2015**, *196*, 39–48. [CrossRef]

30. Sady, W.; Smoleń, S. *The Influence of Pentakeep V and Nitrogen Fertilization on the Yield and Biological Quality of Carrot and Spinach Crop*; Final Report for Cosmo Oil Co., Ltd.; Cosmo Oil Co., Ltd.: Tokio, Japan, 2007.
31. Smoleń, S.; Sady, W. The influence of nitrogen fertilization and Pentakeep V application on contents of nitrates in carrot. *Acta Hortic. Regiotec.* **2009**, *12*, 221–223.
32. Smoleń, S.; Sady, W.; Wierzbińska, J. The effect of plant biostimulation with 'Pentakeep V' and nitrogen fertilization on yield, nitrogen metabolism and quality of spinach. *Acta Sci. Pol. Hortorum Cultus* **2010**, *9*, 25–36.
33. Paul, K.; Sorrentino, M.; Lucini, L.; Rouphael, Y.; Cardarelli, M.; Bonini, P.; Reynaud, H.; Canaguier, R.; Trtílek, M.; Panzarová, K.; et al. Understanding the biostimulant action of vegetal-derived protein hydrolysates by high-throughput plant phenotyping and metabolomics: A case study on tomato. *Front. Plant Sci.* **2019**, *10*, 47. [CrossRef] [PubMed]
34. Paul, K.; Sorrentino, M.; Lucini, L.; Rouphael, Y.; Cardarelli, M.; Bonini, P.; Miras Moreno, M.B.; Reynaud, H.; Canaguier, R.; Trtílek, M.; et al. A combined phenotypic and metabolomic approach for elucidating the biostimulant action of a plant-derived protein hydrolysate on tomato grown under limited water availability. *Front. Plant Sci.* **2019**, *10*, 493. [CrossRef] [PubMed]
35. Re, R.; Pellegrini, N.; Proteggente, A.; Pannala, A.; Yang, M.; Rice-Evans, C. Antioxidant activity applying an improved abts radical cation decolorization assay. *Free Rad. Biol. Med.* **1999**, *26*, 1231–1237. [CrossRef]
36. Fogliano, V.; Verde, V.; Randazzo, G.; Ritieni, A. Method for measuring antioxidant activity and its application to monitoring the antioxidant capacity of wines. *J. Agric. Food Chem.* **1999**, *47*, 1035–1040. [CrossRef] [PubMed]
37. Lichtenthaler, H.K.; Buschmann, C. *Chlorophylls and Carote-Noids–Measurement and Characterisation by UV-VIS.–Cur-rent Protocols in Food Analytical Chemistry (CPFA), (Sup-plement 1), F4.3.1–F4.3.8*; John Wiley: New York, NY, USA, 2001. [CrossRef]
38. Sah, R.N. Nitrate-nitrogen determination: A critical review. *Commun. Soil Sci. Plant Anal.* **1994**, *25*, 2841–2869. [CrossRef]
39. Kampfenkel, K.; Van Montagu, M.; Inzé, D. Extraction and determination of ascorbate and dehydroascorbate from plant tissue. *Anal. Biochem.* **1995**, *225*, 165–167. [CrossRef]
40. Colla, G.; Cardarelli, M.; Bonini, P.; Rouphael, Y. Foliar applications of protein hydrolysate, plant and seaweed extracts increase yield but differentially modulate fruit quality of greenhouse tomato. *HortScience* **2017**, *52*, 1214–1220. [CrossRef]
41. Trevisan, S.; Manoli, A.; Quaggiotti, S.A. Novel biostimulant, belonging to protein hydrolysates, mitigates abiotic stress effects on maize seedlings grown in hydroponics. *Agronomy* **2019**, *9*, 28. [CrossRef]
42. Szczepanek, M.; Siwik-Ziomek, A. P and K accumulation by rapeseed as affected by biostimulant under different NPK and S fertilization doses. *Agronomy* **2019**, *9*, 477. [CrossRef]
43. Kyriacou, M.C.; Rouphael, Y.; Colla, G.; Zrenner, R.M.; Schwarz, D. Vegetable grafting: The implications of a growing agronomic imperative for vegetable fruit quality and nutritive value. *Front. Plant Sci.* **2017**, *8*, 741. [CrossRef]
44. Jannin, L.; Arkoun, M.; Etienne, P.; Laîné, P.; Goux, D.; Garnica, M.; Fuentes, M.; Francisco, S.S.; Baigorri, R.; Cruz, F. *Brassica napus* growth is promoted by *Ascophyllum nodosum* (L.) Le Jol. seaweed extract: Microarray analysis and physiological characterization of N, C, and S metabolisms. *J. Plant Growth Regul.* **2013**, *32*, 31–52. [CrossRef]
45. Ertani, A.; Schiavon, M.; Nardi, S. Transcriptome-wide identification of differentially expressed genes in *Solanum Lycopersicon* L. in response to an alfalfa-protein hydrolysate using microarrays. *Front. Plant Sci.* **2017**, *8*, 1159. [CrossRef] [PubMed]
46. Tsouvaltzis, P.; Koukounaras, A.; Siomos, S.A. Application of amino acids improves lettuce crop uniformity and inhibits nitrate accumulation induced by the supplemental inorganic nitrogen fertilization. *Int. J. Agric. Biol.* **2014**, *16*, 951–955. [CrossRef]
47. Colla, G.; Kim, H.J.; Kyriacou, M.C.; Rouphael, Y. Nitrate in fruits and vegetables. *Sci. Hortic.* **2018**, *237*, 231–238. [CrossRef]
48. Ali, M.; Cheng, Z.H.; Hayat, S.; Ahmad, H.; Ghani, M.I.; LIU, T. Foliar spraying of aqueous garlic bulb extract stimulates growth and antioxidant enzyme activity in eggplant (*Solanum melongena* L.). *J. Integr. Agric.* **2019**, *18*, 1001–1013. [CrossRef]

49. Tadros, M.J.; Omari, H.J.; Turk, M.A. The morphological, physiological and biochemical responses of sweet corn to foliar application of amino acids biostimulants sprayed at three growth stages. *Aust. J. Crop Sci.* **2019**, *13*, 412–417. [CrossRef]
50. Khanam, U.K.S.; Oba, S.; Yanase, E.; Murakami, Y. Phenolic acids, flavonoids and total antioxidant capacity of selected leafy vegetables. *J. Fun. Foods* **2012**, *4*, 979–987. [CrossRef]
51. Kyriacou, M.C.; Rouphael, Y.; Di Gioia, F.; Kyratzis, A.; Serio, F.; Renna, M.; De Pascale, S.; Santamaria, P. Micro-scale food production and the rise of microgreens. *Trends Food Sci. Technol.* **2016**, *57*, 103–115. [CrossRef]
52. Kyriacou, M.C.; El-Nakhel, C.; Graziani, G.; Pannico, A.; Soteriou, G.A.; Giordano, M.; Ritieni, A.; De Pascale, S.; Rouphael, Y. Functional quality in novel food sources: Genotypic variation in the nutritive and phytochemical composition of thirteen microgreens species. *Food Chem.* **2019**, *277*, 107–118. [CrossRef]
53. Kyriacou, M.C.; Rouphael, Y. Towards a new definition of quality for fresh fruits and vegetables. *Sci. Hortic.* **2018**, *234*, 463–469. [CrossRef]
54. Rouphael, Y.; Kyriacou, M.C. Enhancing quality of fresh vegetables through salinity eustress and biofortification applications facilitated by soilless cultivation. *Front. Plant Sci.* **2018**, *9*, 1254. [CrossRef] [PubMed]
55. Wang, Z.H.; Li, S.X.; Malhi, S. Effects of fertilization and other agronomic measures on nutritional quality of crops. *J. Sci. Food Agric.* **2008**, *88*, 7–23. [CrossRef]
56. Vasantharaja, R.; Stanley Abraham, L.; Inbakandan, D.; Thirugnanasambandam, R.; Senthilvelan, T.; Ayesha Jabeen, S.K.; Prakash, P. Influence of seaweed extracts on growth, phytochemical contents and antioxidant capacity of cowpea (*Vigna unguiculata* L. Walp). *Biocatal. Agric. Biotechnol.* **2019**, *17*, 589–594. [CrossRef]

12

Morphological and Biochemical Responses of *Glycine max* (L.) Merr. to the use of Seaweed Extract

Sławomir Kocira [1,*], Agnieszka Szparaga [2], Maciej Kuboń [3], Ewa Czerwińska [4] and Tomasz Piskier [2]

1 Department of Machinery Exploitation and Management of Production Processes, University of Life Sciences in Lublin, Akademicka 13, 20-950 Lublin, Poland
2 Department of Agrobiotechnology, Koszalin University of Technology, Racławicka 15–17, 75-620 Koszalin, Poland; agnieszka.szparaga@tu.koszalin.pl (A.S.); tomasz.piskier@tu.koszalin.pl (T.P.)
3 Institute of Agricultural Engineering and Informatics, Agricultural University Krakow, Balicka 116B, 30-149 Krakow, Poland; maciej.kubon@ur.krakow.pl
4 Department of Biomedical Engineering, Koszalin University of Technology, Śniadeckich 2, 75-453 Koszalin, Poland; ewa.czerwinska@tu.koszalin.pl
* Correspondence: slawomir.kocira@up.lublin.pl

Abstract: Currently, modern agriculture aims to improve the quantity and quality of crop yield, while minimizing the negative impact of treatments on the natural environment. One of the methods to increase plant yield and quality, especially after the occurrence of both abiotic or biotic stress factors, is the application of biostimulants. The aim of the study was to determine the effect of *Ecklonia maxima* extract on plant growth, and the yield, nutritional, and nutraceutical properties of soybean seeds. A field experiment was conducted in three growing seasons (2014–2016). Soybean seeds of Atlanta cultivar were sown in the third 10-day period of April. *Ecklonia maxima* extract was applied in the form of single or double, spraying in the concentrations of 0.7% and 1.0%. Determinations were conducted for: biometric traits, seed yield, seed number, thousand seeds weight, contents of lipids, and proteins in seeds. Further analyses included the contents of total polyphenols, flavonoids, anthocyanins, and reducing power. The number of seaweed extract applications and its concentration modified biometric traits, yield, and quality of crop, while also also altering the nutraceutical and antioxidative potential of soybean. The application of this preparation improved the growth and yield of soybean without any negative effect on the nutritive value of seeds.

Keywords: antioxidant activity; growth; nutrients; nutraceutical potential; soybean; yield

---

1. Introduction

Soya (*Glycine max* (L.) Merrill.) is one of the most important leguminous plants that are cultivated around the world because it is a precious source of both protein and fat [1,2]. Its use for production of food, oil, and fodder means that the demand for this plant is continuously growing [3]. Due to its broad use, it is called a "wonderful crop" [4]. However, this plant is sensitive to unfavourable climatic conditions [5]. Thus, to ensure its effective protection against biotic and abiotic factors, it is recommended to use it in the cultivation of biostimulants, which may improve the biochemical, morphological, and physiological processes that take place in a plant [6–8].

Biostimulants as plant-growth promoters were defined for the first time in the world literature by Kaufman [9]. In turn, Du Jardin [10] claims that "a plant biostimulant is any substance or microorganism, in the form in which it is supplied to the user, applied to plants, seeds or the root environment with the intention to stimulate natural processes of plants to benefit their nutrient use efficiency and/or their tolerance to abiotic stress, regardless of its nutrients content, or any combination of such substances and/or microorganisms intended for this use". The European Biostimulants Industry Council (EBIC) was established in order to develop legal regulations regarding the registration of biostimulants, according to the specificity of their action. However, currently, the registration of these preparations is still based on legal regulations that are set for fertilizers and plant protection products [11–13]. According to Colla et al. [14,15] and Battacharyya et al. [16], among the entire group of those preparations, extracts from seaweed and protein hydrolysates constitute the two most-important categories of substances of natural biostimulants. According to Aguilar, brown algae are the most often used in agriculture [17]. The most popular are, among others, *Ecklonia maxima* (Osbeck) Papenfuss and *Ascophyllum nodosum* (L.) Le Jolis. Brown-algae extracts include various phytohormones, such as auxins, gibberellins, cytokininis, abscisic acid, ethylene, betaine, and polyamins, and other growth promoters as well as trace elements and microelements [18]. Seaweeds also include a varied range of organic compounds, among others, aminoacids, such as asparaginic acid, glutamine acid, and alanine. While, alginic acid, laminarin, and mannitol constitute almost half of the total content of carbohydrates in such biostimulating preparations. Seaweeds also contain a wide range of vitamins that can be used by plants, such as C, B2, B12, D3, E, K, niacin, panthotenic acid, and folic acid. Although, vitamin A does not occur in algae extracts, the presence of its precursor—carotene and another possible precursor, fucoxnathin—was determined [18–20].

According to Ecoforce [21] and Van Oosten et al. [22], seaweed extracts that are used as biostimulants increase the yield and its quality in two ways: a. they stimulate hormone synthesis, influence absorption, and translocation of nutrients; b. they condition the soil, improving its ability to retain moisture and stimulate the activity of favourable microorganisms. Medjdoub estimates that the use of biostimulants, which include extracts from seaweeds, has greater meaning in agriculture. That is because plant growth and development is controlled by plant hormones, which directly or indirectly control the course of various physiological reactions and their integration with the total metabolism [23]. Many studies on seaweed extract indicate that they may increase: a. plant growth, b. activity of photosynthesis, c. resistance to fungi, bacteria and viruses, d. tolerance to ground frost, drought, and salt content, and e. yield and productivity of many cultivations [24–26], mainly by activation of protective mechanisms of plants [27]. Foliar application of biostimulants that are based on seaweeds is an agrotechnical treatment that brought many advantages in numerous cultivations, including grapevine, watermelon, strawberry, apple, tomato, spinach, onion, bean, pepper, carrot, potato, wheat, corn, barley, rice, and turf grass. The results show that plants treated with lower concentrations of extract indicated a stronger growth, higher yield, and higher mineral and nutritive elements content relative to the control [28–33]. Positive reactions also included an improved flowering and fructification ability, product quality and efficiency, and resistance to abiotic stress [31,34,35]. Studies that were performed on a wide group of crops proved that the application of sea-algae-based biostimulants stimulates the primary and secondary metabolisms in plants through the absorption and assimilation of nutrients [36–44]. Growth of productivity of crops induced by the use of such biostimulants in optimal and suboptimal conditions may be related to several direct and indirect mechanisms, including the stimulation of enzymatic activities that are related to carbon, nitrogen metabolism, Krebs cycle, and glycolysis. Such use may also induce activity similar to hormones, especially the one that is assigned to auxins and gibberellins, and improve the nutrition of treated plants by the modulation of the root system [14–16,45].

However, other results indicate that the application of such biostimulants, despite its numerous advantages, like faster germination and earlier growth [46,47], may inhibit the growth and development of many plants. This calls for greater care in the use of seaweeds extracts [48]. Therefore, the inhibition of plant growth after application of biostimulants is a potential problem in plant production. The concentration of these products is an important factor in this regard [49]. This issue may be caused directly by elements that are included in the extract [50], or it might be a consequence of modifications in the regular physiological growth of the plant [51,52]. Improvement of commercial-product formulas, and knowing the mechanisms of active substances in plants and their persistence, should mitigate such negative effects. The exact effect of various elements (e.g., nutrients, betaines, oligomers, polymers) from seaweeds on improving plant growth, vigour, and fractioning of extracts is not fully known. Therefore, a detailed analysis of composition and the fractioning of elements on plant physiology, together with a better ability to monitor the impact of such extracts on those variables and on the expression of genes, would shed light on some of the performance mechanisms [53,54].

Among biostimulants that include seaweeds, Kelpak is particularly interesting. It is extracted from the species *Ecklonia maxima* Osbeck and then harvested along the shores of Africa. Kelpak contains phytohormones, such as auxins (11 mg dm^{-3}) and cytokinins (0.031 mg dm^{-3}), and also alginates (1.5 g L^{-1}), amino acids (total 441.3 mg 100 g^{-1}), mannitol (2261 mg L^{-1}), neutral sugars (1.08 g L^{-1}), and small amounts of macro- (mean composition: N 0.09%, P 90.7 mg kg^{-1}, K 7163.3 mg kg^{-1}, Ca 190.4 mg kg^{-1}, Mg 337.2 mg kg^{-1}, Na 1623.7 mg kg^{-1}) and microelements (mean composition: Mn 17.3 mg kg^{-1}, Fe 40.7 mg kg^{-1}, Cu 13.5 mg kg^{-1}, Zn 17.0 mg kg^{-1}, B 33.0 mg kg^{-1}) [55,56]. Moreira Sisalema [57] indicates that Kelpak, due to the unique extraction process, contains a very-high auxins-to-cytokinins ratio. The dominance of auxins stimulates the dynamic growth of plant roots, which increases the absorption of indispensable nutrients and minerals, and consequently, of plant production. Activity of this biostimulant may also increase plant resistance to drought and enable faster plant regeneration after water stress. Cals [58] shows that Kelpak should be used in leguminous plants before flowering in the dose of 2.0 L ha^{-1}, in relation to the condition of plant nutrition. The concentration of these preparations in foliar applications is usually from 0.2% to 1% and it rarely exceeds these concentrations. Depending on the specific cultivation, cultivar, and climatic conditions, farmers are usually recommended to use these biostimulants in the form of two-week spraying in the stage of intense plant growth [28,59,60].

Because of the varied reactions of many plants to the application of biostimulants from seaweeds, and due to small number of studies on their influence in soya cultivation, a three-year log field study was carried out. Its main aim was to assess the impact of the application of *Ecklonia maxima* extract (Kelpak) on plant growth, yield size, and the quality and nutraceutical potential of genetically non-modified seeds of Atlanta cultivar. The initial hypothesis was that the introduction of agrotechnical treatment to soya cultivation in the form of plant spraying with Kelpak preparation would modify the plant growth, yield, and chemical composition of soya seeds. To test this hypothesis, the yield and structural elements of soya cropping were assessed, as well as the protein, fat, and anti-oxidant potential of seeds in relation with the applied doses and concentrations of the tested preparation. In order to know the morphological and biochemical plant reaction on seaweed extract performance, the responses of treated and untreated control plants in the same environmental conditions were compared. It was expected that the observation of plant reaction would considerably increase knowledge regarding the manner of seaweed extract performance, particularly in leguminous plants cultivation, which are sensitive to biotic and abiotic stresses. The present work is a concrete step towards broadening the understanding of the advantages of the application in agricultural practice of *Ecklonia maxima* seaweed extracts for the improvement of the size and quality of crops.

2. Materials and Methods

2.1. Plant Materials and Growth Conditions

The field experiment was carried out in 2014–2016 in Perespa (50°66′ N; 23°63′ E, Poland). It was established in a randomized block design in four replications on experimental plots with an area of 10 m^2. Soybean was cultivated on the soil belonging to the Gleyic Phaeozems, which was characterized by alkaline pH (pH in 1M KCl: 7.4–7.5). The soil content in the assimilable nutrients was at the medium level, as follows: P (12.6–14.2 mg P_2O_5 in 100 g soil), K (15.3–17.1 mg K_2O in 100 g soil), and Mg (6.2–6.8 mg Mg in 100 g soil). Each year, winter wheat was used as a forecrop. Soybean seeds (*Glycine max* (L.) Merr.) of Atlanta cultivar (Agroyoumis, Poland) were sown on the 25 of April in 2014 and 2015, and 23 of April in 2016 in rows every 30 cm at a raw spacing of

3.5 cm. The weeds were mechanically and manually removed. No pesticides were used (pests did not exceed the thresholds of harmfulness). In the growing season, the plants were sprayed with biostimulant (water solutions) that was based on the *Ecklonia maxima* extract (Kelpak). Kelpak contains phytohormones (mostly auxins 11 mg kg^{-1}, cytokinins 0.03 mg kg^{-1}, and auxin: cytokinin ratio 367:1), carbohydrates (16.9 g kg^{-1}), amino acids (2.5 g kg^{-1}), vitamin B1 (0.9 mg kg^{-1}), B2 (0.1 mg kg^{-1}), C (20 mg kg^{-1}), and E (0.7 mg kg^{-1}). The elemental profile of the biostimulant is: N 3.6 g kg^{-1}, P 8.2 g kg^{-1}, K 7.2 g kg^{-1}, Ca 0.8 g kg^{-1}, Mg 0.2 g kg^{-1}, Fe 13.6 mg kg^{-1}, Mn 8.4 mg kg^{-1}, B 0.24 mg kg^{-1}, Zn 4.2 mg kg^{-1}, and Cu 0.2 mg kg^{-1} [14,61]. The scheme of doses, developmental stages of plants, and terms of spraying are presented in Table 1.

Table 1. Plant developmental stages and dates of biostimulants application.

Biostimulant	Number of Sprays and Plant Developmental Stages in Which the Biostimulants were Applied	Concentration	Volume of Wrking Solution/ Working Pressure	Date of Spraying		
				2014	2015	2016
Kelpak SL	Single spraying BBCH 13-15 (LSS)	0.7%	300 l·ha^{-1}/ 0.30 MPa	June 21	June 20	June 7
	Double spraying BBCH 13-15, BBCH 61 (LDS)	0.7%		June 21, July 5	June 20, July 3	June 7, June 23
	Single spraying BBCH 13-15 (HSS)	1.0%		June 21	June 20	June 7
	Double spraying BBCH 13-15, BBCH 61 (HDS)	1.0%		June 21, July 5	June 20, July 3	June 7, June 23

Plants sprayed with water served as the control. The biostimulant (or water) was sprayed with a GARLAND FUM 12B battery field sprayer (Lechler LU 120-03) at a pressure of 0.30 MPa, using 300 l liquid per hectare. The average temperature and rainfalls in the soybean growing season are shown in Table 2.

Table 2. Temperature (T) and rainfalls during the soybean growing season 2014–2016.

Month	Year						Average from 2002 to 2013	
	2014		2015		2016			
	T (°C) Average (min/max)	Rainfall (mm)	T (°C) Average (min/max)	Rainfall (mm)	T (°C) Average (min/max)	Rainfall (mm)	T (°C)	Rainfall (mm)
IV	9.4 (−6.0/22.7)	36.5	8.2 (−1.7/24.3)	30.1	9.2 (−1.2/22.6)	68.4	8.5	41.2
V	13.7 (0.5/27.7)	208.3	12.7 (1.5/24.9)	108.6	13.8 (2.6/26.7)	61.3	12.7	63.4
VI	16.1 (6.7/28.9)	67.1	17.4 (6.6/30.5)	14.1	18.1 (4.2/31.5)	97.1	17.7	68.6
VII	20.3 (10.0/31.0)	104.2	19.6 (8.4/33.4)	59.2	19.5 (8.8/31.2)	107.6	18.9	79.1
VIII	18.2 (6.3/34.0)	115.4	21.6 (5.6/35.5)	23.4	18.2 (7.1/30.7)	95.3	19.4	71.8
IX	13.7 (3.7/25.8)	89.4	15.1 (4.2/34.5)	137.6	15.2 (1.6/28.7)	41.2	14.1	69.2
Average/Total	15.1	620.9	15.8	373.0	17.1	470.9	15.2	393.3

2.2. Plant Growth, Yield, and Nutritional Value Determination

After the pods have matured, when the seeds have obtained a typical color and hardness (BBCH 89), the plant height, the internode number on the main shoot, and the first pod height were recorded. In addition, after harvesting, the number of pods per plant, the number of seeds per 1 m^2, the weight of seeds, and the weight of thousand seeds were determined. Subsequently, the seeds were dried and then grinded. The flour was used for further analysis.

Protein content was determined with the Kjeldahl method, whereas the content of lipids was based on the acid hydrolysis method [62].

2.3. *Nutraceutical Potential*

Seed extract was prepared following the methodology that was proposed by Świeca et al. [63]. The ground soybean seeds were extracted with a mixture of acetone, water, and hydrochloric acid (70:29:1; v/v/v). Afterwards, the samples were centrifuged for 10 min (6800× *g*) and the resulting supernatant was collected and then used for further analyses.

2.3.1. Phenolics Determination

Determination of Total Phenolic Compounds (TPC)

The content of total phenolic compounds (TPC) was determined with the method of Singleton and Rossi using the Folin–Ciocalteau reagent [64]. Absorbance of the samples was measured with a UV-vis spectrophotometer at a wavelength of 725 nm, then TPC was computed and expressed as gallic acid equivalents (GAE) in mg per g of dry matter (DM).

Determination of Flavonoid Content (TFC)

The total content of flavonoids was determined acc. to the method that was presented by Lamaison and Carnet [65]. The prepared soybean extract was mixed with a methanolic solution of $AlCl_3 \times 6H_2O$. After incubation, the absorbance was measured with a UV-vis spectrophotometer at the wavelength of 430 nm. The total flavonoid content was expressed as quercetin equivalents (QE) in mg per g DM.

Determination of Anthocyanins (TAC)

Using the method that was proposed by Fuleki and Francis using potassium chloride and sodium acetate buffer at two pH values (1.0 and 4.5), the content of anthocyanins was assayed [66]. After 15 min, absorbance of each sample was measured at wavelengths of 520 nm and 700 nm. Subsequently, anthocyanin content was calculated as cynidin-3-glucoside equivalents (Cy3-GE) in mg per g DM.

2.3.2. Reducing Power

Reducing power was measured following the method that was provided by Pulido et al. [67]. The soybean extract was mixed with a phosphate buffer (200 mM, pH 6.6) and 1% solution of $K_3[Fe(CN_6)]$. Next, the samples were incubated at 50 °C for 20 min. The reaction was stopped with trichloroacetic acid and the samples were centrifuged (6800× *g*, 10 min). The resulting supernatant was mixed with distilled water and $FeCl_3$. Afterwards, absorbance was measured at the wavelength of 700 nm. Reducing power was expressed as Trolox equivalents in mg per g DM.

2.4. *The Index of Biostimulant Effect*

The index of biostimulant effect (ABT-C) was determined as the difference between the mean result that was obtained after biostimulant application (ABT) and the control (C), which enabled the evaluation of the effect of biostimulant type on the analyzed traits. The mean value for each treatment has been obtained clustering the means of lower concentration single spraying (LSS), lower concentration double spraying (LDS), single application of the higher concentration (HSS), and higher concentration double spraying (HDS) from different years all together. The standard deviation value (SD) was determined for all reported mean values of ABT-C [5].

2.5. Statistical Analysis

The obtained results were statistically elaborated with Statistica 13 software (StatSoft, Inc.). The materials were collected over three seasons (2014–2016). Laboratory analyses were performed in triplicate. Normality of data distribution was assessed with the Shapiro–Wilk test. The significance of differences between the evaluated mean values was estimated with the Tukey test at a significance level of $p < 0.05$.

3. Results

3.1. Effect of Biostimulants on Biometric Traits

3.1.1. Plant Height

The single application of the higher concentration (HSS) of Kelpak biostimulant ensured better effects in increasing soybean plant height (increased by 35% as compared to the control) (Table 3). The highest plants were obtained in the growing season 2016 after their single spraying with the higher concentration of Kelpak. In contrast, the smallest plants were produced in the 2015 season and their height differed significantly from the values noted in seasons 2014 and 2016. The biostimulant increased the height of plants, which was indicated by a value of the Kelpak effect index (ABT-C) of 28.2 cm for this trait (Table 4).

Table 3. Effect of *Ecklonia maxima* extract (Kelpak) treatment on biometric traits of soybean (average from 2014-2016).

Parameters	*Ecklonia maxima* Treatment	Season			Average from 2014 to 2016
		2014	2015	2016	
Plant height (cm)	C	85.4^{a}	81.9^{a}	88.1^{a}	85.1^{a}
	LSS	114.9^{b}	107.7^{b}	112.6^{b}	111.7^{b}
	LDS	117.5^{b}	108.0^{b}	117.0^{b}	114.1^{b}
	HSS	118.8^{b}	106.4^{b}	120.0^{b}	115.0^{b}
	HDS	114.9^{b}	108.0^{b}	114.4^{b}	112.4^{b}
	AS	110.3^{b}	102.4^{a}	110.3^{b}	
Number of internodes in the main shoot	C	11.2^{a}	10.1^{a}	9.6^{a}	10.3^{a}
	LSS	10.4^{a}	8.6^{a}	10.2^{a}	9.7^{a}
	LDS	9.9^{a}	9.3^{a}	9.0^{a}	9.4^{a}
	HSS	10.0^{a}	9.8^{a}	10.3^{a}	10.0^{a}
	HDS	9.9^{a}	8.8^{a}	11.1^{a}	9.9^{a}
	AS	10.3^{b}	9.3^{a}	10.0^{ab}	
Location height of the first pod (cm)	C	12.5^{a}	11.1^{a}	11.7^{a}	11.7^{a}
	LSS	13.0^{a}	14.2^{a}	12.2^{a}	13.2^{ab}
	LDS	13.8^{a}	14.0^{a}	13.3^{a}	13.7^{b}
	HSS	12.0^{a}	12.5^{a}	12.2^{a}	12.2^{ab}
	HDS	13.0^{a}	12.7^{a}	13.3^{a}	13.0^{ab}
	AS	12.8^{a}	12.9^{a}	12.5^{a}	
Number of pods (per plant)	C	15.2^{a}	14.7^{a}	16.3^{a}	15.4^{a}
	LSS	20.9^{b}	22.5^{cd}	21.0^{b}	21.5^{bc}
	LDS	22.4^{b}	23.4^{d}	21.5^{b}	22.4^{c}
	HSS	19.9^{b}	21.4^{bc}	21.1^{b}	20.8^{b}
	HDS	20.4^{b}	20.3^{b}	20.8^{b}	20.5^{b}
	AS	19.8^{a}	20.4^{a}	20.1^{a}	

Abbreviations: C, control; LSS, lower concentration single spraying; LDS, lower concentration double spraying; HSS higher concentration single spraying; HDS, higher concentration double spraying. Means in the columns, concerning the selected traits, followed by different small letters are significantly different at $p < 0.05$.

Table 4. The index of biostimulant effect (ABT-C).

Parameters	Kelpak
Plant height (cm)	28.2
Number of nodes in the main shoot	−0.5
Location height of the first pod (cm)	1.3
Number of pods (per plant)	5.9
Number of seeds (per m^{-2})	622
Seed yield (t ha^{-1})	0.824
1000 seed weight (g 1000^{-1})	−10.7
Total protein (% DM)	0.35
Total fat (% DM)	−1.56
Total phenols (mg g^{-1} DM)	2.53
Total flavonoids (mg g^{-1} DM)	1.23
Anthocyanins (mg g^{-1} DM)	0.01
Reducing power (mg TE g^{-1} DM)	0.15

3.1.2. Number of Internodes in the Main Shoot

Internode number decreased regardless of the Kelpak concentration and the number of its applications, although the differences were insignificant (Table 3). The highest number of internodes on the main shoot was obtained in the first and third season. The highest number of internodes on the main shoot was obtained in the first season and it differed significantly from the number that was determined in 2015. The value of the ABT-C index computed for Kelpak was negative (Table 4).

3.1.3. Location Height of the First Pod

Biostimulant treatment increased the height of the first pod as compared to the control. Significant differences were observed between the double application of the lower concentration of Kelpak and the control (increased by 17%) (Table 3). The tallest heights of the first pods were observed in the 2015 season, however they did not significantly differ from the values that were reported in the two other seasons. Values of the ABT-C index demonstrate that the height of the first pod was larger with the application of Kelpak preparation (Table 4).

3.1.4. Number of Pods per Plant

Double foliar application of the lower concentration of Kelpak permitted achieving the highest number of pods per plant (increased by 45% as compared to the control) (Table 3). The study demonstrated that the mean number of pods determined in particular growing seasons was at a similar level and did not significantly differ among seasons. In turn, biostimulant increased the pod number per plant because the value of Kelpak effect index was 5.9 pods/plant after spraying with this preparation (Table 4).

3.2. Effect of Biostimulants on Soybean Yield

3.2.1. Number of Seeds

Double spraying soybean plants with the higher concentrations (HDS) of Kelpak had the largest effect on the increase in seed number per m2 (increased by 43% as compared to the control) (Table 5). The analysis of growing seasons demonstrated the largest value of this trait in 2016 and the smallest one in 2015 (lower by 5% than that noted in 2016). The application of seaweed extract increased this number, which was indicated by values of the ABT-C index that were calculated for this trait (Table 4).

Table 5. Effect of *Ecklonia maxima* extract (Kelpak) treatment on yield and nutritional properties of soybean (average from 2014–2016).

Parameters	*Ecklonia maxima* Treatment	Season			Average from 2014 to 2016
		2014	2015	2016	
Number of seeds (per m^{-2})	C	1793^{a}	1581^{a}	1907^{a}	1760^{a}
	LSS	2255^{b}	2210^{b}	2337^{b}	2267^{b}
	LDS	2340^{b}	2376^{bc}	2406^{bc}	2374^{b}
	HSS	2344^{b}	2350^{b}	2401^{bc}	2365^{b}
	HDS	2466^{c}	2528^{c}	2576^{c}	2524^{c}
	AS	2240^{a}	2209^{a}	2326^{b}	
Seed yield (t ha^{-1})	C	3.267^{a}	2.664^{a}	3.262^{a}	3.064^{a}
	LSS	3.677^{b}	3.636^{b}	3.767^{b}	3.693^{b}
	LDS	3.805^{b}	3.876^{b}	3.852^{bc}	3.844^{b}
	HSS	3.758^{b}	3.874^{b}	3.907^{bc}	3.846^{b}
	HDS	4.137^{c}	4.171^{c}	4.198^{c}	4.169^{c}
	AS	3.729^{a}	3.644^{a}	3.797^{b}	
1000 seed weight (g)	C	182.2^{b}	168.5^{a}	171.0^{b}	173.9^{b}
	LSS	163.1^{a}	164.6^{a}	161.2^{a}	162.9^{a}
	LDS	162.6^{a}	163.2^{a}	160.1^{a}	161.9^{a}
	HSS	160.3^{a}	164.9^{a}	162.7^{a}	162.3^{a}
	HDS	167.8^{a}	165.0^{a}	163.1^{a}	165.3^{a}
	AS	167.2^{b}	165.2^{ab}	163.6^{a}	
Total protein (% DM)	C	36.8^{a}	46.5^{d}	35.9^{a}	39.7^{a}
	LSS	37.7^{b}	45.7^{c}	38.6^{d}	40.7^{a}
	LDS	37.7^{b}	47.4^{e}	36.3^{b}	40.5^{a}
	HSS	38.0^{b}	42.7^{b}	38.9^{d}	39.9^{a}
	HDS	39.1^{c}	40.9^{a}	38.1^{c}	39.4^{a}
	AS	37.9^{b}	44.6^{c}	37.6^{a}	
Total fat (% DM)	C	17.5^{d}	15.0^{d}	16.6^{c}	16.4^{b}
	LSS	14.5^{a}	15.5^{e}	14.5^{a}	14.8^{a}
	LDS	15.4^{b}	12.8^{a}	15.1^{b}	14.4^{a}
	HSS	15.5^{b}	13.8^{c}	15.1^{b}	14.8^{a}
	HDS	15.7^{c}	13.3^{b}	16.4^{c}	15.2^{ab}
	AS	15.7^{c}	14.1^{a}	15.5^{b}	

Abbreviations: C, control; LSS, lower concentration single spraying; LDS, lower concentration double spraying; HSS higher concentration single spraying; HDS, higher concentration double spraying. Means in the columns, concerning the selected traits, followed by different small letters are significantly different at $p < 0.05$.

3.2.2. Seed Yield

The most positive response of plants to the use of biostimulant was observed after double spraying with the higher concentration of Kelpak preparation, as indicated by their seed yield increase by 36% when compared to the control (Table 5). The highest mean seed yield for Atlanta cv. was obtained in 2016. In contrast, the seed yield of 2015 season turned out to be the lowest among the studied seasons (lower by 4% than that noted in 2016). Foliar application of Kelpak increased the seed yield of soybean of Atlanta cv., which was indicated by positive values of the ABT-C index that were calculated for this trait (Table 4).

3.2.3. Thousand Seed Weight

Foliar application of Kelpak decreased 1000 seed weight. Its lowest value was determined after double application of Kelpak in the lower concentration (decrease by 7% as compared to the control) (Table 5). The least decrease of 1000 seed weight was achieved after double plant spraying with the higher concentration of Kelpak biostimulant. The highest mean 1000 seed weight was reported in the 2014 growing season. The values of the biostimulant effect index calculated for this trait were negative, which points to the negative impact of Kelpak preparation on 1000 seed weight (Table 4).

3.3. Effect of Biostimulant on the Nutritional Properties

3.3.1. Total Protein in Soybean Seeds

Depending on concentration and number of applications, Kelpak increased or decreased the protein content in a dry matter of seeds. However, the statistical analysis demonstrated that differences in the effects of biostimulant on this trait were insignificant. Increased protein content was determined in seeds of plants single-sprayed with the lower concentrations of Kelpak (Table 5). Concerning growing seasons, the highest protein content of seeds was noted in 2015. Values of the ABT-C index that were calculated for this trait were positive for this preparations (Table 4).

3.3.2. Total Fat in Soybean Seeds

Regardless of the number of sprayings and concentration of biostimulant, its use decreased the fat content in dry matter of soybean seeds, with the greatest decrease (by 14% as compared to the control) being noted after double spraying the plants with the lower concentration of Kelpak (Table 5). In contrast, the smallest decrease in fat content of the seeds as compared to the control was determined after double spraying with the higher concentrations of Kelpak. The highest fat content of soybean seeds was noted in season 2014 and the lowest in 2015. The values of the ABT-C index that were calculated for this preparation were negative (Table 4), which is indicative of its negative effect on fat content in of Atlanta cv. soybean seeds.

3.4. Effect of Biostimulants on the Antioxidant Potential in Soybean Seeds

3.4.1. Total Phenolic Content

The use of Kelpak in soybean cultivation caused changes in contents of total polyphenols (TPC) in seeds (Table 6), which varied depending on both the number of applications and the concentration of this preparation. The use of the biostimulant based on *Ecklonia maxima* extract caused an increase in phenolics compounds content in soybean seeds. However, significant differences were only demonstrated in plants that were single-sprayed with 1% Kelpak (HSS). The TPC content that was determined for this combination was over twofold higher, when compared to the control combination. This nutraceutical property of soybean was influenced by meteorological conditions that occurred in a given growing season. The highest significant differences were observed in 2014 and 2016. A positive value of the difference between contents of phenolics in combinations that were treated with Kelpak biostimulant and the control samples (ABT-C) was calculated for soybean seeds (Table 4).

3.4.2. Total Anthocyanins Content

The presence of anthocyanins was detected in seven out of the 15 analyzed combinations of Kelpak biostimulant use in soybean cultivation. These compounds were not detected in the control samples in any of the growing seasons studied.

The use of Kelpak affected the content of anthocyanins in soybean seeds. However, their presence was only detected in 17% of the analyzed combinations. The number of applications and concentration of the biostimulant were the factors that determined anthocyanins content. The highest value of which was noted after plants spraying with the higher concentration of Kelpak. In this case, significant differences were also observed as influenced by conditions that occurred during the plant growth stage (Table 6). The values of biostimulant effect ABT-C index calculated for this trait were positive (Table 4).

Table 6. Effect of *Ecklonia maxima* extract (Kelpak) treatment on the antioxidant potential in soybean seeds (average from 2014–2016).

Parameters	*Ecklonia maxima* Treatment	Season			AA
		2014	2015	2016	
Total phenols (mg g^{-1} DM)	C	5.77^{a}	4.50^{a}	5.77^{b}	5.35^{a}
	LSS	7.36^{b}	5.85^{e}	7.74^{c}	6.98^{a}
	LDS	8.56^{c}	4.70^{b}	8.40^{d}	7.22^{a}
	HSS	15.02^{d}	5.05^{c}	15.20^{e}	11.76^{b}
	HDS	5.78^{a}	5.26^{d}	5.54^{a}	5.53^{a}
	AS	8.50^{b}	5.07^{a}	8.53^{b}	
Total flavonoids (mg g^{-1} DM)	C	1.99^{a}	1.44^{a}	1.99^{a}	1.81^{a}
	LSS	1.87^{a}	1.92^{c}	1.92^{a}	1.90^{a}
	LDS	2.64^{b}	1.84^{b}	2.59^{b}	2.36^{a}
	HSS	5.10^{d}	2.93^{e}	5.15^{d}	4.39^{b}
	HDS	4.18^{c}	2.08^{d}	4.21^{c}	3.49^{b}
	AS	3.16^{b}	2.04^{a}	3.17^{b}	
Anthocyanins (mg g^{-1} DM)	C	0.00^{a}	0.00^{a}	0.00^{a}	0.000^{a}
	LSS	0.00^{a}	0.00^{a}	0.00^{a}	0.000^{a}
	LDS	0.00^{a}	0.02^{b}	0.00^{a}	0.007^{a}
	HSS	0.00^{a}	0.04^{c}	0.00^{a}	0.013^{ab}
	HDS	0.04^{b}	0.00^{a}	0.05^{b}	0.030^{b}
	AS	0.008^{a}	0.012^{b}	0.010^{ab}	
Reducing power (mg TE g^{-1} DM)	C	0.15^{a}	0.10	0.15^{a}	0.13^{a}
	LSS	0.30^{bc}	0.22	0.33^{c}	0.28^{b}
	LDS	0.21^{ab}	0.14	0.28^{b}	0.21^{ab}
	HSS	0.45^{d}	0.08	0.42^{e}	0.31^{b}
	HDS	0.38^{cd}	0.16	0.37^{d}	0.30^{b}
	AS	0.30^{b}	0.14^{a}	0.31^{b}	

Abbreviations: C, control; LSS, lower concentration single spraying; LDS, lower concentration double spraying; HSS higher concentration single spraying; HDS, higher concentration double spraying. Means in the columns, concerning the selected traits, followed by different small letters are significantly different at $p < 0.05$.

3.4.3. Total Flavonoid Content

Flavonoid content analysis showed a significant effect of the application of biostimulant on its values. The foliar application of Kelpak resulted in the increased content of flavonoids in seeds. Significantly, the highest content of these compounds was noted after plant spraying with 1% solution of this preparation, regardless of the number of applications.

The analysis of the effect of biostimulants with different composition revealed that their foliar application resulted in an increased content of flavonoids when compared to the control samples (a positive value of the ABT-C difference) (Table 4).

3.4.4. Reducing Power

The evaluation of the effect of applying biostimulants with different compositions on the antioxidant activity of soybean included the determination of the reducing power, the value of which was increased by almost all combinations of this biostimulant.

Significant differences in reducing power values were observed upon the application of Kelpak biostimulant (Table 6). A tendency for an increase of reducing potential was noted after the application of this preparation in the higher concentration and after single spraying the plants with its 0.7% solution. In the second study year, the value of reducing power was the lowest when compared to the other analyzed years (over twofold decrease of RP value). Foliar application of *Ecklonia maxima* extract increased its reducing power values, which was indicated by positive values of the ABT-C index that were calculated for this trait (Table 4).

4. Discussion

Biostimulants induce the growth and development of plants, from seed germination throughout the entire ontogenesis. They affect the metabolic processes that occur in the plant by enhanced activity and synthesis of phytohormones, by stimulating the growth of the root system, and by improving the uptake, translocation, and retention of nutrients, which determines quantity and quality of crop yield [6,68].

Our study demonstrates a significant increase in the growth of soybean plants after the foliar application of biostimulant that is based on *Ecklonia maxima* extract. An earlier study also showed the growth stimulation of soybean treated with a biostimulant (Fylloton) containing *Ascophyllum nodosum* extract and free amino acids [8]. The marked growth responses in soybean plants are possibly due to *Ecklonia maxima* extract (Kelpak) composition, especially the PGRs (Plant Growth Regulator) that were identified (cytokinins, auxins, polyamines, gibberellins, brassinosteroids) and the mineral content in this biostimulant [55,69,70]. Additionally, the stimulatory role of Kelpak in the production of phytohormones has been demonstrated. For example, it increased the content of cytokinins in *Eucomis autumnalis* [70].

However, despite the observed favourable effects related to the application of biostimulants, including seaweeds, the precise mechanism of their activity still remains mostly unknown [26]. It should be emphasised that a full explanation of the principle or principles of their operation may cause a potential increase in the use of these preparations. According to Crouch and Van Staden [71] and Craigie [72], a wide scope of reported physiological responses of plants in cultivations where seaweed extracts were used is related to the fact that those products include numerous active compounds. Cytokinins, auxins, gibberellins, brassinosteroids, and other activating particles, like, for example, oligomers and polysaccharides are included [73]. According to Depuydt and Hardtke [74], cytokinines, together with auxins, function as regulators of various physiological processes, including those that are related to plant growth and development [75–79]. Thus, each change in the concentration of endogenic cytokinines influences the regulation of many physiological processes and as a result impacts the growth of the entire plant [80,81]. Studies by Aremu et al. [70] proved that the total content of cytokinines increases in plants after the application of the Kelpak preparation. The qualitative composition of the listed compounds is changed, which is related to their functional and physiological role, in particular, during plant morphogenesis. According to Strnad [76], isoprenic cytokinines determine the growth processes that include a continuation of the cell cycle. On the other hand, aromatic cytokinines model growth processes, such as morphogenesis and ageing. Aremu et al. [70] even assumed that the quantification of the endogenic content of cytokinins might provide information regarding possible physiological mechanisms that are related to the application of Kelpak biostimulant. However, researchers stress that, due to numerous active substances and compounds that are contained in Kelpak, the observed, favourable impact on the growth and development of plants may not only be assigned to cytokinins, but instead be the result of possible cross reactions of those compounds with other biologically active particles that are included in seaweed biostimulants. Therefore, further research concerning those fields is indispensable in order to obtain a full explanation for the Kelpak performance [70].

Still, in the literature, the prevailing hypothesis is that the majority of responses of plants that were cultivated with biostimulants' application, including seaweeds, results from the presence of compounds from the group of plant hormones, namely cytokinins [18]. The assumption stems from the fact that these compounds, isolated from seaweed extracts and individually tested in cultivations, mitigates the stress that is caused by free radicals through direct capturing and the prevention of reactive oxygen forms (ROS). This is done through the inhibition of xanthine oxidation [18,34,82–84].

Khan et al. [85] and Panda et al. [18] additionally indicate that extracts from seaweeds, such as Kelpak, support plant tolerance to stress, influencing the increase of K^+ capture in plants.

Additionally, in the literature, we may find hypotheses regarding modelling the growth and development of plants through the application of biostimulants from seaweeds as an effect of the presence of substances that are similar to gibberellins [86]. Research by Stephenson [19] indicates that these extracts include at least two compounds that behave like gibberellins (GA3 and GA7). However, they also show the presence of terpenoids and α-tokopherol, the performance of which may imitate gibberellins' activity on plants [18,87,88].

Recent theories indicate that the activity of seaweed extracts may be the result of their content of betains. Panda et al. shows that these compounds have a similar impact on plants as the aforementioned cytokinins [18]. It was proved that seaweed extracts include, among others, gamma aminobutyric acid betaine, 6-amino valeric acid betaine, and glyco betaine. Mancuso et al. [89] suggested that, due to the presence of those active compounds, extracts influence the mitigation of the osmotic and oxidization stress in plants, which may lead to the damage of DNA lipids, carbohydrates, and proteins, and also disturb correct cell signalling. Genard et al. [90] and Blunden et al. [91] even assigned an improvement of plant yield to the presence of betaines in extracts, since that led to the increased concentration of chlorophyll. According to Naidu et al. [92], betaines constitute a source of nitrogen when they are provided in low doses, or act as osmolites in higher concentrations. Many studies also show that betaines play a role in the correct formation of somatic germs from cotyledons tissues and mature seeds [18,93,94].

The stimulation of plant growth and development may also result from the occurrence of polyamins in biostimulants based on seaweed extracts, since these compounds may act as plant growth regulators. However, it should be emphasised that they are not classified as plant hormones. Several amino groups that usually replace hydrogen in the alkaline chain (putrescine, spermidine, and spermine) are characteristic of polyamins' structure. Research by Haman et al. [95] proves that polyamins determine the stability of various RNA and DNA conformation states. These compounds are often related to important stages of the cell division cycle. They also ensure the stability of a membrane to various cell membranes. Thus, due to the fact that polyamins affect a wide scope of physiological growth processes, their occurrence in biostimulating products that are made of seaweed may influence plant growth [18].

In our study, the use of biostimulant increased the fat and protein content in soybean seeds. The stimulating effect of biostimulants on the nutritional composition of various plants is mainly due to the number of PGR contained in the solutions [89]. In addition, the increased nutritional content in *C. triloba* is probably related to the ability of biostimulators to improve the slow release of nutrients and their uptake by plants [68,96]. A stimulating activity of seaweeds extract is also found in the presence of abscisic acid (ABA) [97,98]. However, the ABA function remains not fully characterized. Nevertheless, it is known that this acid induces protein synthesis, which are needed by plants in dealing with stress factors during water deficiencies [99,100]. Davies [101] shows that, during drought, this compound in plants caused numerous physiological reactions, including the closing of stoma, increase of a trend for accumulation of protein in seeds, gene transcription for proteinase inhibitors, as well as inhibition of sprouts growth or initiation of some states of seeds dormancy.

Unfortunately, the precise mechanisms that are activated by those biostimulants are still difficult to be identified, despite even greater knowledge on the composition of extracts from seaweeds. This is also due to the fact that these preparations constitute an abundance of many biologically active chemical compounds. They also include bioactive secondary metabolites, vitamins, and vitamin precursors [102,103]. Many authors underline the meaning of their synergetic cooperation, which stimulates that growth of plants assuming a mechanism that has not been fully known yet [24,72,104]. One of the main components of seaweed extracts are polysaccharides, including alginians, fucoidans, and laminarans [85,105]. Fucoidans have various structures due to a varied degree of methylation, sulphurization, and branching [72]. Alginians are polymers of mannuronic and galuronic acid, with a confirmed activity of plant-growth promotion [106]. Finally, laminarins that are included in extracts are registered compounds that increase plant resistance to fungi and bacteria pathogens [107].

Improvement of a nutritive value of soya seeds observed in our research, as expressed in protein and fat content, could also be caused by the fact that seaweeds are rich in phenolic compounds (complex chloroclucinol, eckol, and dieckol polymers). Phenols belong to secondary metabolites synthesized in plants under the influence of stress. Their task is to protect cells and the components of cell nuclei [108,109]. The ability to chelate metal ions [110] is a significant role of these compounds, besides their antioxidant activity. Research by Raj et al. [60] confirms that phenolic compounds with dihydroxybenzene or trihydroxybenzene groups show strong chelating activity. Rengasamy et al. [111] also shows that eckol belonging to phenolic compounds proves to have a strong auxinosimilar activity. According to the authors, the impact of these polyphenols, included in seaweed extracts, on the endogenic content of auxins is a key element that is necessary for understanding basic mechanisms of these preparations. Korasick et al. [112] reached similar conclusions. They conclude that auxins have a strong impact on many important stages of physiological growth in the life cycle of a plant. Researchers emphasise that maintaining proper concentration of active auxin in cells is of key significance for controlling almost all aspects that are related to the plant growth. The concentration of cell auxin is affected by the speed of anabolism, catabolism, transport, and conjugation [113]. In relation to the type of concentration, polyphenols may inhibit or stimulate the development of vegetative plants. It mainly takes place due to their abilities to modulate the metabolism and concentration of active auxin forms in plants [114–117]. Gaspar et al., [118] proves that the phenolic inhibitors of oxidase IAA, such as chlorogenic acid, influence the activity of auxins. Some of the mentioned compounds even constitute alternative substrates for the oxidizing enzyme, which in turn is related to the protection of auxins before oxygen decomposition. Wilson and Van Staden [119] prove that some of the phenolic acids that protect auxins before decarboxilation increase the concentration of active forms of auxins, which are indispensable for the stimulation of growth and development of crop roots. However, according to Arem et al., attempts to explain mechanisms that are responsible for a positive response of plants to application of extracts from seaweeds should take into account the possible cross reactions between phytohormones included therein and the quantitative concentration of auxins, which may justify the observed morphological differences [120].

In the literature, one may find hypotheses that assume that the increased growth and yield of plants that are treated with seaweed extracts resulted from a positive impact of those preparations on the activity of esterase enzymes. This enzyme is considered to be a marker of plant growth processes due to its role in organogenesis. It also works as an index of somatic embryogenesis [121–124]. According to Aremu et al., a higher activity of esterase in plants that were treated with seaweed extracts indicated their stimulating impact on the increase of plant biomass production [120].

Plant metabolism may be modelled through the use of biostimulants. According to Nardi et al., this group of active preparations affects most of all carbon and nitrogen metabolism, which is associated with an enhanced activity of enzymes participating in, among others, the process of glycolysis, Krebs cycle, or nitrogen assimilation [125]. Oboh at al. and Ertani et al. demonstrated that biostimulants application yielded metabolic pathways that are linked with secondary metabolites, like e.g. phenolic compounds [126,127]. It should also be emphasized that the synthesis of secondary metabolites proceeds as an element of chemical defense [128]. Already, in 1959, these compounds were no longer treated as ballast substances [129]. Today, they are believed to play a significant role in plant protection against adverse factors [130]. The most common indicator of plants resistance to biotic factors is the content of phenolic compounds [131], which are precursors of more complex phenolic structures, like flavonoids or lignins [132].

In our study, the foliar application of Ecklonia maxima extract (Kelpak) caused a significant increase in polyphenols content. Ertani et al. and Lakhdar et al. showed that the application of biostimulants in plant cultivation enhanced the synthesis of antioxidative compounds, which are indicators of increased plant resistance to biotic and abiotic stress factors [38,133]. The physiological response of plants to the use of biostimulants results from the presence of active substances in them, such as phytohormones, amino acids, proteins, phenols, or triacontanol [38,127,134].

A positive impact of compounds that are included in seaweed extracts on the total content of phenolic compounds in soya seeds has a significant meaning in the attempt to explain the mechanisms of operation of those biostimulants. The antioxidant potential of plants is inseparable from the amount and quality of phenolic compounds [135]. In real environmental conditions, the regulation of phytochemical synthesis includes a range of advanced mechanisms that enable a precise control of production of specific particles in a suitable place and time, and also in response to outside signals [136]. Such compounds include those in seaweed extracts that may activate specific biochemical pathways that are responsible for the synthesis of secondary metabolites in plants [136–138]. According to Cheynier et al., eckol that is contained in biostimulants influences the phenylpropanoid pathway in the biochemical synthesis of phenolic acids [136]. Researchers assume that this impact is caused by the regulation of the enzymatic activity of ammonia lyase of pheyloalanin and chalconic sythase. However, only the approach that is based on the genetic analysis will enable the observation of gene regulation engaged in those pathways [16,120]. Research results that were carried out by Jannin et al. proved that cysteine protease, related to the process of synthesis of phenolic compounds, were regulated downwards, while the expression of genes that are related to photosynthesis, cell metabolism, response to stress, and nitrogen metabolism were significantly raised in the case of plants treated with seaweed extracts [139]. Roupahel et al. observed an increased concentration of phenolic compounds after the application of such biostimulants assigned to their main components, such as polysaccharides (alginians, fucoidans, and laminarins) [140]. These compounds influence endogenic hormonal homeostasis [63,64]. Additionally, processes of synthesis and accumulation of secondary metabolites may be related to the activity of enzymatic groups that are engaged in phytochemical homeostasis (the so-called direct effect) [127,140]. They also depend on the plant nutrition condition and potassium and magnesium concentration (direct effect) [41]. Roupahel et al. [140] also search for the growth of the concentration of bioactive compounds in the activation of key enzymes, such as chalkone isomerase, which is engaged in the biosynthesis of flavon precursors [141].

According to Azcona et al. [142] and Ertani et al. [127], the high effectiveness of biostimulants in plant crops is also influenced by the number of treatments at the appropriate stages of plant development. The first treatment of plant with these preparations resulted most of all in the increased number and weight of leaves, which is referred to as "short-time effect". Another dose of biostimulants, applied at the plant blooming stage, led to the long-term effect, which was manifested by changes in crop size and quality. In the case of fruits, it results in, among others, an increase in their number and weight when compared to control samples that were not treated with biostimulants [125,127]. The increased content of polyphenols in the crop may indeed result from the use of biostimulants at the appropriate growth stages of plants. Experiments that were conducted by Oboh et al. [126] and by Zhang and Hamauzu [143] confirmed that the first application of these preparations led to an increased content of phenolic compounds in leaves, and that this increase was smaller after the second application of biostimulants. According to Ertani et al. [127], changes in the total polyphenolics content resulting from different numbers of applications of biostimulants are also linked with changes in contents of individual phenolic acids.

This was since the increasing total content of phenolic acids led to an increased number of their functional groups, which are sequesters of free radicals [144]. It must be emphasized that the increased content of polyphenols in plant tissues, as evoked by the action of biostimulants, is a beneficial phenomenon, not only because of the increased plant resistance to stress factors, but also because of significant importance to consumers, since such plant products are rich sources of antioxidative compounds being valuable to the human body [145,146]. Phenolic acids, such as caffeic, gallic, and ferulic, are claimed to exhibit anticarcinogenic and antimicrobial activities [147,148].

To sum up, biostimulants that contain seaweed extracts enable many opportunities to improve plant growth. Their use in agriculture is considered to be favourable for cropping. However, the operation mechanism of such products is not completely described [149]. That is because the impact of the application of growth regulators on plants is not only a consequence of their direct ability to control metabolic pathways, since their activity may be multidirectional. Limited knowledge on the mechanisms of the preparations' activity are still mainly based on assumptions and hypotheses highlights the need of further research within this scope [5]. So far it has been proved that seaweed extracts influence the plant physiology through changes in their general profile of transcriptome, and also in metabolism [139,150]. Research by Fan et al., concerning the analysis of gene expression, expanded the understanding of the possible mechanisms that regulate the activity of these preparations [141]. Researchers indicate that, after the application of extracts from seaweed, increases in the amount of transcripts of regulatory enzymes that are related to the nitrogen metabolism (cytolase glutamine synthesis), antioxidant ability (glutathione reductase), and glycine betaine synthesis (betaine aldehyde dehydrogenesis and choline monooxigenesis) were observed [16,18,22,72].

Although biostimulants are extensively used in agricultural practice, presently the most significant research on those preparations requires a better understanding of the mechanism of their influence [22,151]. According to Van Osten et al. [22] and Povero et al. [152], only after obtaining a complete explanation of those mechanisms, can the design and production of new generation biostimulants can take place. Due to their complex composition and interactions between particular compounds, mechanisms of operation of preparations based on seaweed extracts are slowly and successively discovered, with applications of molecular biology, metabolomics, and genomics techniques. However, according to many researchers, observed favourable biological effects of extracts activity is caused by the activity of small organic particles, as well as polymers that are included in products that have an ability to regulate genes' operation responsible for ensuring and modelling plant resistance systems [16,71,139].

5. Conclusions

The number of biostimulant applications and its concentration modified the biometric traits, crop size, and yield, as well as the nutraceutical and antioxidative potential of soybean seeds. The study demonstrated that the foliar application of *Ecklonia maxima* extract improved the growth and yield of soybean without any negative effect on the nutritive value of its seeds. Our experiment showed a positive effect of double foliar application of the higher concentrations of this biostimulant on soybean seed number and yield. The application of *Ecklonia maxima* extract increased the antioxidative activity of soybean seeds, and content of total phenolic compounds, flavonoids, and anthocyanins. The results of our study indicate the need for continuing investigations and extending their scope with the aim to identify responses of different cultivable plants on the use of biostimulants that are based on various biologically-active compounds.

Author Contributions: S.K. and A.S. conceived and designed the research. S.K., A.S. and E.C. performed the experiments. S.K., A.S., E.C., M.K. and T.P. prepared the materials. S.K., A.S., and M.K. analyzed the data. S.K. and A.S. wrote the paper. M.K. and T.P. revised the manuscript. All authors read and approved the final manuscript.

References

1. Bezerra, A.R.G.; Sediyama, T.; Borém, A.; Soares, M.M. Importância econômica. In *Soja: Do plantio à colheita*; Sediyama, T., Silva, F., Borém, A., Eds.; UFV: Viçosa, Brazil, 2015; pp. 9–26.
2. Kocira, S. Effect of amino acid biostimulant on the yield and nutraceutical potential of soybean. *Chil. J. Agric. Res.* **2019**, *79*, 17–25. [CrossRef]

3. Stein, H.H.; Berger, L.L.; Drackley, J.K.; Fahey, G.C., Jr.; Hernot, D.C.; Parsons, C.M. Nutritional properties and feeding values of soybeans and their coproducts. In *Soybeans, Chemistry, Production, Processing, and Utilization*; Johnson, L.A., White, P.J., Galloway, R., Eds.; AOCS Press: Urbana, IL, USA, 2008; pp. 613–660.
4. Lodhi, K.K.; Choubey, N.K.; Dwivedi, S.K.; Pal, A.; Kanwar, P.C. Impact of seaweed saps on growth, flowering behavior and yield of soybean [*Glycine max* (L.) Merrill.]. *Bioscan* **2015**, *10*, 479–483.
5. Szparaga, A.; Kocira, S.; Kocira, A.; Czerwińska, E.; Świeca, M.; Lorencowicz, E.; Kornas, R.; Koszel, M.; Oniszczuk, T. Modification of growth, yield, and the nutraceutical and antioxidative potential of soybean through the use of synthetic biostimulants. *Front. Plant Sci.* **2018**, *9*. [CrossRef] [PubMed]
6. Basak, A. Biostimulators. Definitions, classification and legislation. In *Biostimulators in Modern Agriculture, General Aspects*; Gawronska, H., Ed.; Editorial Housen Wies Jutra: Warszawa, Poland, 2008; pp. 7–17.
7. Paradiković, N.; Vinkovic, T.; Vinkovic, V.I.; Zuntar, I.; Bojic, M.; Medic, M. Effect of natural biostimulants on yield and nutritional quality: An example of sweet yellow pepper (*Capsicum annuum* L.) plants. *Sci. Food Agric.* **2011**, *91*, 2146–2152. [CrossRef] [PubMed]
8. Kocira, S.; Szparaga, A.; Kocira, A.; Czerwińska, E.; Wójtowicz, A.; Bronowicka-Mielniczuk, U.; Koszel, M.; Findura, P. Modeling biometric traits, yield and nutritional and antioxidant properties of seeds of three soybean cultivars through the application of biostimulant containing seaweed and amino acids. *Front. Plant Sci.* **2018**, *9*, 388. [CrossRef] [PubMed]
9. Kaufmann, G.L., III; Kneivel, D.P.; Watschke, T.L. Effects of a biostimulant on the heat tolerance associated with photosynthetic capacity, membrane thermostability, and polyphenol production of perennial ryegrass. *Crop Sci.* **2007**, *47*, 261–267. [CrossRef]
10. Du Jardin, P. The Science of Plant Biostimulants—A bibliographic analysis. Ad hoc Study Report to the European Commission DGENTR. 2012. Available online: https://orbi.uliege.be/bitstream/2268/169257/1/Plant_Biostimulants_final_report_bio_2012_en.pdf (accessed on 1 February 2019).
11. Traon, D.; Amat, L.; Zotz, F.; Du Jardin, P. *A Legal Framework for Plant Biostimulants and Agronomic Fertiliser Additives in the EU. Report for the European Commission Enterprise & Industry Directorate—General*; Publications Office of the European Union: Luxembourg, 2014.
12. Matyjaszczyk, E. The introduction of biostimulants on the Polish market. The present situation and legal requirements. *Przem. Chem.* **2015**, *10*, 1841–1844. [CrossRef]
13. Du Jardin, P. Plant biostimulants: Definition, concept, main categories and regulation. *Sci. Hortic.* **2015**, *196*, 3–14. [CrossRef]
14. Colla, G.; Cardarelli, M.; Bonini, P.; Rouphael, Y. Foliar applications of protein hydrolysate, plant and seaweed extracts increase yield but differentially modulate fruit quality of greenhouse tomato. *HortScience* **2017**, *52*, 1214–1220. [CrossRef]
15. Colla, G.; Hoagland, L.; Ruzzi, M.; Cardarelli, M.; Bonini, P.; Canaguier, R.; Rouphael, Y. Biostimulant action of protein hydrolysates: Unraveling their effects on plant physiology and microbiome. *Front. Plant Sci.* **2017**, *8*, 2202. [CrossRef] [PubMed]
16. Battacharyya, D.; Babgohari, M.Z.; Rathor, P.; Prithiviraj, B. Seaweed extracts as biostimulants in horticulture. *Sci. Hortic.* **2015**, *196*, 39–48. [CrossRef]
17. Aguilar, J. Algas marinas para la agricultura de alto rendimiento. Available online: http://www.interempresas.net/Horticola/Articulos/136576-Algas-marinas-para-la-agricultura-de-alto-rendimiento.html (accessed on 1 February 2019).
18. Panda, D.; Pramanik, K.; Nayak, B.R. Use of sea weed extracts as plant growth regulators for sustainable agriculture. *Int. J. Bioresour. Stress Manag.* **2012**, *3*, 404–411.
19. Stephenson, W.A. *Seaweed in Agriculture and Horticulture*; Faber and Faber: London, UK, 1968.
20. Crouch, I.J.; van Staden, J. Evidence for the presence of plant growth regulators in commercial seaweed products. *Plant Growth Regul.* **1993**, *13*, 21–29. [CrossRef]
21. Ecoforce, A. Extracto de algas para agricultura ecológica—Fertilizantes ecoforce. Available online: http://fertilizantesecoforce.es/es/blog/index/list/cat/agricultura-ecologica/?p=19 (accessed on 1 February 2019).
22. Van Oosten, M.J.; Pepe, O.; De Pascale, S.; Silletti, S.; Maggio, A. The role of biostimulants and bioeffectors as alleviators of abiotic stress in crop plants. *Technol. Agric.* **2017**, *4*, 5. [CrossRef]
23. Medjdoub, R. *La algas marinas y la agricultura*; Adiego: Zaragoza, Spain; pp. 1–3. Available online: http://catsaigner.adiego.com/sites/default/files/las_algas_marinas.pdf (accessed on 1 February 2019).

24. Norrie, J.; Keathley, J.P. Benefits of *Ascophyllum nodosum* marine-plant extract applications to 'Thompson seedless' grape production. *Acta Hortic.* **2006**, *727*, 243–247. [CrossRef]
25. Gajc-Wolska, J.; Spiżewski, T.; Grabowska, A. The effect of seaweed extracts on the yield and quality parameters of broccoli (*Brassica oleracea* var. *cymosa* L.) in open field production. *Acta Hortic.* **2013**, *1009*, 83–89. [CrossRef]
26. Sharma, H.S.S.; Fleming, C.; Selby, C.; Rao, J.R.; Martin, T. Plant biostimulants: A review on the processing of microalgae and use of extracts for crop management to reduce abiotic and biotic stresses. *J. Appl. Phycol.* **2014**, *26*, 465–490. [CrossRef]
27. Guerrero, J. La aplicación de las algas marinas para la fertilización Hortalizas. Available online: http://www.hortalizas.com/nutricion-vegetal/la-aplicacion-de-las-algas-marinas-para-la-fertilizacion/ (accessed on 1 February 2019).
28. Jayaraj, J.; Norrie, J.; Punja, Z.K. Commercial extract from the brown seaweed *Ascophyllum nodosum* reduces fungal diseases in greenhouse cucumber. *J. Appl. Phycol.* **2011**, *23*, 353–361.
29. Araujo, I.B.; Peruch, L.A.M.; Stadnik, M.J. Efeito do extrato de alga e da argila silicatada na severidade da alternariose e na produtividade da cebolinha comum (*Allium fistulosum* L.). *Trop. Plant Pathol.* **2012**, *37*, 363–367. [CrossRef]
30. Uppal, A.K.; El Hadrami, A.; Adam, L.R.; Tenuta, M.; Daayf, F. Biological control of potato Verticillium wilt under controlled and field conditions using selected bacterial antagonists and plant extracts. *Biol. Control* **2008**, *44*, 90–100. [CrossRef]
31. Zodape, S.T.; Gupta, A.; Bhandari, S.C.; Rawat, U.S.; Cahudhary, D.R.; Eswara, K.; Chikara, J. Foliar application of seaweed sap as biostimulant for enhancement of yield and field quality of tomato (*Lycopersicon esculentum* Mill.). *J. Sci. Ind. Res.* **2011**, *70*, 215–219.
32. El Modafar, C.; Elgadda, M.; El Boutachfaiti, R.; Abouraicha, E.; Zehhar, N.; Petit, E.; El Alaoui-Talibi, Z.; Courtois, B.; Courtois, J. Induction of natural defence accompanied by salicylic acid-dependant systemic acquired resistance in tomato seedlings in response to bioelicitors isolated from green algae. *Sci. Hortic.* **2012**, *138*, 55–63. [CrossRef]
33. Shah, M.T.; Zodape, S.T.; Chaudhary, D.R.; Eswaran, K.; Chikara, J. Seaweed sap as an alternative to liquid fertilizer for yield and quality improvement of wheat. *J. Plant Nutr.* **2013**, *36*, 192–200. [CrossRef]
34. Zhang, X.Z.; Ervin, E.H. Impact of seaweed extract-based cytokinins and zeatin riboside on creeping bentgrass heat tolerance. *Crop Sci.* **2008**, *48*, 364–370. [CrossRef]
35. Jayaraj, J.; Wan, A.; Rahman, M.; Punja, Z.K. Seaweed extract reduces foliar fungal diseases on carrot. *Crop Prot.* **2008**, *10*, 1360–1366. [CrossRef]
36. Botta, A. Enhancing plant tolerance to temperature stress with amino acids: An approach to their mode of action. *Acta Hortic.* **2013**, *1009*, 29–35. [CrossRef]
37. Colla, G.; Svecova, E.; Rouphael, Y.; Cardarelli, M.; Reynaud, H.; Canaguier, R.; Rouphael, Y. Effectiveness of a plant-derived protein hydrolysate to improve crop performances under different growing conditions. *Acta Hortic.* **2013**, *1009*, 175–179. [CrossRef]
38. Ertani, A.; Schiavon, M.; Muscolo, A.; Nardi, S. Alfalfa plant-derived biostimulant stimulate short-term growth of salt stressed *Zea mays* L. plants. *Plant Soil* **2013**, *364*, 145–158. [CrossRef]
39. Lucini, L.; Rouphael, Y.; Cardarelli, M.; Canaguier, R.; Kumar, P.; Colla, G. The effect of a plant-derived protein hydrolysate on metabolic profiling and crop performance of lettuce grown under saline conditions. *Sci. Hortic.* **2015**, *182*, 124–133. [CrossRef]
40. Xu, C.; Leskovar, D. Effects of *A. nodosum* seaweed extracts on spinach growth, physiology and nutrition valued under drought stress. *Sci. Hortic.* **2015**, *183*, 39–47. [CrossRef]
41. Rouphael, Y.; Colla, G.; Giordano, M.; El-Nakhel, C.; Kyriacou, M.C.; De Pascale, S. Foliar applications of a legume-derived protein hydrolysate elicit dose-dependent increases of growth, leaf mineral composition, yield and fruit quality in two greenhouse tomato cultivars. *Sci. Hortic.* **2017**, *226*, 353–360. [CrossRef]
42. Rouphael, Y.; Cardarelli, M.; Bonini, P.; Colla, G. Synergistic action of a microbial-based biostimulant and a plant derived-protein hydrolysate enhances lettuce tolerance to alkalinity and salinity. *Front. Plant Sci.* **2017**, *8*, 131. [CrossRef] [PubMed]
43. Rouphael, Y.; De Micco, V.; Arena, C.; Raimondi, G.; Colla, G.; De Pascale, S. Effect of *Ecklonia maxima* seaweed extract on yield, mineral composition, gas exchange and leaf anatomy of zucchini squash grown under saline conditions. *J. Appl. Phycol.* **2017**, *29*, 459–470. [CrossRef]

44. Di Stasio, E.; Rouphael, Y.; Colla, G.; Raimondi, G.; Giordano, M.; Pannico, A.; El-Nakhel, C.; De Pascale, S. The influence of *Ecklonia maxima* seaweed extract on growth, photosynthetic activity and mineral compositionof *Brassica rapa* L. subsp. *sylvestris* under nutrient stress conditions. *Eur. J. Hortic. Sci.* **2017**, *82*, 286–293. [CrossRef]
45. Colla, G.; Nardi, S.; Cardarelli, M.; Ertani, A.; Lucini, L.; Canaguiere, R.; Rouphael, J. Protein hydrolysates as biostimulants in horticulture. *Sci. Hortic.* **2015**, *196*, 28–38. [CrossRef]
46. Möller, M.; Smith, M.L. The significance of the mineral component of seaweed suspension on lettuce (*Lactuca sativa* L.) seedling growth. *J. Plant Physiol.* **1998**, *153*, 658–663. [CrossRef]
47. Farooq, M.; Aziz, T.; Basra, S.M.A.; Cheema, M.A.; Rehman, H. Chilling tolerance in hybrid maize induced by seed treatments with salicylic acid. *J. Agron. Crop Sci.* **2008**, *194*, 161–168. [CrossRef]
48. Sivritepe, N.; Sivritepe, H.Ö.; Türkan, I.; Bor, M.; Özdemir, F. NaCl pre-treatments mediate salt adaptation in melon plants through antioxidative system. *Seed Sci. Technol.* **2008**, *36*, 360–370. [CrossRef]
49. Zodape, S.T.; Kawarkhe, V.J.; Patolia, J.S.; Warade, A.D. Effect of liquid seaweed fertilizer on yield and quality of okra (*Abelmoschus esculentus* L.). *J. Sci. Ind. Res. India* **2008**, *67*, 1115–1117.
50. Aitken, J.E.; Senn, J.I. Seaweed products as fertilizers and soil conditioners. *Bot. Mar.* **1965**, *8*, 144–147. [CrossRef]
51. Rayorath, P.; Narayanan, J.M.; Farid, A.; Khan, W.; Palanisamy, R.; Hankins, S.; Critchley, A.T.; Prithiviraj, B. Rapid bioassays to evaluate the plant growth promoting activity of *Ascophyllum nodosum* (L.) Le Jol. using a model plant, *Arabidopsis thaliana* (L.) Heynh. *J. Appl. Phycol.* **2008**, *20*, 423–429. [CrossRef]
52. Subramanian, S.; Sangha, J.S.; Gray, B.A.; Singh, R.P.; Hiltz, D.; Critchley, A.T.; Prithiviraj, B. Extracts of the marine brown macroalga, *Ascophyllum nodosum*, induce jasmonic acid dependent systemic resistance in *Arabidopsis thaliana* against *Pseudomonas syringae* pv. *Tomato* DC3000 and *Sclerotinia sclerotiorum*. *Eur. J. Plant Pathol.* **2011**, *131*, 237–248. [CrossRef]
53. Pereira, L.; Gheda, S.F.; Ribeiro-Claro, P.J.A. Analysis by vibrational spectroscopy of seaweed polysaccharides with potential use in food, pharmaceutical, and cosmetic industries. *Int. J. Carbohydr. Chem.* **2013**, *7*. [CrossRef]
54. Ordog, V.; Stirk, W.A.; van Staden, J.; Novak, O.; Strand, M. Endogenous cytokinins in the three genera of microalgae from the Chlorophyta. *J. Phycol.* **2004**, *40*, 88–95. [CrossRef]
55. Stirk, W.A.; Tarkowská, D.; Turečová, V.; Strnad, M.; van Staden, J. Abscisic acid, gibberellins and brassinosteroids in Kelpak®, a commercial seaweed extract made from Ecklonia maxima. *J. Appl. Phycol.* **2014**, *26*, 561–567. [CrossRef]
56. Lötze, E.; Hoffman, E.W. Nutrient composition and content of various biological active compounds of three south African-based commercial seaweed biostimulants. *J. Appl. Phycol.* **2016**, *28*, 1379–1386. [CrossRef]
57. Moreira Sisalema, J.M. Comportamiento agronómico del cultivo de soya (*Glycine max* L.), a la aplicación de tres extractos de algas marinas, en la zona de Puebloviejo. Available online: http://dspace.utb.edu.ec/bitstream/49000/5048/1/TE-UTB-FACIAG-ING%20AGROP-000018.pdf (accessed on 1 February 2019).
58. Ficha Técnica. Available online: https://www.calstiendavirtual.cl/html5/Fichas/6CASA%20JARD%C3%8DN/0126756%20KELPAK.pdf (accessed on 1 February 2019).
59. Ali, N.; Farrell, A.; Ramsubhag, A.; Jayaraman, J. The effect of *Ascophyllum nodosum* extract on the growth, yield and fruit quality of tomato under tropical conditions. *J. Appl. Phycol.* **2016**, *28*, 1353–1362. [CrossRef]
60. Raj, T.; Nishanthi, P.; Hane Graff, K.; Ann Suji, H. Seaweed extract as a biostimulant and a pathogen controlling agent in plants. *Int. J. Trop. Agric.* **2018**, *36*, 563–580.
61. Rouphael, Y.; Giordano, M.; Cardarelli, M.; Cozzolino, E.; Mori, M.; Kyriacou, M.C.; Bonini, P.; Colla, G. Plant- and seaweed-based extracts increase yield but differentially modulate nutritional quality of greenhouse spinach through biostimulant action. *Agronomy* **2018**, *8*, 126. [CrossRef]
62. AOAC. *Official Methods of Analysis of AOAC International*, 17th ed.; Horwitz, W., Ed.; AOAC International: Gaithersburg, MD, USA, 2000.
63. Świeca, M.; Gawlik-Dziki, U.; Kowalczyk, D.; Złotek, U. Impact of germination time and type of illumination on the antioxidant compounds and antioxidant capacity of Lens culinaris sprouts. *Sci. Hortic.* **2012**, *140*, 87–95. [CrossRef]
64. Singleton, V.; Rossi, J. Colorimetry of total phenolics with phosphomolybdic-phosphotungstic acid reagents. *Am. J. Enol. Viticult.* **1965**, *16*, 144–158.

65. Lamaison, J.L.C.; Carnet, A. Teneurs en principaux flavonoids des fleurs de *Crataegeus monogyna* Jacq et de *Crataegeus laevigata* (Poiret D. C) en fonction de la vegetation. *Pharm. Acta Helv.* **1990**, *65*, 315–320. [CrossRef]
66. Fuleki, T.; Francis, F.J. Quantitative methods for anthocyanins. 1. Extraction and determination of total anthocyanin in cranberries. *J. Food Sci.* **1968**, *33*, 72–77. [CrossRef]
67. Pulido, R.; Bravo, L.; Saura-Calixto, F. Antioxidant activity of dietary polyphenols as determined by a modified ferric reducing/antioxidant power assay. *J. Agric. Food Chem.* **2000**, *48*, 3396–3402. [CrossRef] [PubMed]
68. Calvo, P.; Nelson, L.; Kloepper, J.W. Agricultural uses of plant biostimulants. *Plant Soil* **2014**, *383*, 3–41. [CrossRef]
69. Papenfus, H.B.; Stirk, W.A.; Finnie, J.F.; Van Staden, J. Seasonal variation in the polyamines of *Ecklonia maxima*. *Bot. Mar.* **2012**, *55*, 539–546. [CrossRef]
70. Aremu, A.O.; Plačková, L.; Gruz, J.; Bíba, O.; Novák, O.; Stirk, W.A.; Doležal, K.; Van Staden, J. Seaweed-derived biostimulant (Kelpak®) influences endogenous cytokinins and bioactive compounds in hydroponically grown Eucomis autumnalis. *J. Plant Growth Regul.* **2016**, *35*, 151–162. [CrossRef]
71. Crouch, I.J.; van Staden, J. Commercial seaweed products as biostimulants in horticulture. *J. Home Consum. Hortic.* **1994**, *1*, 19–76. [CrossRef]
72. Craigie, J.S. Seaweed extract stimuli in plant science and agriculture. *J. Appl. Phycol.* **2011**, *23*, 371–393. [CrossRef]
73. Stadnik, M.J.; de Freitas, M.B. Algal polysaccharides as source of plant resistance inducers. *Trop. Plant Pathol.* **2014**, *39*, 111–118. [CrossRef]
74. Depuydt, S.; Hardtke, C.S. Hormone signalling crosstalk in plant growth regulation. *Curr. Biol.* **2011**, *21*, 365–373. [CrossRef] [PubMed]
75. Horgan, R. Present and future prospects for cytokinin research. In *Physiology and Biochemistry of Cytokinins in Plants*; Kamĭnek, M., Mok, D.W.S., Eds.; SPB Academic Publishing: Hague, The Netherlands, 1992; pp. 3–13.
76. Strnad, M. The aromatic cytokinins. *Physiol. Plant* **1997**, *101*, 674–688. [CrossRef]
77. Sakakibara, H. Cytokinins: Activity, biosynthesis, and translocation. *Ann. Rev. Plant Biol.* **2006**, *57*, 431–449. [CrossRef] [PubMed]
78. Santner, A.; Calderon-Villalobos, L.I.A.; Estelle, M. Plant hormones are versatile chemical regulators of plant growth. *Nat. Chem. Biol.* **2009**, *5*, 301–307. [CrossRef] [PubMed]
79. Frébort, I.; Kowalska, M.; Hluska, T.; Frébortová, J.; Galuszka, P. Evolution of cytokinin biosynthesis and degradation. *J. Exp. Bot.* **2011**, *62*, 2431–2452. [CrossRef] [PubMed]
80. Stirk, W.A.; Van Staden, J. Flow of cytokinins through the environment. *Plant Growth Regul.* **2010**, *62*, 101–116. [CrossRef]
81. Zalabák, D.; Pospısilová, H.; Šmehilová, M.; Mrızová, K.; Freébort, I.; Galuszka, P. Genetic engineering of cytokinin metabolism: Prospective way to improve agricultural traits of crop plants. *Biotechnol. Adv.* **2013**, *31*, 97–117. [CrossRef] [PubMed]
82. McKersie, B.D.; Leshem, Y.Y. *Stress and Stress Coping in Cultivated Plants*; Kluwer Academic Publishers: Dordrecht, The Netherlands, 1994.
83. Fike, J.H.; Allen, V.G.; Schmidt, R.E.; Zhang, X.; Fontenot, J.P.; Bagley, C.P.; Ivy, R.L.; Evans, R.R.; Coelho, R.W.; Wester, D.B. Tasco-Forage: I. Influence of a seaweed extract on antioxidant activity in tall fescue and in ruminants. *J. Anim. Sci.* **2001**, *79*, 1011–1021. [CrossRef] [PubMed]
84. Ervin, E.H.; Zhang, X.; Fike, J. Ultraviolet-B Radiation Damage on Kentucky Bluegrass II: Hormone Supplement Effects. *HortScience* **2004**, *39*, 1471–1474. [CrossRef]
85. Khan, W.; Rayirath, U.P.; Subramanian, S.; Jithesh, M.N.; Rayorath, P.; Hodges, D.M.; Critchley, A.T.; Craigie, J.S.; Norrie, J.; Prithiviraj, B. Seaweed extracts as biostimulants of plant growth and development. *J. Plant Growth Regul.* **2009**, *28*, 386–399. [CrossRef]
86. Jennings, R.C. Gibberellins as endogenous growth regulators in green and brown algae. *Planta* **1968**, *80*, 34–42. [CrossRef]
87. Gopala, R.P. Gibberellin-like behaviour of α-tocopherol in green gram *Vigna radiata*. *Geobios* **1984**, *11*, 21–25.
88. Jensen, A. Tocopherol content of seaweed and seaweed meal. I. Analytical methods and distribution of tocopherols in benthic algae. *J. Sci. Food Agric.* **1969**, *20*, 449–453. [CrossRef]
89. Mancuso, S.; Azzarello, E.; Mugnai, S.; Briand, X. Marine bioactive substances (IPA extract) improve foliar iron uptake and water tolerance in potted *Vitis vinifera* plants. *Adv. Hort. Sci.* **2006**, *20*, 156–161.

90. Genard, H.; Le Saos, J.; Billard, J.P.; Tremolieres, A.; Boucaud, J. Effect of salinity on lipid composition, glycine betaine content and photosynthetic activity in chloroplasts of *Suaeda maritima*. *Plant Physiol. Biochem.* **1991**, *29*, 421–427.
91. Blunden, G.; Jenkins, T.; Liu, Y. Enhanced leaf chlorophyll levels in plants treated with seaweed extract. *J. Appl. Physiol.* **1997**, *8*, 535–543. [CrossRef]
92. Naidu, B.P.; Jones, G.P.; Paleg, L.G.; Poljakoff-Mayber, A. Proline analogues in *Melaleuca* species: Response of *Melaleuca lanceolata* and *M. uncinata* to water stress and salinity. *Aust. J. Plant Physiol.* **1987**, *14*, 669–677. [CrossRef]
93. Wachira, F.; Ogada, J. In vitro regeneration of *Camellia sinensis* (L.) O. Kuntze by somatic embryogenesis. *Plant Cell Rep.* **1995**, *14*, 463–466. [CrossRef] [PubMed]
94. Akula, A.; Akula, C.; Bateson, M. Betaine: A novel candidate for rapid induction of somatic embryogenesis in tea (*Camellia sinensis* [L.] O. Kuntze). *Plant Growth Regul.* **2000**, *30*, 241–246. [CrossRef]
95. Hamana, K.; Matsuzaki, S.; Niitsu, M.; Samejima, K.; Nagashima, H. Polyamines of inicellular thermoacidophillic red alga, *Cyanidium caldarium*. *Phytochemistry* **1990**, *29*, 377–380. [CrossRef]
96. Rathore, S.S.; Chaudhary, D.R.; Boricha, G.N.; Ghosh, A.; Bhatt, B.P.; Zodape, S.T.; Patolia, J.S. Effect of seaweed extract on the growth, yield and nutrient uptake of soybean (*Glycine max*) under rainfed conditions. *S. Afr. J. Bot.* **2009**, *75*, 351–355. [CrossRef]
97. Hirsch, R.; Hartung, W.; Gimmler, H. Abscisic acid content of algae under stress. *Bot. Acta* **1989**, *102*, 326–334. [CrossRef]
98. Boyer, G.L.; Dougherty, S.S. Identification of abscisic acid in the seaweed *Ascophyllum nodosum*. *Phytochemistry* **1988**, *27*, 1521–1522. [CrossRef]
99. Rensing, S.A.; Lang, D.; Zimmer, A.D.; Terry, A.; Salamov, A.; Shapiro, H.; Nishiyama, T.; Perroud, P.F.; Lindquist, E.A.; Kamisugi, Y.; et al. The *Physcomitrella* genome reveals evolutionary insight into the conquest of land by plants. *Science* **2008**, *319*, 64–69. [CrossRef] [PubMed]
100. Verslues, P.E.; Agarwal, M.; Katiyar-Agarwal, S.; Zhu, J.; Zhu, J.K. Methods and concepts in quantifying resistance to drought, salt and freezing, abiotic stresses that affect water plant status. *Plant J.* **2006**, *45*, 523–539. [CrossRef] [PubMed]
101. Davies, P.J. *Plant Hormones. Biosynthesis, Signal Transduction, Action!* 3rd ed.; Kluwer Academic Publishers: Dordrecht, The Netherlands, 2004; p. 750.
102. Berlyn, G.P.; Russo, R.O. The use of organic biostimulants to promote root growth. *Belowground Ecol.* **1990**, *2*, 12–13.
103. Blunden, G.; Currie, M.; Mathe, J.; Hohmann, J.; Critchley, A.T. Betaine yields from marine algal species utilized in the preparation of seaweed extracts used in agriculture. *Nat. Prod. Commun.* **2010**, *5*, 581–585. [PubMed]
104. Craigie, J.; MacKinnon, S.; Walter, J. Liquid seaweed extracts identified using H NMR profiles. *J. Appl. Phycol.* **2008**, *20*, 665–671. [CrossRef]
105. Rayirath, P.; Benkel, B.; Hodges, D.M.; Allan-Wojtas, P.; MacKinnon, S.; Critchley, A.; Prithiviraj, B. Lipophilic components of the brown seaweed, *Ascophyllum nodosum*, enhance freezing tolerance in *Arabidopsis thaliana*. *Planta* **2009**, *230*, 135–147. [CrossRef] [PubMed]
106. Yabur, R.; Bashan, Y.; Hernandez-Carmona, G. Alginate from the macroalgae *Sargassum sinicola* as a novel source for microbial immobilization material in waste water treatment and plant growth promotion. *J. Appl. Phycol.* **2007**, *19*, 43–53. [CrossRef]
107. Mercier, L.; Laffite, C.; Borderies, G.; Briand, X.; Esquerre-Tugaye, M.T.; Fournier, J. The algal polysaccharide carrageenans can act as an elicitor of plant defence. *New Phytol.* **2011**, *149*, 43–51. [CrossRef]
108. Nakamura, T.; Nagayama, K.; Uchida, K.; Tanaka, R. Antioxidant activity of phlorotannins isolated from the brown alga *Eisenia bicydis*. *Fish. Sci.* **1996**, *62*, 923–926. [CrossRef]
109. Wang, T.; Jonsdottir, R.; Olafsdottir, G. Total phenolic compounds, radical scavenging and metal chelation of extracts from Icelandic seaweeds. *Food Chem.* **2009**, *116*, 240–248. [CrossRef]
110. Andjelkovic, M.; Van Camp, J.; De Meulenaer, B.; Depaemelaere, G.; Socaciu, C.; Verloo, M.; Verhe, R. Iron-chelation properties of phenolic acids bearing catechol and galloyl groups. *Food Chem.* **2006**, *98*, 23–31. [CrossRef]

111. Rengasamy, K.R.R.; Kulkarni, M.G.; Stirk, W.A.; Van Staden, J. Eckol Improves Growth, Enzyme Activities, and Secondary Metabolite Content in Maize (*Zea mays* cv. Border King). *J. Plant Growth Regul.* **2015**, *34*, 410–416. [CrossRef]
112. Korasick, D.A.; Enders, T.A.; Strader, L.C. Auxin biosynthesis and storage forms. *J. Exp. Bot.* **2013**, *64*, 2541–2555. [CrossRef] [PubMed]
113. Ljung, K. Auxin metabolism and homeostasis during plant development. *Development* **2013**, *140*, 943–950. [CrossRef] [PubMed]
114. Peer, W.A.; Murphy, A.S. Flavonoids and auxin transport: Modulators or regulators? *Trends Plant Sci.* **2007**, *12*, 556–563. [CrossRef] [PubMed]
115. Wu, H.C.; Du Toit, E.S.; Reinhardt, C.F.; Rimando, A.M.; Van der Kooy, F.; Meyer, J.J.M. The phenolic, 3,4-dihydroxybenzoic acid, is an endogenous regulator of rooting in *Protea cynaroides*. *Plant Growth Regul.* **2007**, *52*, 207–215. [CrossRef]
116. De Klerk, G.J.; Guan, H.; Huisman, P.; Marinova, S. Effects of phenolic compounds on adventitious root formation and oxidative decarboxylation of applied indoleacetic acid in Malus 'Jork 9'. *Plant Growth Regul.* **2011**, *63*, 175–185. [CrossRef]
117. Teixeira da Silva, J.; Dobranszki, J.; Ross, S. Phloroglucinol in plant tissue culture. *In Vitro Cell. Dev. Biol. Plant* **2013**, *49*, 1–16. [CrossRef]
118. Gaspar, T.; Kevers, C.; Penel, C.; Greppin, H.; Reid, D.; Thorpe, T. Plant hormones and plant growth regulators in plant tissue culture. *In Vitro Cell. Dev. Biol. Plant* **1996**, *32*, 272–289. [CrossRef]
119. Wilson, P.J.; Van Staden, J. Rhizocaline, rooting co-factors, and the concept of promoters and inhibitors of adventitious rooting—A review. *Ann. Bot.* **1990**, *66*, 479–490. [CrossRef]
120. Aremu, A.O.; Nqobile, A.; Kannan, M.; Rengasamy, R.R.; Amoo, S.O.; Gruz, J.; Bĭba, O.; Šubrtová, M.; Pênčĭk, A.; Novák, O.; et al. Physiological role of phenolic biostimulants isolated from Brown seaweed *Ecklonia maxima* on plant growth and development. *Planta* **2015**, *241*, 1313–1324. [CrossRef] [PubMed]
121. Chibbar, R.N.; Schyluk, J.; Georges, F.; Mallard, C.S.; Constabel, F. Esterase isozymes as markers of somatic embryogenesis in cultured carrot cells. *J. Plant Physiol.* **1988**, *133*, 367–370. [CrossRef]
122. Pedersen, S.; Andersen, S.B. Developmental expression of isoenzymes during embryiogenesis in barley anther culture. *Plant Sci.* **1993**, *91*, 75–86. [CrossRef]
123. Krsnik-Rasol, M.; Cipcic, H.; Hagege, D. Isoesterases related to cell differentiation in plant tissue culture. *Chem. Biol. Interact.* **1999**, *120*, 587–592. [CrossRef]
124. Balen, B.; Krsnik-Rasol, M.; Simeon-Rudolf, V. Isoenzymes of peroxidase and esterase related to morphogenesis in *Mammillaria gracillis* Pfeiff. tissue culture. *J. Plant Physiol.* **2003**, *160*, 1401–1406. [CrossRef] [PubMed]
125. Nardi, S.; Carletti, P.; Pizzeghello, D.; Muscolo, A. Biological activities of humic substances, in biophysico-chemical processes involving natural nonliving organic matter in environmental systems. In *Fundamentals and Impact of Mineral-Organic-Biota Interactions on the Formation, Transformation, Turnover, and Storage of Natural Nonliving Organic Matter (NOM)*, 3rd ed.; John Wiley: Hoboken, NJ, USA, 2009; pp. 305–339.
126. Oboh, G.; Puntel, R.L.; Rocha, J.B.T. Hot pepper (*Capsicum annuum*, Tepin and *Capsicum chinese*, Habanero) prevents Fe^{2+}-induced lipid peroxidation in brain-*in vitro*. *Food Chem.* **2007**, *102*, 178–185. [CrossRef]
127. Ertani, A.; Pizzeghello, D.; Francioso, O.; Sambo, P.; Sanchez-Cortes, S.; Nardi, S. *Capsicum chinensis* L. growth and nutraceutical properties are enhanced by biostimulants in a long-term period: Chemical and metabolomic approaches. *Front. Plant Sci.* **2014**, *5*, 375. [CrossRef] [PubMed]
128. Złotek, U.; Świeca, M.; Jakubczyk, A. Effect of abiotic elicitation on main health-promoting compounds, antioxidant activity and commercial quality of butter lettuce (*Lactuca sativa* L.). *Food Chem.* **2014**, *148*, 253–260. [CrossRef] [PubMed]
129. Karolewski, P.; Jagodziński, A.M. Udział węgla w związkach obronnych przed czynnikami biotycznymi u roślin drzewiastych. *Sylwan* **2013**, *157*, 831–841.
130. Czerwińska, E.; Szparaga, A. Antibacterial and antifungal activity of plant extracts. *Rocznik Ochrona Środowiska* **2015**, *17*, 209–229.
131. Stevens, M.T.; Lindroth, R.L. Induced resistance in the interminate growth of aspen (*Populus tremuloides*). *Oecologia* **2005**, *145*, 298–306. [CrossRef] [PubMed]

132. Kocira, S.; Szparaga, A.; Kocira, A.; Czerwińska, E.; Depo, K.; Erlichowska, B.; Deszcz, E. Effect of applying a biostimulant containing seaweed and amino acids on the content of fiber fractions in three soybean cultivars. *Legum. Res.* **2018**. [CrossRef]
133. Lakhdar, A.; Iannelli, M.A.; Debez, A.; Massacci, A.; Jedidi, N.; Abdelly, C. Effect of municipal solid waste compost and sewage sludge use on wheat (Triticum durum): Growth, heavy metal accumulation, and antioxidant activity. *J. Sci. Food Agric.* **2010**, *90*, 965–971. [CrossRef] [PubMed]
134. Jindo, K.; Martim, S.A.; Navarro, E.C.; Pérez-Alfocea, F.; Hernandez, T.; Garcia, C.; Aguiar, N.O.; Canellas, L.P. Root growth promoting by humic acids from composted and non-composted urban organic wastes. *Plant Soil* **2012**, *353*, 209–220. [CrossRef]
135. Gülçin, I. Antioxidant activity of food constituents: An overview. *Arch. Toxicol.* **2012**, *86*, 345–391. [CrossRef] [PubMed]
136. Cheynier, V.; Comte, G.; Davies, K.M.; Lattanzio, V.; Martens, S. Plant phenolics: Recent advances on their biosynthesis, genetics, and ecophysiology. *Plant Physiol. Biochem.* **2013**, *72*, 1–20. [CrossRef] [PubMed]
137. Pavarini, D.P.; Pavarini, S.P.; Niehues, M.; Lopes, N.P. Exogenous influences on plant secondary metabolite levels. *Anim. Feed Sci. Technol.* **2012**, *176*, 5–16. [CrossRef]
138. Moyo, M.; Amoo, S.O.; Aremu, A.O.; Gruz, J.; Šubrtová, M.; Doležal, K.; Van Staden, J. Plant regeneration and biochemical accumulation of hydroxybenzoic and hydroxycinnamic acid derivatives in *Hypoxis hemerocallidea* organ and callus cultures. *Plant Sci.* **2014**, *227*, 157–164. [CrossRef] [PubMed]
139. Jannin, L.; Arkoun, M.; Etienne, P.; Laîné, P.; Goux, D.; Garnica, M.; Fuentes, M.; San Francisco, S.; Baigorri, R.; Cruz, F.; et al. *Brassica napus* growth is promoted by *Ascophyllum nodosum* (L.) Le Jol. Seaweed extract:microarray analysis and physiological characterization of N, C, and S metabolisms. *J. Plant Growth Regul.* **2013**, *32*, 31–52. [CrossRef]
140. Rouphael, Y.; Kyriacou, M.C.; Petropoulos, S.A.; De Pascale, S.; Colla, G. Improving vegetable quality in controlled environments. *Sci. Hortic.* **2018**, *234*, 275–289. [CrossRef]
141. Fan, D.; Hodges, D.M.; Critchley, A.T.; Prithiviraj, B. A commercial extract of brown macroalga (*Ascophyllum nodosum*) affects yield and the nutritional quality of Spinach in vitro. *Commun. Soil Sci. Plant Anal.* **2013**, *44*, 1873–1884. [CrossRef]
142. Azcona, I.; Pascual, I.; Aguirreolea, J.; Fuentes, M.; García-Mina, J.M.; Sánchez-Díaz, M. Growth and development of pepper are affected by humic substances derived from composted sludge. *J. Plant Nutr. Soil Sci.* **2011**, *174*, 916–924. [CrossRef]
143. Zhang, D.; Hamauzu, Y. Phenolic compounds, ascorbic acid, carotenoids and antioxidant properties of green, red and yellow bell peppers. *J. Food Agric. Environ.* **2003**, *1*, 22–27.
144. Pantelidis, G.E.; Vasilakakis, M.; Manganaris, G.A.; Diamantidis, G.R. Antioxidant capacity, phenol, anthocyanin and ascorbic acid contents in raspberries, blackberries, red currants, gooseberries and cornelian cherries. *Food Chem.* **2007**, *102*, 777–783. [CrossRef]
145. Świeca, M.; Sęczyk, Ł.; Gawlik-Dziki, U. Elicitation and precursor feeding as tools for the improvement of the phenolic content and antioxidant activity of lentil sprouts. *Food Chem.* **2014**, *161*, 288–295. [CrossRef] [PubMed]
146. Złotek, U.; Szychowski, K.A.; Świeca, M. Potential in vitro antioxidant, anti-inflammatory, antidiabetic, and anticancer effect of arachidonic acid-elicited basil leaves. *J. Funct. Foods* **2017**, *36*, 290–299. [CrossRef]
147. Raina, K.; Rajamanickam, S.; Deep, G.; Singh, M.; Agarwal, R.; Agarwal, C. Chemopreventive effects of oral gallic acid feeding on tumor growth and progression in TRAMP mice. *Mol. Cancer Ther.* **2008**, *7*, 1258–1267. [CrossRef] [PubMed]
148. Chen, J.Y.; Wen, P.F.; Kong, W.F.; Pan, Q.H.; Zhan, J.C.; Li, J.M.; Wan, S.B.; Huang, W.D. Effect of salicylic acid on phenylpropanoids and phenylalanine ammonia-lyase in harvested grape berries. *Postharvest Biol. Technol.* **2006**, *40*, 64–72. [CrossRef]
149. Polo, J.; Mata, P. Evaluation of a biostimulant (Pepton) based in enzymatic hydrolyzed animal protein in comparison to seaweed extracts on root development, vegetative growth, flowering, and yield of gold cherry tomatoes grown under low stress ambient field conditions. *Front. Plant Sci.* **2018**, *8*, 2261. [CrossRef] [PubMed]
150. Nair, P.; Saveetha, K.; Zhang, J.; Ji, X.; Chris, K.; Bernhard, B.; Mark, D.H.; Alan, T.C.; David, H.; Balakrishnan, P. Transcriptional and metabolomics analysis of *Ascophyllum nodosum* mediatel freezing tolerance in *Arabidopsis thaliana*. *BMC Genom.* **2012**, *13*, 643. [CrossRef] [PubMed]

151. Biesaga-Kościelniak, J.; Kościelniak, J.; Filek, M.; Marcińska, I.; Krekule, J.; Machackova, I.; Kuboń, M. The effect of plant growth regulators and their interaction with electric current on winter wheat development. *Acta Physiol. Plant* **2010**, *32*, 987–995. [CrossRef]
152. Povero, G.; Mejia, J.F.; Tommaso, D.D.; Piaggesi, A.; Warrior, P. A systematic approach to discover and characterize natural plant biostimulants. *Front. Plant Sci.* **2016**, *7*, 435. [CrossRef] [PubMed]

13

Chemical Traits of Fermented Alfalfa Brown Juice: Its Implications on Physiological, Biochemical, Anatomical and Growth Parameters of Celosia

Nóra Bákonyi [1], Szilvia Kisvarga [2], Döme Barna [1], Ibolya O. Tóth [1], Hassan El-Ramady [1,3], Neama Abdalla [1,4], Szilvia Kovács [1], Margaréta Rozbach [5], Csaba Fehér [5], Nevien Elhawat [1,6], Tarek Alshaal [1,3,*] and Miklós Gábor Fári [1]

1 Department of Agricultural Botany, Plant Physiology and Biotechnology (MEK), Debrecen University, Böszörményi Street 138, 4032 Debrecen, Hungary; nbakonyi@agr.unideb.hu (N.B.); barnadoeme@gmail.com (D.B.); olahne@agr.unideb.hu (I.O.T.); ramady2000@gmail.com (H.E.-R.); neama_ncr@yahoo.com (N.A.); szkovacs@agr.unideb.hu (S.K.); nevienelhawat@gmail.com (N.E.); miklos0810@gmail.com (M.G.F.)
2 Research Institute for Fruitgrowing and Ornamentals, National Agricultural Research and Innovation Center (NARIC) Szent-Györgyi Street 4., 2100 Gödöllő, Hungary; fullerina@gmail.com
3 Soil and Water Department, Faculty of Agriculture, Kafrelsheikh University, Kafr El-Sheikh 33516, Egypt
4 Plant Biotechnology Department, Genetic Engineering Division, National Research Center, 33-El-Behouth Street, Cairo 12622, Egypt
5 Department of Applied Biotechnology and Food Science (BME-ABÉT), Szent Gellért Square 4, 1111 Budapest, Hungary; rozbachgreti@gmail.com (M.R.); fehercsaba87@gmail.com (C.F.)
6 Department of Biological and Environmental Sciences, Faculty of Home Economic, Al-Azhar University, Sperpay Number 4 Rd, Tanta, 31732, Egypt
* Correspondence: alshaaltarek@gmail.com

Abstract: Brown juice is a byproduct of fractionated green biomass during leaf protein isolation. It represents approximately 45%–50% of the total pressed fresh biomass. Disposal of brown juice is a serious issue in leaf protein production due to its high biological oxygen demand and carbohydrates content. The current study aimed to find a possible potential use of brown juice. Therefore, chemical and biochemical properties of brown juice—derived from alfalfa green biomass—were determined before and after fermentation by lactic acid bacteria. Additionally, the growth stimulation potential of fermented brown juice on plumed cockscomb (*Celosia argantea* var. plumose 'Arrabona') plants were tested. Celosia seedlings were sprayed at different rates of fermented brown juice (i.e., 0.5%, 1%, 2.5%, 5%, and 10%) and tap water was applied as control. The results revealed that lactic acid bacteria successfully enhanced the stabilization of brown juice via reducing sugars content and increasing organic acids content. After fermentation, contents of glucose monomers were 15 times lower; while concentrations of lactic and acetic acids increased by 7- and 10-fold, respectively. This caused a reduction in the pH of fermented brown juice by 13.9%. Treating Celosia plants at lower rates of fermented brown juice (up to 1.0%) significantly induced their growth dynamics and antioxidant capacity. Higher values of vegetative parameters were measured in treated plants compared to control. The brown juice treatments caused significant changes in histological parameters as well. The activity of catalase and peroxidase increased in plants that received fermented brown juice especially at low rates. Moreover, an increase in water-soluble protein and phenol was measured in different tissues of plants sprayed with fermented brown juice. Malondialdehyde content was lowered in treated plants compared to control. Fermented brown juice at high rates slightly reduced the amount of photosynthetic pigments; however, this reduction was not reported for low rates of fermented brown juice. These results surely illustrate the potential use of fermented alfalfa brown juice as a growth stimulator for crops particularly at rates below 2.5%.

Keywords: deproteinized leaf juice; fermentation; lactic acid bacteria; plant nutrition; antioxidant capacity; ornamental plants

---

1. Introduction

Due to the continuous growth in the global population (7.2 billion) and malnutrition, the global demand for the protein will increase in the next years [1]. The lack of protein supply has existed as a health problem for many years and is considered as one of the main types of malnutrition in developing and developed countries [2]. Over the next decades, a dramatic increase in the global protein demand is expected and overall protein consumption is predicted to nearly double by 2050. These rapid changes will create serious and accelerated pressure on land and water resources and their scarcity [3]. To meet the increased protein demand there are several approaches to introduce novel protein sources or alternatives [4,5]. The extraction of proteins from forage crops such as alfalfa, clover or grass is a potential process for the production of leaf protein concentrates (LPC), which can be utilized as feed or food but also hydrolyzed into amino acids for the cosmetics or pharma industries [6]. Alfalfa or lucerne plant is well known as the king of forage. It is a perennial flowering plant belonging to the legume family Fabaceae. This plant has several advantages including high-quality leaf protein (50%–60%), strong adaptability, high nutritional value, good taste, wide distribution, and stable productivity [7]. It can also yield crude protein 2-, 3-, and 4-fold higher than peas, soybean, and wheat, respectively [8]. Therefore, alfalfa nowadays is considered as the most promising crop for LPC. Isolation of leaf protein in form of LPC aims to extract solid or insoluble proteins (i.e., the protein of mitochondria, chloroplasts, nucleoprotein, and cell wall) and soluble proteins (i.e., the soluble fraction of mitochondrial proteins, chloroplast matrix, and cytoplasm proteins). Therefore, the thermal treatment of green juice obtained by pressing the fresh biomass is needed to coagulate these types of proteins. During coagulation of leaf protein, a brown liquid byproduct is produced, and it is referred to as "brown juice". One kilogram of fresh alfalfa biomass can produce up to 500 g of brown juice [9]. These large amounts of brown juice are rich in protein and phenols as well as micronutrients. Plant phenolic compounds are known to be able to modulate important physiological routes like signal transduction and transcriptional regulation. Phenolic compounds in brown juice associated with auxin bioregulators [10] prove that the disposal of these amounts of valuable brown juice is a waste. Disposal is high in its costs and will waste the nutritional value of this byproduct, which would be easily adaptable to the circular economy concept; a technology that generates no further waste by utilizing all the produced renewable resources [11–13]. The main product, the leaf protein produced by coagulation, is widely studied [14]; however, the brown liquid, also known as whey or brown juice [15], has limited literature especially in the case of the plant nutrition aspect. Brown juice is mentioned in some articles as DPJ (Deproteinized Plant Juice) [16] or deproteinized leaf extracts or leaf juice, deproteinized whey [17] as a byproduct of plant protein-producing technologies. DPJ can be applied for several purposes; for instance, as a fertilizer for plants, milk for calves, excellent fodder for cattle and rabbits, medium for microbial growth, and also for in vitro rhizogenesis [18–21]. The dry matter and protein content of brown juice range from 13% to 15% and 16%–20%, respectively, whereas the cellulose content is 25%–30% [18]. The alfalfa brown juice has a dry matter content of 4%–8% which is influenced by the species, varieties, weather conditions, phenophase, methods of harvest, and processing.

Several microorganisms like lactic acid bacteria (LAB) are useful, having advantageous features, and can be found in a range of locations from soil and natural water, to the surface of plants up to the human intestinal tract [22]. These microorganisms have been applied for decades in the fermentation processes of raw materials because of their beneficial effects. It has been validated that ferments containing lactic acid bacteria (or other PGPB—plant growth-promoting bacteria) (isolated from different sources) have plant growth-promoting properties [23]. Lactic acid bacteria containing ferments were proven to be effective biofertilizers, biocontrol agents, and biostimulants because they promote plant health, growth, and resilience as they improve nutrient availability [24], however, the functional roles of these bacteria in the phytomicrobiome have not been discovered yet [25]. *Celosia* genus is native to tropical America and Africa. *Celosia argentea* is a food crop in West Africa as well as a medicinal plant in China and India with considerable pharmacological properties [26]. Among 13 green leafy vegetables, *Celosia argentea* was one of the few that had exceptionally high iron (13.5 mg 100 g^{-1}), calcium (188 mg 100 g^{-1}), sodium (240.6 mg 100 g^{-1}), ascorbic acid (26 mg 100 g^{-1}), and β-carotene (4.42 mg 100 g^{-1}) content. The edible portion of *Celosia argentea* was found to be 55 g 100 g^{-1} fresh weight which was one of the highest, while its moisture and protein content was found to be 87.6 and 3.2 g 100 g^{-1}, respectively [27]. Plumosa Group of *Celosia argentea* is an attractive ornamental plant characterized by a wide range in size and color of flowers. Plumosa cultivars can grow from dwarf to tall. The inflorescence of narrow pyramidal, plume-like, is consistent with tiny, vivaciously colored (e.g., orange, red, purple, yellow) flowers.

This research aimed to enhance the stability of stored brown juice through fermentation by lactic acid bacteria; assess physiochemical traits of alfalfa brown juice before and after fermentation; determine whether the different fermented brown juice concentrations have any impact on the formation of the stem's anatomy; and evaluate the potential of fermented brown juice as a growth stimulator using *Celosia argantea* var. *plumosa* as a model plant.

2. Materials and Methods

2.1. Brown Juice Production and Its Characteristics

2.1.1. Source of Alfalfa Biomass

A field experiment of alfalfa (*Medicago sativa* L. var. Hunor-40) was carried out during 2017 and 2018, under the GINOP (2.2.1-15-2017-00051) project labeled Proteomill [28], at the experimental farm of Tedej Zrt., Hajdúnánás, Hungary. The seeds were sown on chernozem soil at the rate of 25 kg ha^{-1}. All recommended agronomic practices such as irrigation, weed control, and fertilization were done. The alfalfa fresh biomass was used as a source for brown juice. The first cut of alfalfa plants was carried out in the middle of May 2018 directly before the flowering stage since at this time protein in alfalfa biomass is at its highest content. Plants were harvested early morning and directly transferred in special boxes to the laboratory to avoid the degradation of protein by protease enzyme.

2.1.2. Extraction of Brown Juice

Alfalfa fresh biomass was fractionated into the fiber, leaf protein concentrate (LPC), and deproteinized plant juice (DPJ, brown juice) as follows: fresh biomass was pressed and pulped mechanically using Angel Juicer (5500, Angel Ltd., Praha, Czech Republic) into fiber and green juice fractions. Later, the green juice was thermally treated at 80 °C in order to coagulate mainly the chloroplastic and cytoplasmic proteins. After thermal coagulation, the mixture was left at room temperature for approximately 10 min, then the coagulant was separated from brown juice using moistened 100% natural unbleached cotton cloth filter (pore size = 10 microns).

2.1.3. Fermentation of Brown Juice

Fermentation of brown juice was necessary to increase the stability of brown juice and its storage period because fresh brown juice rapidly spoils due to high sugar and protein content. After cooling, the brown juice was transferred into a 20-L container and inoculated with AdiSil LG-100 Perfect (Fides Agro, Šardice, Czech Republic) containing heterofermentative lactic acid bacterial cultures (10^{11} CFU g^{-1}, *Pediococcus acidilactici, Lactobacillus paracasei, Lactobacillus plantarum*) at the rate of 0.01 g L^{-1}. The inoculated samples were kept at 35 °C for 48 h.

2.1.4. Determination of Lactic Acid Bacteria

The qualitative measurement of lactic acid bacteria in the fermented brown juice was determined at the end of the fermentation process by methylene blue test [29]. Briefly, 1 mL methylene blue reagent was added to 10 mL fermented brown juice, and then the samples were incubated at 37 °C for 48 h. The time needed for the disappearing of blue color is an indication of lactic acid bacteria density in the solution.

2.1.5. Chemical Properties of Brown Juice

The pH of brown juice was measured by pH-meter (Mettler Toledo S20 Seven Easy, Switzerland). Electrical conductivity (EC) was determined using EC-meter (Thermo Scientific, Orion Model $209A^+$ type, Germany). Degree Brix was recorded manually by a refractometer (RBR32-ATC, Germany). The content of macro- and micro-elements in brown juice before and after fermentation was measured using HNO_3-H_2O_2 wet digestion method as described by Kovács et al. [30]. Briefly, 1 g lyophilized brown juice was weighed into a Kjeldahl digestion tube, then 10 mL HNO_3 (99%, VWR International, USA) was added. The mixture was placed on the heater at 100 °C for 45 min; after cooling 5 mL H_2O_2 (30%, Sigma-Aldrich, St. Louis, MO, USA) was added for complete oxidation of organic materials and samples were kept on the heater for additional 45 min at 120 °C. After cooling the sample volume was brought to 50 mL using distilled water and then filtered using MN 640 W filter paper. The elemental content of brown juice was measured by ICP-OES spectrometer (Perkin Elmer made OPTIMA 3300 DV, Pittsboro, NC, USA).

Total phenol content in brown juice was determined spectrophotometerically using Ultrospec spectrophotometer (2100 pro, Amersham BioSciences, Amersham, United Kingdom) as previously described by Boór and Bélafiné Bakó [31]. Determination of total N content was carried out by Kjeldahl method [32] (Sparks et al., 1996). Concentrations of glucose and organic acids were determined by HPLC using BioRad (Hercules, CA, USA) Aminex HPX-87H (300 × 7.8 mm) column at 65 °C, and a refractive index detector. The eluent was 5 mmol L^{-1} H_2SO_4 at a flow rate of 0.5 mL min^{-1}. The injection volume was 40 μL. Concentrations of fructose, xylose, and arabinose were determined by HPLC using Phenomenex (Torrance, CA, USA) Rezex RPM-Monosaccharide Pb^{+2} (300 × 7.8 mm) column at 80 °C, and a refractive index detector. The eluent was ultrapure (milli-Q) water at a flow rate of 0.5 mL min^{-1}. The injection volume was 40 μL. Total sugars include monomer sugars and sugar oligomers solubilized. Monomer sugar concentrations were determined by HPLC after a sample preparation of 5 min boiling followed by centrifugation (5000 rpm) to eliminate residual proteins. To determine the oligomer sugar content of the samples, weak acid hydrolysis was performed. The samples were mixed with 8 w/w % H_2SO_4 at a volume ratio of 1:1 and treated at 120 °C in the autoclave for 15 min to decompose sugar oligomers into monomers, which were determined by HPLC.

2.2. *Celosia Experiment*

This experiment was carried out to assess the potential use of brown juice as a plant growth stimulator. In the present study, Celosia (*Celosia argantea* var. *plumosa* 'Arrabona') was used as a model plant for examining physiological, biochemical, and anatomical responses to fermented brown juice in our department and the National Agricultural Research and Innovation Center (NARIC, Budapest, Hungary). Celosia seeds were obtained from NARIC.

2.2.1. Experimental Design

A greenhouse pot experiment was carried out at the NARIC. The experimental layout was the Randomized Complete Block design (RCB) with 15 replicates. A polyethylene pot (7 × 7 × 8 cm) was filled with potting soil for horticultural crops (Klassman-Deilmann TS 3 FINE type, Geeste, Germany). The physical and chemical properties of potting soil are structure fine, pH (H_2O) 6, N 140 mg L^{-1}, P (P_2O_5) 100 mg L^{-1}, K (K_2O) 180 mg L^{-1}, Mg 100 mg L^{-1}, S 150 mg L^{-1}. Seeds of Celosia were sown in nursery substrate on 16th July 2018 and 4 days later germinated seeds were fertilized using different rates of brown juice. After two weeks, identical and healthy seedlings were transferred to the pots. Fermented brown juice was applied as a foliar application at rates of 0.5%, 1.0%, 2.5%, 5%, and 10%. The final application volume was 250 mL and equally shared among all replicates of the same treatment. The control plants were sprayed with tap water. Brown juice was applied once a week from starting the experiment on 16th July until 14th August, then we applied brown juice twice a week (Tuesdays and Fridays) until the end of the experiment on 11th September. At the end of the experiment, the following vegetative parameters were measured: root and stem length, root and stem volume, root and stem fresh and dry mass, and the number of leaves.

2.2.2. Determination of Water-Soluble Protein and Antioxidant Enzymes

Water-soluble protein fraction of lyophilized root, stem, and leaf tissues was determined using Coomassie Brilliant Blue G-250 according to Bradford [33] in triplicate with bovine serum albumin as standard. Briefly, 20 mg plant tissue was ground into homogenate in the mortar with quartz sand, then transferred into a volumetric flask, and then suspended in 100 mL distilled water to extract water-soluble protein fraction. The solution was centrifuged at 3000 rpm for 5 min. The supernatant was used for the assay of water-soluble protein content using UV-160A spectrophotometer (Shimadzu, Japan) at 595 nm. Peroxidase activity was determined in lyophilized roots, stems, and leaves of Celosia plants according to Roxas et al. [34]. Briefly, 100 mg plant tissue was macerated in 4 mL of phosphate buffer 0.01 M (pH 6.0). The homogenate was centrifuged at 13,000 rpm for 10 min to collect the supernatant. The supernatant was used to measure peroxidase activity using UV-160A spectrophotometer (Shimadzu, Japan) at 460 nm for 1 min. The unit of peroxidase activity was defined with the increase of one unit of absorbance per mL^{-1} min^{-1} g^{-1} of dry matter. Catalase (CAT) activity in lyophilized Celosia leaves was measured by following the decomposition of hydrogen peroxide at 240 nm according to Woodbury et al. [35]. The reaction included 0.2 mL supernatant, 1.5 mL phosphate buffer (pH 7.8, 0.2 M), and 1 mL distilled water. The colorimetric determination of CAT was conducted by the model UV-160A spectrophotometer (Shimadzu, Japan) at 240 nm. The biochemical reaction was initiated by adding 0.3 mL 0.1 M H_2O_2. The activity of CAT was expressed as μmol H_2O_2 consumed/mg protein/min.

2.2.3. Malondialdehyde Measurement

The malondialdehyde (MDA) content was determined from roots, stems, and leaves of Celosia plants by the method of Zhang and Huang [36]. Briefly, 100 mg lyophilized sample was homogenized in 1 mL 0.1% (*w/v*) TCA solution using cold mortar and pestle. The homogenates were centrifuged at 10,000× *g* for 10 min. Then, 4 mL of 0.5% thiobarbituric acid (TBA) in 20% TCA solution was added into 1 mL of supernatant and incubated at 96 °C for 30 min. The tubes were cooled by transferring into an ice bath. The absorbance of the supernatant was recorded at 532 nm. The standard curve was generated from MDA standard. The concentration of MDA of samples was calculated from the absorbance knowing calibration curve.

2.2.4. Photosynthetic Pigment

The photosynthetic pigment content of Celosia leaves was measured spectrophotomertically based on methods described by Porra et al. [37]. For the sample preparation, the leaf disc was cut and the chlorophyll content was extracted by N'N dimethyl-formamide overnight. The absorbance was measured by spectrophotometer (Amersham Biosciences Ultrospec 2100 Pro UV/Visible) on 663 and 645 nm wavelengths and from these data, the chl a, b, a + b, and a/b ratio were calculated.

2.2.5. Histology

We used three specimens per treatment for the stem's histological examination. Each plant was cut into smaller pieces and the third internodes (from beneath) fixed separately in a mixture of glycerin:alcohol:water (1:1:1) for a week. Then, several cross-sections were prepared using blades, after clarification, they were stained with Toluidin-blue. All analyses were performed under a light microscope (Zeiss Axioscope 2+; Zeiss International, Oberkochen, Ostalbkreis, Germany) with a compatible camera, and the Scope Photo software (Scopetek, München, Germany) was used for processing the images. For the measurement, we used at least 15 different cross-sections per internodes. The measured parameters were thick at the epidermis, primary cortex, pith, primary and secondary vascular tissue.

2.3. Statistical Analysis

Before the ANOVA test, Levene's Test for Equality of Variances was performed. The Levene's test for different variables at the six treatments of brown juice (i.e., 0%, 0.5%, 1%, 2.5%, 5%, and 10%) was negative, $p < 0.05$, and then the variances showed homogeneity. Results of the experiments were subjected to one-way (for fresh and dry weight, chlorophyll pigments, protein, MDA, POD, and catalase) and two-way (for root and shoot lengths, root and shoot volumes, and number of leaves) ANOVA by 'SigmaPlot 12.0' software and the means were compared by Duncan's Multiple Range Test [38] at $p < 0.05$.

3. Results

3.1. Characteristics of Brown Juice

3.1.1. Chemical Traits of Brown Juice

The fermentation of brown juice significantly changed its chemical properties (Table 1). Inoculation of fresh brown juice by lactic acid bacteria under anaerobic conditions caused a 13.9% reduction in pH. The degree Brix slightly increased after fermentation as it changed from 7.03 to 7.20. Total phenolic content dropped down after fermentation by almost 33.4%. Moreover, EC of fermented brown juice was 25.2% lower than fresh brown juice. Additionally, the density of brown color, that brown juice has, was reduced as its absorbance at 430 nm was diminished by 35.9%.

Table 1. Physiochemical characteristics of alfalfa brown juice before and after fermentation using lactobacillus.

Parameter	Before	After
pH	4.54 ± 0.03	3.91 ± 0.05
Brix [†] (%)	7.03 ± 0.02	7.20 ± 0.01
Total phenolic content (μg mL^{-1})	36.5 ± 1.19	24.26 ± 0.55
Electrical conductivity (dS m^{-1})	11.13 ± 0.11	8.47 ± 0.06
Color-absorbance (at 430 nm)	0.594 ± 0.006	0.381 ± 0.004
Lactic acid bacteria (CFU × 10^8 per mL)	11.33 ± 4.04	8.00 ± 4.36
Sugars content (g L^{-1})		
Glucose monomer *H*	21.19 ± 0.64	1.33 ± 0.03
Glucose oligomer *H*	2.80 ± 0.58	Nd [‡]
Xylose monomer *Pb*	12.0 ± 0.06	nd
Xylose oligomer *Pb*	1.90 ± 0.03	0.60 ± 0.02
Arabinose monomer *Pb*	nd	0.10 ± 0.01
Arabinose oligomer *Pb*	1.50 ± 1.16	0.64 ± 0.01
Fructose monomer *Pb*	3.70 ± 0.02	0.89 ± 0.01
Fructose oligomer *Pb*	nd	nd
Acids content (g L^{-1})		
Acetic acid *H*	1.5 ± 0.02	10.4 ± 0.03
Lactic acid *H*	5.0 ± 0.25	50.1 ± 0.68
Propionic acid *H*	nd	1.2 ± 0.02

Notes: [†] Degree Brix = water-soluble sugar content (one degree Brix means 1 g of sucrose in 100 mL aqueous solution); [‡] nd = not detected; sample size (n = 6); *H*-samples run on Aminex HPX 87 H column; *Pb*-samples run on Aminex HPX 87 Pb column.

3.1.2. Contents of Sugars and Organic Acids in the Brown Juice

Furthermore, the effect of lactic acid bacteria was not only reflected in the chemical characteristics of brown juice but also was noticed in sugars content. Interestingly, contents of monomer and oligomer forms of glucose, xylose, arabinose, and fructose reduced after fermentation, except arabinose monomer which was below the detected limit in fresh brown juice and became 0.1 g L^{-1} after fermentation; also, no fructose oligomer was detected either in fresh or fermented brown juice samples (Table 1). The highest decrease was found for glucose monomer as it lowered by 16 times in fermented brown juice compared to fresh brown juice. Fructose monomer, also, was four times lower in fermented brown juice, while arabinose oligomer recorded a decrease of 57.3% (Table 1). In contrast to sugars content, organic acids such as acetic, lactic, and propionic acids were considerably increased after fermentation by lactic acid bacteria. The content of lactic acid was 10-fold higher in fermented brown juice, as the highest recorded increase for any measured organic acid, while acetic acid content changed by seven times higher. Propionic acid content was below the detected limit in fresh brown juice; however, after fermentation it increased, recording 1.2 g L^{-1} (Table 1).

3.1.3. Macro- and Microelements Content of Brown Juice

Content of macro-and microelements of brown juice meaningfully changed due to fermentation by lactic acid bacteria (Table 2). Fermentation of brown juice resulted in a substantial reduction in the concentration of N, P, K, and S by 11%, 32%, 38%, and 21%, respectively. Otherwise, the contents of other elements displayed in Table 2 were found to be considerably higher after treating brown juice with lactic acid bacteria under anaerobic conditions. Interestingly, concentrations of Ca, Mg, Mo, Sr, and Ba were increased by 55%, 63%, 36%, 54%, and 109%, respectively. Furthermore, Na, Mn, Fe, Zn, B, and Al contents were 14.5-, 2.0-, 11.0-, 2.5-, 1.5-, and 5.7-fold higher, respectively, in fermented brown juice than fresh brown juice. No Cu was detected in brown juice either fresh or fermented.

Table 2. Content of macro- and microelements (mg L^{-1}) in alfalfa brown juice before and after fermentation.

Elements	Before	After
N	18.24 ± 0.66 †	16.19 ± 0.01
P	286 ± 30	238 ± 5.98
K	6090 ± 571	5276 ± 153
Ca	1270 ± 70	2326 ± 66.58
Mg	379 ± 16	739 ± 24.83
Na	31.03 ± 8.70	452 ± 15.02
S	352 ± 22	425 ± 6.57
Mn	1.61 ± 0.13	4.66 ± 0.09
Mo	0.29 ± 0.12	0.45 ± 0.01
Fe	2.04 ± 0.45	32.20 ± 0.86
Cu	0.06 ± 0.05	nd ‡
Zn	2.03 ± 0.19	8.59 ± 0.21
Sr	5.46 ± 0.24	8.80 ± 0.23
B	3.69 ± 0.33	11.61 ± 0.33
Al	0.24 ± 0.27	1.67 ± 0.13
Ba	0.39 ± 0.02	0.94 ± 0.02

Notes: † Standard deviation; ‡ not detected; sample size (n = 6).

3.2. Fermented Brown Juice as A Growth Stimulator

The possible utilization of fermented brown juice as a growth stimulator was evaluated. Celosia seedlings were treated with different doses of fermented brown juice through foliar application.

3.2.1. Growth Dynamic of Celosia

Spraying of Celosia seedlings with fermented brown juice significantly induced the development of stems (Figure 1). The application of brown juice at low concentrations had better effects on plant growth than higher concentrations. Spraying Celosia plants with 0.5% of fermented brown juice resulted in the tallest stem (26.0 cm); however, higher concentrations drastically diminished stem length. For instance, at the rate of 10% fermented brown juice stem length was 15.6 cm (Figure 1A). The root system of Celosia plants responded to fermented brown juice differently to the shoot part. All rates of brown juice resulted in very similar lengths of root systems except the rate of 10% which caused a significant reduction in root systems (16.9 cm). However, the tallest root system was found in control plants sprayed with tap water (Figure 1A).

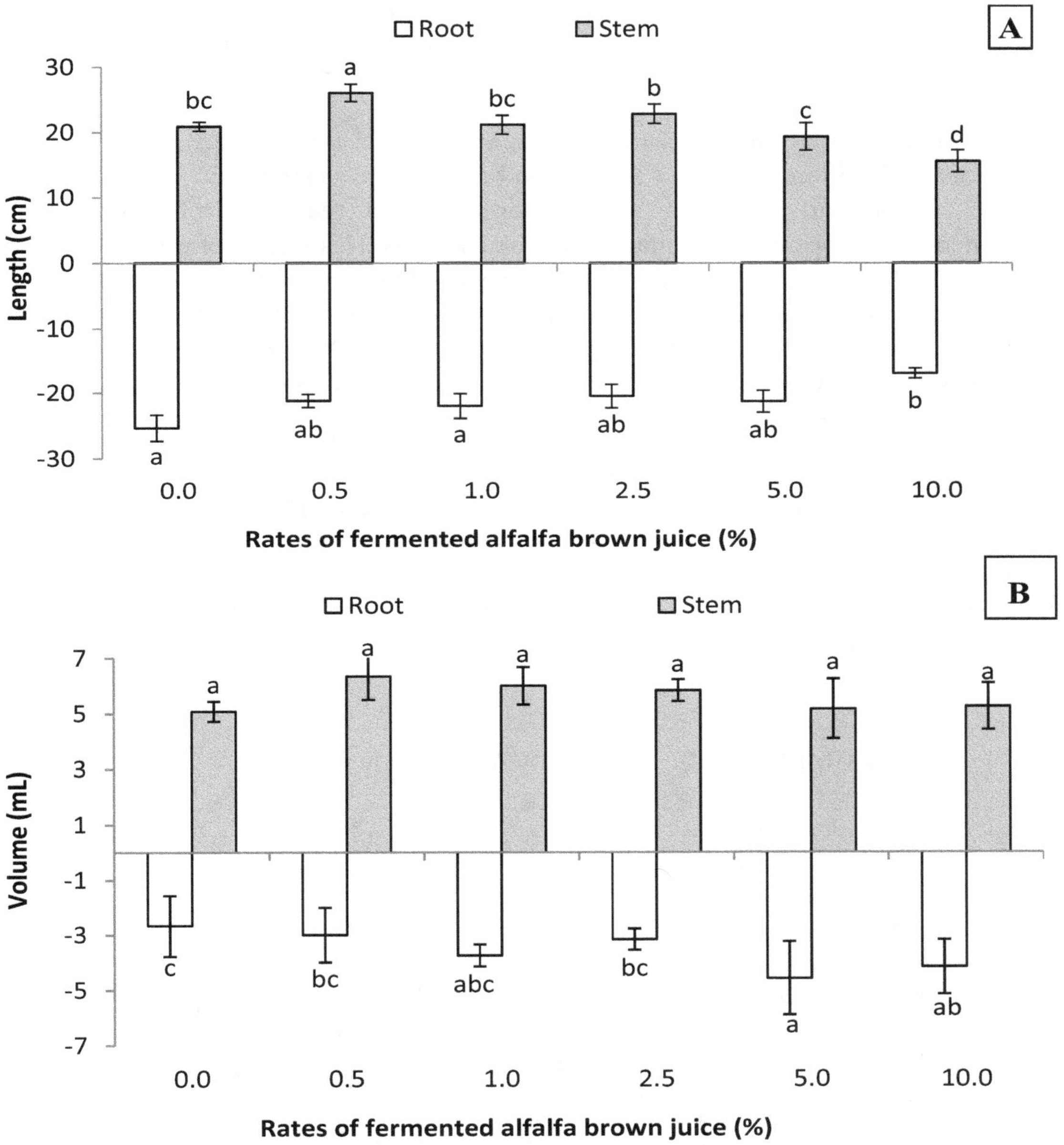

Figure 1. Length (**A**) and volume (**B**) of root and stem systems of Celosia plants fertilized at different rates of fermented alfalfa brown juice applied as a foliar application. Sample size ($n = 6$). Different letters above the same columns show significant differences at the level of $p < 0.05$.

Although, length of shoot and root systems is considered as a good indicator for plant growth and its response to the newly added fertilizers and/or stimulators, alone it does not precisely describe the real status of plant health. Therefore, to have a comprehensive description of the shoot and root systems, their volumes should be also measured. This is very essential particularly to describe the root system and its architecture as shoot parts respond to growth conditions in a proportional way. Concerning stem volume, similar findings as for its length were reported. At lower rate of fermented brown juice (0.5%) the highest volume of stem (6.0 cm^3) was measured while increasing the rate of fermented brown juice gradually and significantly declined the stem volume and lowest volume (2.3 cm^3) was measured for plants sprayed at 10% fermented brown juice (Figure 1B).

Results of root volume presented in Figure 1B displayed that although control plants had the tallest root length, its volume was the lowest among all the treatments. This means that control plants had long roots but unbranched ones with few lateral roots. All treated Celosia plants with fermented brown juice showed higher root volume compared to control plants. The highest root volume was noticed at plants sprayed with 5% of fermented brown juice. Additionally, results show that higher rates of fermented brown juice, i.e., 5% and 10% resulted in higher measured root volumes (Figure 1B).

Fresh mass of different Celosia tissues (roots, stems, and leaves) significantly responded to spraying the plants with different rates of fermented brown juice as shown in Figure 2. Fresh mass of roots, stems, and leaves of all plant parts was higher for plants treated with fermented brown juice compared to control plants sprayed with tap water. The highest fresh mass of roots, stems, and leaves was 4.30, 8.87, and 8.49 g plant^{-1}, respectively, that measured at rates of 2.5%, 0.5%, and 5% fermented brown juice, respectively (Figure 2A). Control plants displayed the lowest dry mass of roots, stems, and the number of leaves 0.17, 0.31, and 0.55 g plant^{-1}, respectively; while sprayed plants with 2.5% fermented brown juice showed the highest determined dry mass 0.44, 0.64, and 0.86 g plant^{-1}, for roots, stems, and leaves, respectively (Figure 2B).

All rates of fermented brown juice, except 10%, significantly increased the number of leaves per plant (Figure 2C). Applying fermented brown juice at the rate of 10% significantly decreased the number of leaves not only compared to other fermented brown juice rates but also control plants. The highest number of leaves per plant was 18 and was counted for plants treated with 1% fermented brown juice. However, the differences between treatments of 0.5%, 1%, 2.5%, and 5% of fermented brown juice were not significant.

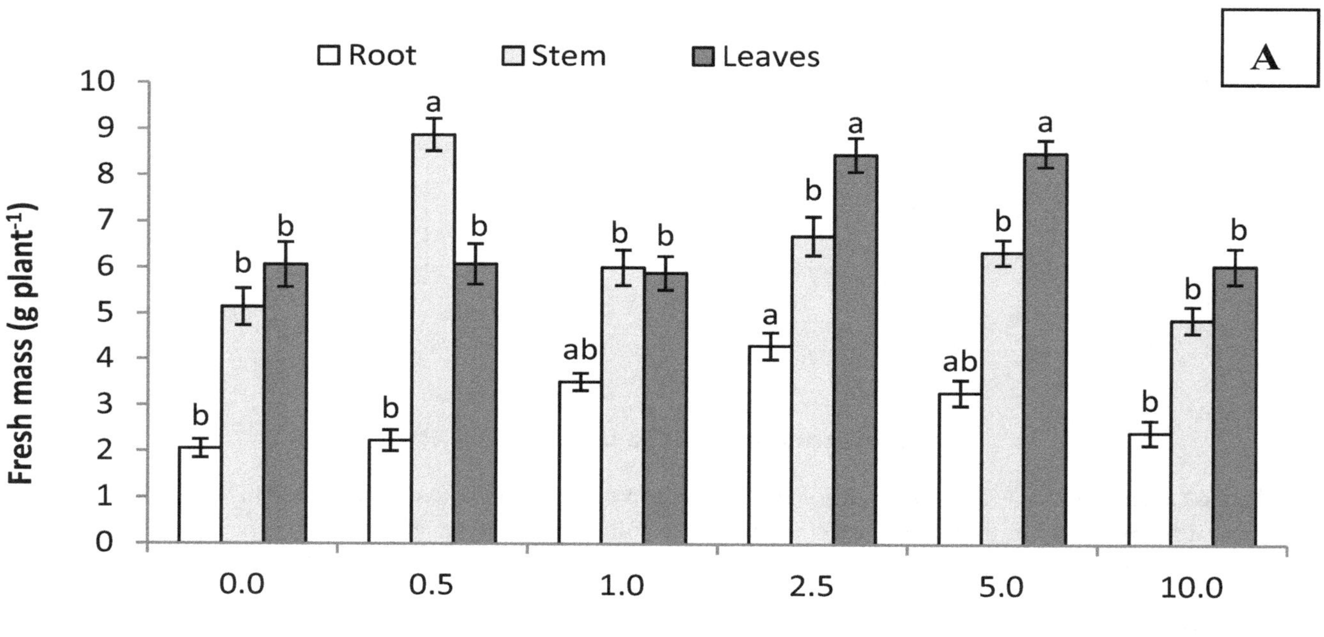

Figure 2. *Cont.*

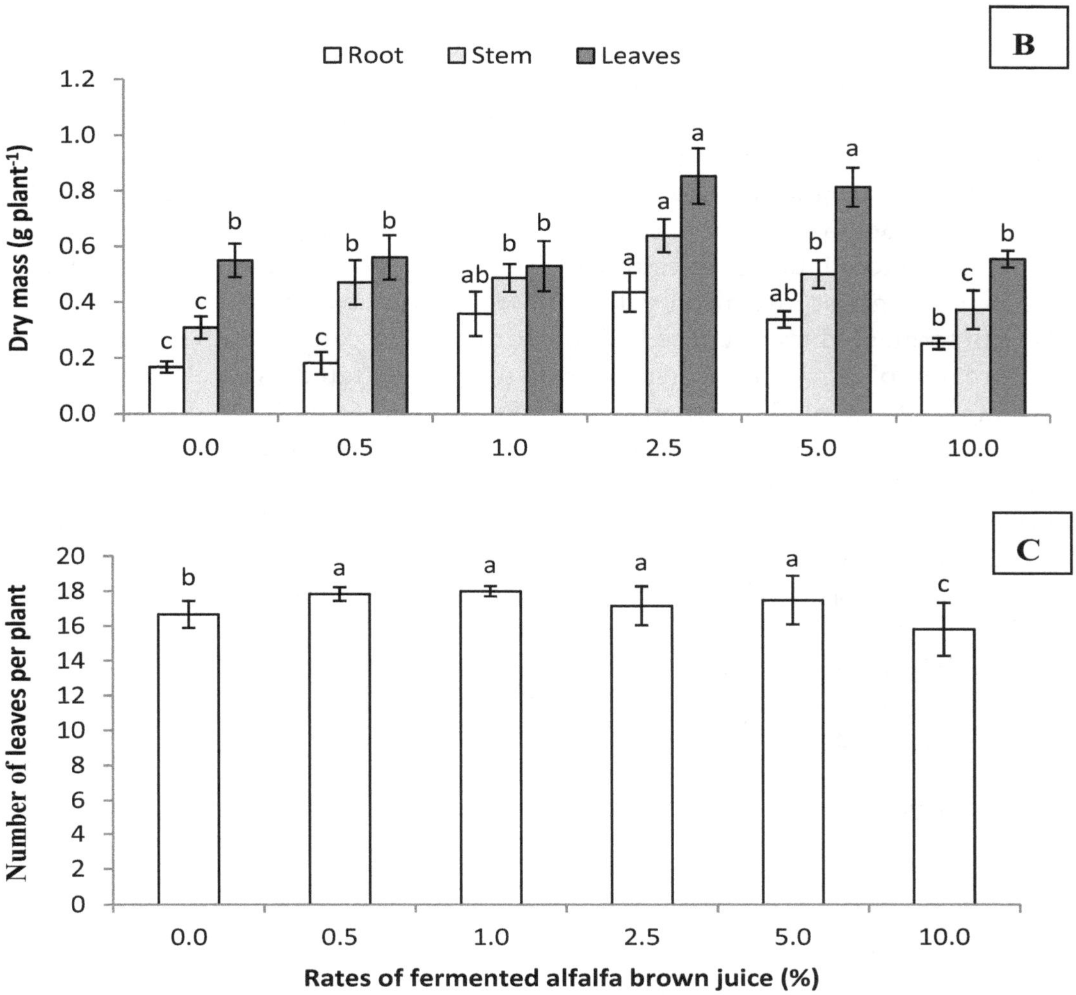

Figure 2. Fresh (**A**) and dry (**B**) masses and the number of leaves (**C**) of different plant tissues (roots, stems, and leaves) of Celosia plants sprayed at different rates of fermented alfalfa brown juice. Sample size ($n = 6$). Different letters above the same columns show significant differences at the level of $p < 0.05$.

3.2.2. Antioxidant Capacity of Celosia Plants Treated with Fermented Brown Juice

Spraying Celosia plants with fermented brown juice significantly induced the activity of catalase (CAT) enzyme in the leaves (Figure 3A). All treated plants had higher activities of CAT enzyme compared to control plants (sprayed with tap water). However, increasing the rate of applied brown juice gradually reduced the CAT activity up to 5%, but this reduction was still higher than the control. Treated Celosia plants at the rate of 10% achieved the highest CAT activity among all treatments (0.290 μmoL H_2O_2 consumed mg^{-1} protein min^{-1}).

Different Celosia plant tissues (i.e., root, stem, and leaf) showed a significant response of peroxidase enzyme activity (POD) to added fermented brown juice (Figure 3B). Higher rates of fermented brown juice above 1.0% resulted in higher POD activity in the root system than both lower rates and control plants. The root POD activity in treatments of 2.5%, 5%, and 10% of fermented brown juice was higher than lower rates and the control; however, no statistically significant differences were calculated among these treatments. Interestingly, applying fermented brown juice at the rate of 1% resulted in the lowest determined activity of POD in the root system among all treatments including the control plants. The activity of POD in the stem tissue of Celosia plants was totally in contrast to POD activity in the root system (Figure 3B). The high rates of fermented brown juice above 1% showed lower POD activity in the stem than low rates (i.e., 0.5% and 1%) and control plants. The lowest POD activity in stems was noticed when plants were sprayed at 10% fermented brown juice, while the highest measured POD activity in the stem was found for plants that received 0.5% fermented brown juice (Figure 3B). Except for treatments of 2.5% and 5% fermented brown juice, all other treatments including control plants showed similar POD activity in leaf tissue without significant differences. The highest leaf POD activity was measured in the leaves of treated plants with 2.5% fermented brown juice, while at the rate of 5% fermented brown juice the lowest leaf POD activity was determined (Figure 3B).

Malondialdehyde (MDA) content in different tissues of Celosia plants was measured as a marker for the degree of lipid peroxidation of unsaturated fatty acids due to oxidative stress. In the root system of Celosia plants, the highest measured value of MDA content was denoted in control plants. All treated plants with fermented brown juice had lower root MDA content than control plants. However, the response of treated plants with fermented brown juice hesitated as no clear trend was seen. The lowest applied rate 1% fermented brown juice showed the lowest root MDA content, while the highest root MDA content was measured in the root system of plants sprayed with 2.5% fermented brown juice (Figure 3C). In contrast to the root system, stem MDA content was found to increase as the rate of fermented brown juice increased up to 5% then reduced at the rate 10% recording the lowest MDA content in stem tissue among all treated plants with fermented brown juice. However, the lowest MDA content in the stem was displayed in control plants. Leaves of control plants showed higher MDA content than plants that received different rates of fermented brown juice. No significant differences were found in the MDA content of leaves of plants sprayed at the rates of 1%, 2.5%, and 5% fermented brown juice (Figure 3C). However, the lowest leaf MDA content was determined in the leaves of plants treated with 0.5% fermented brown juice.

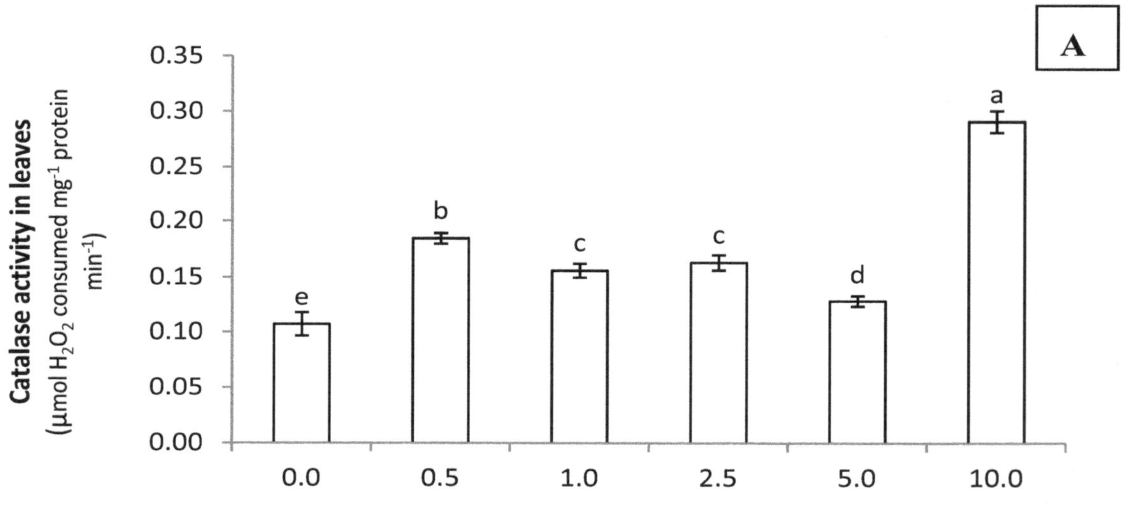

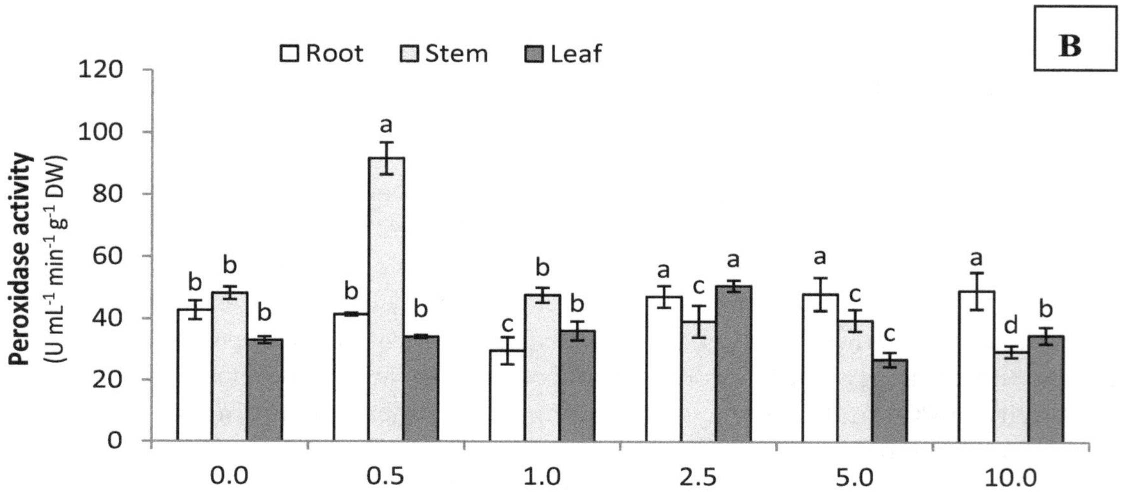

Figure 3. *Cont.*

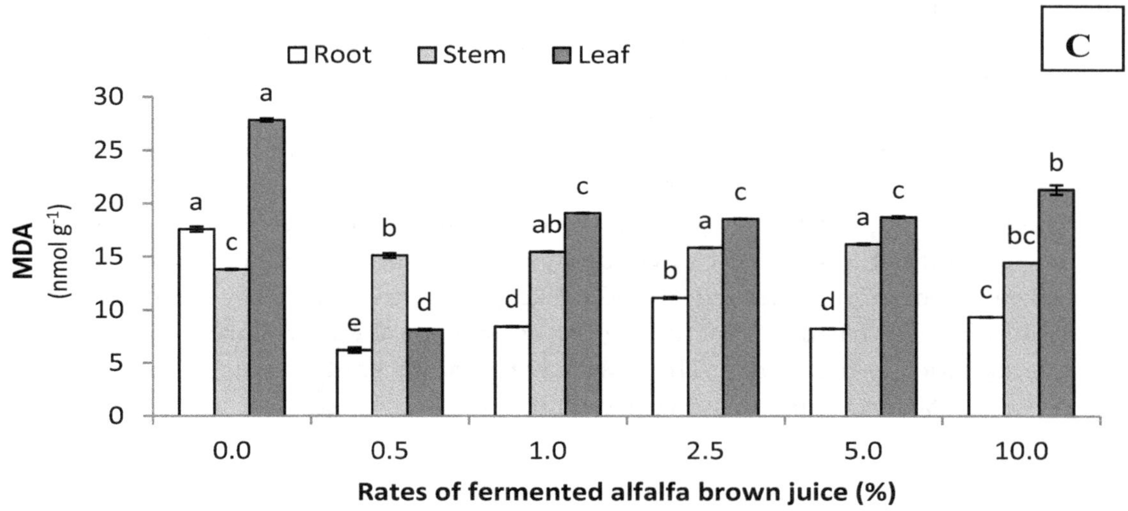

Figure 3. The activity of catalase (**A**) and peroxidase (**B**) and malondialdehyde content (**C**) in different plant tissues (roots, stems, and leaves) of Celosia plants sprayed at different rates of fermented alfalfa brown juice. Sample size (n = 6). Different letters above the same columns show significant differences at the level of $p < 0.05$.

3.2.3. Phenolic, Protein, and Photosynthetic Pigments Contents

The results of water-soluble phenol content are depicted in Figure 4A. Different plant tissues of Celosia plants possessed different water-soluble phenol contents as the root system showed the lowest content, while the highest water-soluble phenol content was measured in leaves. The addition of fermented brown juice as a foliar application to Celosia plants significantly affected the water-soluble phenol content in the root system. The highest water-soluble phenol content (46.7 μg g^{-1}) was measured in the root system of plants that received 0.5% fermented brown juice, while, when plants were allowed to grow in the presence of 10% fermented brown juice, the water-soluble phenol content was 12.8 μg g^{-1} (Figure 4A). The root system of the control plant displayed 18.6 μg g^{-1} water-soluble phenol content. In stem tissues, water-soluble phenol content in plants treated with fermented brown juice showed lower water-soluble phenol content than control plants. Increasing the rate of fermented brown juice up to 2.5% gradually increased the content of water-soluble phenol in stem tissues, then a linear increase was recorded when rates of fermented brown juice were increased up to 10%. The highest stem water-soluble phenol content (56.8 μg g^{-1}) was measured for control plants (Figure 4A). Except for treatment of 0.5% fermented brown juice, all fermented brown juice rates showed higher water-soluble phenol content in leaf tissues. The highest water-soluble phenol content (μg g^{-1}) was measured in leaves of plants that received 1% fermented brown juice; then a gradual decrease was noticed with increasing the rate of fermented brown juice up to 10% (Figure 4A).

The content of water-soluble protein was higher in leaf tissue followed by the root system, while the lowest content was denoted in stem tissue. Significantly, the application of fermented brown juice improved water-soluble protein content in the root system. Treatments of 1%, 2.5%, and 10% fermented brown juice displayed higher water-soluble protein content than control and treatments of 0.5% and 5% (Figure 4B). The lowest water-soluble protein content (2.18 mg g^{-1}) was measured in the root system of plants treated with 0.5% of fermented brown juice. Similar results were found in stem tissue of Celosia plants, where all treated plants with fermented brown juice had higher water-soluble protein content than control except treatment of 0.5% fermented brown juice. Although the content of water-soluble protein in leaves was higher than measured in the root system, the trend in which roots and leaves responded to spraying with fermented brown juice was almost the same. Leaf water-soluble protein contents in plants of treatments of 0.5% and 5% were the lowest among all treatments including the control. Other fermented brown juice rates enhanced the water-soluble protein content in leaf tissue over the control plants (Figure 4B).

Significant differences were noticed in a few cases among treatments for chlorophyll pigments content (Figure 4C). Content of *chl a* was reduced gradually with increasing the rate of applied fermented brown juice. Application of fermented brown juice at low rates (i.e., 0.5%) significantly improves the *chl a* content recording the highest value among all other treatments but was similar to control plants. On the other hand, the *chl b* content was found to respond negatively to increasing the rate of applied fermented brown juice as a gradual significant reduction was noticed. Content of total *chl a + b* displayed a similar tendency as it slightly decreased with increasing the rate of fermented brown juice. The low rate of fermented brown juice showed a slightly higher content than the control, but this increase was not significant (Figure 4C). Except treatment of 0.5% fermented brown juice, carotenoids content in all treatments including the control showed similar values as no significant differences were statistically measured. The lowest carotenoids content was determined in leaves of plants that received 0.5% fermented brown juice (Figure 4C).

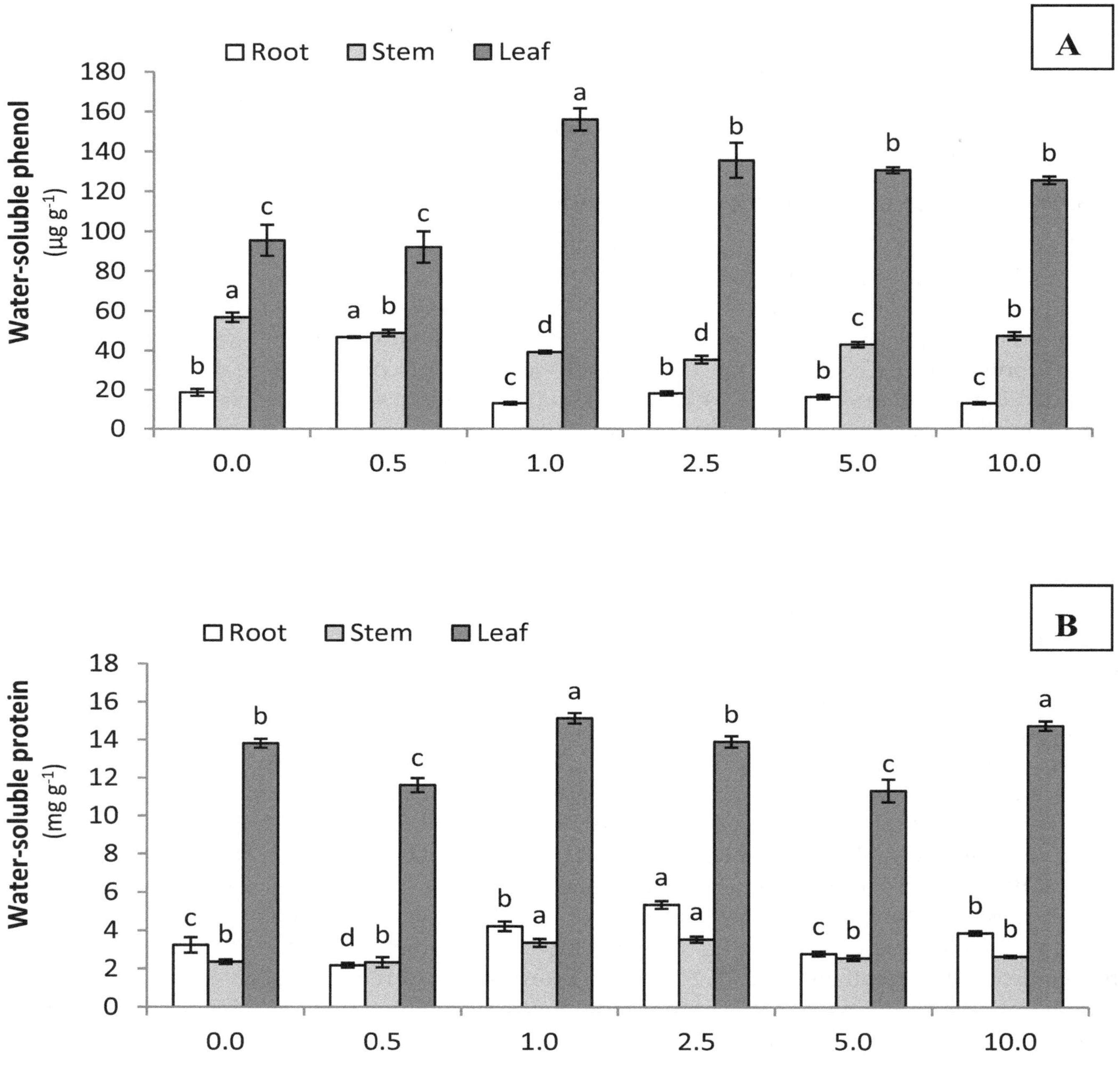

Figure 4. *Cont.*

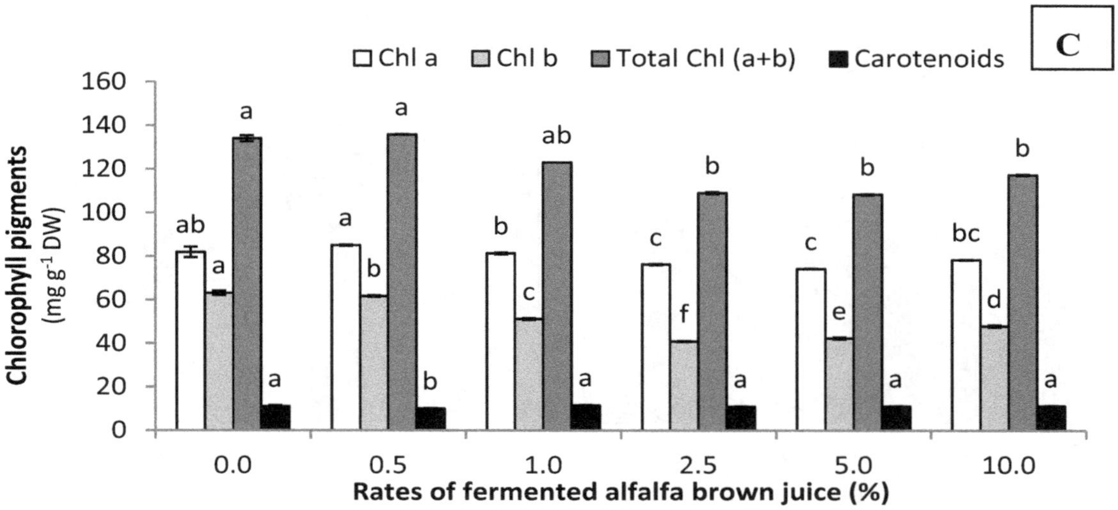

Figure 4. Water-soluble phenol (**A**), water-soluble protein (**B**), and chlorophyll pigments contents (**C**) in different plant tissues of Celosia plants sprayed at different rates of fermented alfalfa brown juice. Sample size ($n = 6$). Different letters above the same columns show significant differences at the level of $p < 0.05$.

3.2.4. Anatomical Features of Celosia Stem after Brown Juice Application

Regarding the cross-sections, 10–15 cm from the apex were analyzed, and the tissue structure was representative of an older Celosia's stem anatomy, with successive cambia [39,40]. Stems were covered by the epidermis (single row); beneath its primer cortex containing angular collenchyma (four to six cells thick) was visible. In the pith primary vascular bundles were located surrounded by a cylinder of anomalous cambium. Secondary and primary vascular tissues were separated by the conjunctive tissue [41]. Both the conjunctive tissue and the inner part of the pith were composed of parenchymatous cells (Figure 5).

There is no fundamental difference in the tissue structure in connection with the treatment, but there are significant differences in the thickness of the tissues, which support the differences that are visible to the naked eye too (e.g., thicker, stiffer stem). All levels of concentration caused a reduction in the thickness of the epidermis, while it was the 1% treatment that caused a reduction to a greater extent. The thickness of the primary cortex reinforced with angular collenchyma was decreased by most treatments, except the 0.5% and 10% treatments, where statistically verified thickening was observed. The proportion of pith involved vascular tissues increased for all treatments. The more concentrated brown juice treatments resulted in significantly thicker primary vascular tissues, except the 1% and the 10% treatments. Growing of the secondary vascular tissue was the highest at the 0.5% treatment, where the new successive cambium formed almost entirely closed xylem and phloem, significantly contributing to the strength of the stem (Table 3).

Table 3. Impact of different concentrations of fermented brown juice on stem tissue of *Celosia argentea* var. *plumosa* (μm) (mean ± SD, $n = 45$).

	Epidermis	Cortex	Pith	Primary Vascular Bundle	Secondary Vascular Tissue
Cont.	31.44 ± 5.28 a	356.14 ± 57.69 ab	1873.30 ± 295.29 b	236.47 ± 79.43 b	221.47 ± 51.79 b
0.5%	30.81 ± 6.72 a	374.10 ± 99.50 ab	1903.79 ± 187.39 b	271.65 ± 73.59 ab	295.59 ±76.55 a
1%	25.83 ± 3.59 b	322.42 ± 61.76 bc	2033.30 ± 205.45 b	242.23 ± 41.22 b	212.56 ± 51.56 b
2.5%	28.50 ± 4.42 ab	261.69 ± 19.64 d	2011.11 ± 198.88 b	303.98 ± 88.94 a	230.18 ± 73.69 b
5%	31.43 ± 3.81 a	339.46 ± 68.14 b	2227.77 ± 310.33 a	348.30 ±122.47 a	256.93 ± 66.78 a
10%	28.69 ± 3.49 ab	392.07 ± 91.05 a	1915.50 ± 209.11 b	310.12 ± 102.47 a	274.80 ± 89.62 a

Notes: Different letters in each column indicate statistically significant differences ($p < 0.05$).

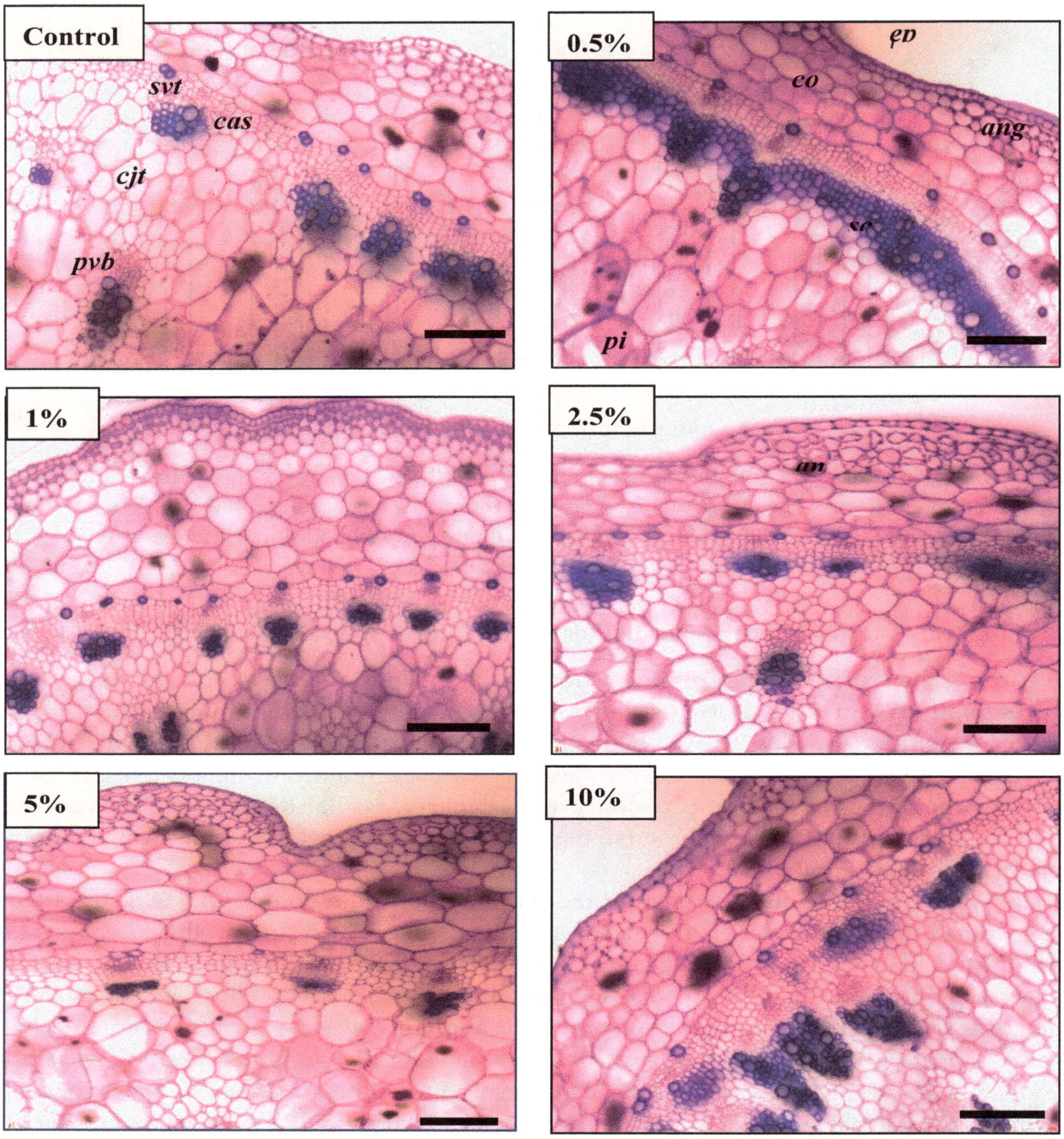

Figure 5. Anatomical sections of *Celosia argentea* var. *plumosa* stem. ***ep*** epidermis, ***co*** cortex, ***ang*** angular collenchyma, ***pi*** pith, ***sc*** successive cylinder (***svt*** secondary vascular tissue), ***ca*** cambium, ***cjt*** conjunctive tissue, ***pvb*** primary vascular bundle after spraying Celosia plants with different rates of fermented brown juice (i.e., control, 0.5%, 1%, 2.5%, 5%, and 10%). *Scale bar is* 200 μm.

To sum up, it can be stated that the brown juice treatments (applied as foliar) influence the proportions of the Celosia stem's tissue. As a result of the treatments, the thinning of the epidermis and the intense growth of the vascular tissues (especially the secondary vascular tissue) can be projected. The growth rate of secondary tissues within the pith is the highest at 0.5% treatment.

4. Discussion

Recently, isolation of protein from plant green leaves has gained increasing attention as an attempt to bridge the gap between protein production and demand due to the dramatic increase in population and the increase in living standards. Brown juice (referred to as deproteinized plant juice or DPJ as well) is a byproduct generated during the coagulation of soluble protein in green juice through thermal treatment. Brown juice has gained less attention than LPC and press cake. It represents nearly 50% of pressed and pulped fresh biomass [9]. Therefore, these huge amounts could be an obstacle facing the acceleration of this approach and its acceptance by both politicians and the public. Disposal of alfalfa brown juice is a serious issue in LPC production due to its high biological oxygen demand (BOD) and carbohydrates content [42]. Due to its richness in free amino acids, peptides, soluble sugars, vitamins, and many macro- and microelements, it can be directed towards animal feeding and production of many chemicals [43]. Additionally, it can be used as a fertilizer, growth stimulator and/or growth medium for microorganisms [42]. Although some pieces of literature have been reporting the possible utilization of brown juice as a ruminant feed [44], few studies have been focusing on brown juice as a fertilizer [42,45].

During our recent experiments on LPC production from alfalfa biomass, it became clear that the storage of the brown juice at room conditions leads to fast spoiling. Therefore, we had to store it below 4 °C. This may be due to its high carbohydrate content, which represents a suitable environment for bacteria to grow [42]. Therefore, converting sugars into organic acids and subsequent decrease in the pH of brown juice through fermentation using lactic acid bacteria seemed to be an ideal solution since fermented brown juice is stable and this facilitates its handling.

Lactic acid bacteria have long been known for their role in the fermentation of carbohydrates. Consequently, it has wide applications in medicine and food processing. Nowadays, lactic acid bacteria have been found to play an important role in agriculture, bioenergy production, and bioremediation of the environment [46]. In the present study we, firstly, aimed to stabilize alfalfa brown juice through reducing its water-soluble sugars content and pH using lactic acid bacteria. Accordingly, sugars content in brown juice was reduced after fermentation, because lactic acid bacteria use sugars as energy and carbon sources [47]. As shown in Table 1, most of the sugars in brown juice were found to be below the quantification limits after fermentation indicating that lactic acid bacteria consumed them. Comparable results have been previously presented by many researchers [48,49]. On the other hand, organic acids (e.g., lactic, acetic, and propionic acids) were increased in fermented brown juice causing a subsequent reduction in pH (Table 1). Novik, et al. [46] reported lactic acid as the main acid produced after fermentation of water-soluble sugars such as glucose and fructose either monomer or oligomer by lactic acid bacteria. Similar findings were cited by Bautista-Trujillo et al. [48], who observed a decrease in pH of maize silage after inoculation by lactic acid bacteria due to the increase in the production of organic acids, mainly lactic and acetic acids. Moreover, they reported an increase of 46.3% in lactic acid content. However, lactic acid content was found to increase by 8-fold after fermentation compared to unfermented brown juice (Table 1). This high increase in lactic acid content may be attributed to the initial low pH of fresh brown juice (4.54), which helps to hydrolyze the oligo- and polysaccharides, therefore they become available for lactic acid bacteria [48]. Additionally, another possible reason for high lactic acid content could be attributed to the high Mn content in brown juice. Cheng et al. [49] stated that applying Mn to Jerusalem artichoke juice enhanced the lactic acid production by lactic acid bacteria up to 12 g L^{-1}. Moreover, Dimitrovski et al. [50] stated that the fermentation of Jerusalem artichoke tuber juice by lactic acid bacteria reduced its pH from 6.5 to 4.7 after 30 h. In the current study, at the end of the fermentation process, the pH of brown juice was 3.91. Lactic acid bacteria significantly reduced the absorbance of brown juice by 35.9%. This result was supported by that previously cited by Kwaw et al. [51]. They studied the effect of different strains of lactic acid bacteria on colorimetric properties of mulberry juice, reporting a 6.9% reduction in the color. They referred this reduction to the increase in content of the monomeric anthocyanin. Although an increase in total phenolic content has been previously reported for fermented mulberry juice [51] and pomegranate juice [52], our results displayed a decrease of 33.4% after inoculation of brown juice by lactic acid bacteria. Except N, P, K, and S other macro- and microelements were higher in fermented alfalfa brown juice (Table 2). The reduction in concentrations of N, P, K, and S could be attributed to the fact that they are essential elements for the growth of lactic acid bacteria. However, similar findings were described by Kim [53], who stated that fermented kale juice had higher elemental composition than unfermented juice. Moreover, he cited significant differences between kale juice fermented by different lactic acid bacterial strains. The increase in the concentration of microelements, in particular, may be due to the increase of brown juice acidity. Although, the concentration of macronutrients (e.g., N, P, and K) was reduced after fermentation, the content of macro- and microelements is still high, and this makes the fermented brown juice a potential growth stimulator. On the whole, these results are supported by earlier findings of Ream et al. [45]. They reported that alfalfa brown juice contains a relatively high content of N and K, in addition to small amounts of P, Ca, Mg, and other microelements.

In the present study, fermented alfalfa brown juice as a growth stimulator was evaluated using the Celosia plant as a model. Brown juice was applied at different rates by foliar application. The foliar application of fermented brown juice showed a significant potential on the development of Celosia seedlings in comparison with control. Noticeably, increasing the application rate of brown juice sprayed on Celosia seedlings led to a considerable reduction in shoot parts, particularly the stem length. From a horticultural point of view, this seems to be a good result since the target is the flower not the vegetative growth of Celosia. Application of brown juice at low rates such as 0.5% and 1.0% enhanced the growth and resulted in high values of stem length, the volume of stem and root, fresh masses of stem and root, and number of leaves. Shorter but more branched root systems were observed when Celosia seedlings were sprayed with brown juice (Figure 2). This phenomenon is supported by data of length and the volume of roots (Figure 1). The beneficial effect of alfalfa brown juice could be attributed to its high content of macro- (i.e., N, P, K, Ca, and Mg) and microelements (i.e., S, Mn, Fe, Cu, Zn, and Mo); all in phyto-available forms (Table 2). Similarly, Ream et al. [45] observed that using brown juice as a fertilizer added at an annual rate of 1.25 cm induced the growth and yield of alfalfa, corn, and bromegrass; while, at the higher rate (2.5 cm) a reduction in yield and plant damage were noticed in all crops. However, they referred to the damage in plant growth caused by high rates of brown juice to unknown reasons; moreover, they considered it a not serious problem since the added amount of brown juice can be controlled. These results are supported by findings of Reddy et al. [42], who earlier stated that the application of alfalfa brown juice at low rates enhanced germination and growth of cowpea, mung bean, and groundnut; while high rates inhibited the germination and reduced the plant growth. They reported that alfalfa brown juice can be used as a fertilizer if it would be added at a lower level than 10%. Additionally, they could not give a reason for such damaging effects of high rates of brown juice, except what previously was mentioned by Pirie [54], who stated that alfalfa brown juice contains some phytotoxic organic compounds.

In our experiment, the reduction in plant growth of treated Celosia plants at high rates of fermented brown juice can be explained by high EC value and low pH of brown juice solutions. Increasing the rates of brown juice caused a gradual increase in EC and a decrease in pH as shown in Table 4. At treatment of 0.5% brown juice, EC (dS m^{-1}) and pH were 0.12 and 4.21, respectively; while, at the highest applied rate 10% they were 1.99 and 4.38, respectively. Low pH is not favorable for the development of plants; it reduces photosynthesis due to a reduction in stomatal conductance [55]. This might explain why we found diminished growth of Celosia plants treated at high rates of brown juice. However, in the current study, the lowest pH was 4.38 when Celosia plants received 10% fermented brown juice. Long et al. [56] had earlier reported a reduction in citrus growth below pH 4, while higher pH did not inhibit the growth and seedlings reached their maximum growth at pH 5. The reduction in growth may be attributed to H^+-toxicity which damages leaves. Absorption of nutrients applied as foliar application depends on the pH of the solution. Extreme pH (below 2 and above 12) was reported to burn the leaves. Moreover, some elements prefer different pH values for their optimum absorption by plant leaves.

Antioxidant enzymes such as CAT and POD are among the most important antioxidant enzymes which play a vital role in scavenging reactive oxygen species generated in cells due to different biotic and abiotic stresses [57]. Thus, enhancing the activity of these enzymes is considered an important step in improving the plants' tolerance to different kinds of stress [58,59]. The results abstracted from this research showed that the application of alfalfa brown juice after fermentation by lactic acid bacteria significantly increased the activity of CAT and POD in different Celosia tissues. However, the low rates of brown juice seemed to be more effective than higher ones, as a reduction in the activities was noticed. These results are confirmed by results of MDA, as treated plants with fermented brown juice had lower MDA content than control plants (untreated plants) regardless of the type of plant tissues. These results demonstrate that fermented brown juice can potentially be exploited as a growth stimulator, particularly at low rates. Besides, fermented brown juice had a significant effect on water-soluble phenol and protein contents, as they were higher in treated plants in comparison to control ones. The high rates of brown juice were found to reduce the photosynthetic pigment content. This could be attributed to low pH at high rates of fermented brown juice. This result was in accordance with that cited by Solati et al. [60], who reported a decrease in chlorophyll content due to low pH.

Brown juice could be very useful as a soil fertilizer/conditioner particularly in alkaline soils and/or sandy soil due to its rich composition in macro- and microelements and sugars. These could induce the microbial growth in soil increasing soil fertility. Additionally, sugars play an important role in soil stabilization through maintaining soil aggregation, which subsequently leads to better water holding capacity [42,61]. While delivering deeper insights into the possible use of alfalfa brown juice as a growth stimulator and trying to precisely determine the most effective rate and application method, there are many issues that should be addressed in the future [62]. Results, undoubtedly, suggest that brown juice tolerance can be plant species dependent; therefore, more studies on different plant species at different rates of brown juice are crucially needed. Additionally, phytotoxicity of brown juice should be the focus of future studies.

Table 4. pH and electrical conductivity (EC, dS m^{-1}) values of alfalfa brown juice solutions before and after fermentation at the beginning of the experiment.

Rates of Brown Juice (%)	pH		EC	
	Before	*After*	*Before*	*After*
0.5	4.65 ± 0.02 †	4.21 ± 0.07	0.15 ± 0.01	0.12 ± 0.03
1.0	4.67 ± 0.02	4.16 ± 0.00	0.28 ± 0.01	0.54 ± 0.58
2.5	4.68 ± 0.01	4.16 ± 0.01	0.67 ± 0.07	0.46 ± 0.02
5.0	4.72 ± 0.01	4.18 ± 0.00	1.20 ± 0.04	0.81 ± 0.05
10.0	4.72 ± 0.01	4.19 ± 0.00	2.26 ± 0.04	1.44 ± 0.12

Notes: † Standard deviation.

5. Conclusions

The present study highlights the possible use of alfalfa brown juice as a growth stimulator. Brown juice is a serious problem in LPC production, where its disposal represents a threat to the environment due to its high content of water-soluble sugars as well as macro- and microelements. Fermentation of brown juice using lactic acid bacteria significantly improved its nutritional value and stability, because these bacteria—as our data showed—produce a significant amount of organic acids i.e., lactic, acetic, and propionic acids through their metabolism making the nutrients more available and the pH of row material (brown juice) lower, thereby stabilizing it. Most water-soluble sugars were under the detectable level after fermentation as the bacteria used them as carbon source. Moreover, the concentration of nutrients increased—showing the effect of bacteria for nutrient availability—after fermentation except N, P, K, and S showed a slight decrease. In this study, treating *Celosia argentea*, a valuable ornamental species with significant food and medical uses, with low rates of fermented brown juice through foliar application significantly improved the growth, as all of the vegetative parameters such as stem and root length, shoot and root volume, fresh mass of stem and root, and the number of leaves increased. The brown juice treatments in low (0.5%) concentration caused positive changes in histological parameters, in the growth rate of secondary tissues. Additionally, fermented brown juice showed a considerable impact on the antioxidant capacity of Celosia plants, as CAT and

POD activities increased while MDA content decreased. Moreover, both water-soluble phenol and protein were found to increase in treated plants with fermented brown juice compared to the control showing the beneficial effect of lactic acid fermentation and chemical properties of brown juice. These results conclude and state the potential use of fermented alfalfa brown juice as a sustainable growth stimulator for crops with a particular interest in horticultural crops. Our data regarding the chemical and microbiological properties of brown juice and the effects (listed plant responses) it triggered confirm the scientific investigations where plant growth-promoting properties of lactic acid bacteria contribute greatly to the maintenance of the health of plants (also strengthening disease resistance) and by consuming these plants, they are also beneficial in human digestive processes. It should be noted that the sample size number was modest in our study that is why it was difficult to draw strong far-going conclusions, however preliminary conclusions support the fact that phytomicrobiome engineering can be a promising strategy for sustainable agriculture, but the data available is limited to understand properly these complex symbiotic relationships. Therefore, examination of fermented alfalfa brown juice's effect on physiological, biochemical, and anatomical parameters of other horticultural and agricultural crops is in progress.

Author Contributions: Conceptualization, N.B., S.K. and M.G.F.; methodology, D.B., N.E. and I.O.T.; software, D.B.; validation, D.B. and I.O.T.; formal analysis, N.A. and S.K.; investigation, M.R. and C.F.; resources, H.E.-R. data curation, T.A.; writing—original draft preparation, N.B. and D.B.; writing—review and editing, N.B., T.A., H.E.-R. and D.B.; visualization, S.K.; supervision, M.G.F.; project administration, N.B.; funding acquisition, N.B. All authors have read and agreed to the published version of the manuscript.

Acknowledgments: The research was financed by the Higher Education Institutional Excellence Programme (NKFIH-1150-6/2019) of the Ministry of Innovation and Technology in Hungary, within the framework of the Biotechnology thematic programme of the University of Debrecen."; "Complex Rural Economic and Sustainable Development, Elaboration of its Service Networks in the Carpathian Basin (Project ID: EFOP-3.6.2-16-2017-00001, Hungary)" research project. Also, support was given by Tempus Public Foundation (TPF), Hungary and Széchenyi 2020 under the GINOP-2.2.1-15-2017-00051.

References

1. Tripathi, A.D.; Mishra, R.; Maurya, K.K.; Singh, R.B.; Wilson, D.W. Estimates for world population and global food availability for global health. In *The Role of Functional Food Security in Global Health*; Academic Press: Cambridge, MA, USA.
2. Alexandratos, N.; Bruinsma, J. *World Agriculture towards 2030/2050: The 2012 Revision*; ESA Working Paper No. 12-03; FAO: Rome, Italy, 2012.
3. Henchion, M.; Hayes, M.; Mullen, A.M.; Fenelon, M.; Tiwari, B. Future Protein Supply and Demand: Strategies and Factors Influencing a Sustainable Equilibrium. *Foods* **2017**, *6*, 53. [CrossRef] [PubMed]
4. Boer, J.; Aiking, H. Strategies towards healthy and sustainable protein consumption: A transition framework at the levels of diets, dishes, and dish ingredients. *Food Qual. Prefer.* **2019**, *73*, 171–181. [CrossRef]
5. Tenorio, A.T.; Kyriakopoulou, K.E.; Suarez-Garcia, E.; van den Berg, C.; van der Goot, A.J. Understanding differences in protein fractionation from conventional crops, and herbaceous and aquatic biomass—Consequences for industrial use. *Trends Food Sci. Technol.* **2018**, *71*, 235–245. [CrossRef]
6. Kromus, S.; Wachter, B.; Koschuh, W.; Mandl, M.; Krotscheck, C.; Narodoslawsky, M. The green biorefinery Austria-development of an integrated system for green biomass utilization. *Chem. Biochem. Eng. Q.* **2004**, *18*, 7–12.
7. Zhang, W.; Grimi, N.; Jaffrin, M.Y.; Ding, L.; Tang, B. A short review on the research progress in alfalfa leaf protein separation technology. *J. Chem. Technol. Biotechnol.* **2017**, *2*, 2894–2900.
8. Tenorio, A.T.; Gieteling, J.; de Jong, G.A.H.; Boom, R.M.; van der Goot, A.J. Recovery of protein from green leaves: Overview of crucial steps for utilisation. *Food Chem.* **2016**, *203*, 402–408. [CrossRef]
9. Manwatkar, V.G.; Gogle, D.P. The Effect of Deproteinised Juice (DPJ) on Seed Germination and Seedling Growth of Different Plants (by Paper Towel Method). *Int. J. Life Sci.* **2014**, *2*, 65–68.

10. Cheyniera, V.; Comte, G.; Davies, K.M.; Lattanzio, V.; Martens, S. Plant phenolics: Recent advances on their biosynthesis, genetics, and ecophysiology. *Plant. Physiol. Biochem.* **2013**, *72*, 1–20. [CrossRef]
11. Salah, E.H. *Sustainability of Agricultural and Rural Waste Management*, 1st ed.; Cradle-to-Cradle for Sustainable Development eBook; Academic Press: New York, NY, USA, 2007; 424p, ISBN 9780080550145.
12. Popp, J.; Lakner, Z.; Harangi-Rakos, M.; Fari, M. The effect of bioenergy expansion: Food, energy, and environment. *Renew. Sustain. Energy Rev.* **2014**, *32*, 559–578. [CrossRef]
13. Obi, F.O.; Ugwuishiwu, B.O.; Nwakaire, J.N. Agricultural waste concept, generation, utilization and management. *Niger. J. Technol.* **2016**, *35*, 957–964. [CrossRef]
14. Grela, E.R.; Pietrzak, K. Production technology, chemical composition and use of alfalfa protein-xanthophyll concentrate as dietary supplement. *J. Food Process. Technol.* **2014**, *5*, 10.
15. Pirie, N.W. *Leaf Protein and its By-products in Human and Animal Nutrition*; Cambridge University Press: Cambridge, UK, 1987; p. 209.
16. Iliyas, S. Effect of different flours on α -amylase production. *Int. J. Bioassays* **2014**, *3*, 3011–3015.
17. Jadhav, R.K.; Chhaya, G. Influence of deproteinised foliage fluid liquid biofertilizer on nitrate reductase activity of Eleusine coracana plants. *Curr. Bot.* **2018**, *9*, 33–36. [CrossRef]
18. Zanin, V. A New Nutritional Idea for Man: Lucerne Leaf Concentrate. APEF, Association Pour la Promotion des Extraits Foliaires en Nutrition. 1998. Available online: http://www.nutrition-luzerne.org/anglais/pdf/EtudeZaninenglish.pdf (accessed on 29 October 2019).
19. Thomsen, M.H.; Bech, D.; Kiel, P. Manufacturing of Stabilised Brown Juice for L-lysine production—From University Lab Scale over Pilot Scale to Industrial Production. *Chem. Biochem. Eng. Q.* **2003**, *18*, 37–46.
20. Shende, G.C.; Gogle, D.P. To study the effect of various concentration of deproteinized leaf juice (DPJ) of selected plants on growth of Aspergillus niger. *Int. J. Life Sci.* **2016**, *6*, 186–188.
21. Jadhav, R.K.; Kadam, R.; Joshi, D. Physiology of Gramineae Crops by Deproteinised Foliage Extract and its Influence on Photosynthetic Chemistry. *J. Phytochem.* **2019**, *110*, 149–163.
22. Hegyi, F. Examination of Lactobacilli by Advanced Colorimetric Method. In Hungarian. Laktobacilluszok Vizsgálata továbbfejlesztett Kolorimetriás Módszerrel. Ph.D. Thesis, Corvinus University, Budapest, Hungary, 2014; pp. 8–93.
23. Lamont, J.R.; Wilkins, O.; Bywater-Ekegard, M.; Smith, D.L. From yogurt to yield: Potential applications of lactic acid bacteria in plant production. *Soil Biol. Biochem.* **2017**, *111*, 1–9. [CrossRef]
24. Smith, D.L.; Subramanian, S.; Lamont, J.R.; Bywater-Ekeg€ard, M. Signaling in the phytomicrobiome: Breadth and potential. *Front. Plant Sci.* **2015**, *6*, 709. [CrossRef]
25. Smith, D.L.; Praslickova, D.; Ilangumaran, G. Inter-organismal signaling and management of the phytomicrobiome. *Front. Plant. Sci.* **2015**, *6*, 722. [CrossRef]
26. Tang, Y.; Xin, H.-L.; Guo, M.-L. Review on research of the phytochemistry and pharmacological activities of Celosia argentea. *Rev. Bras. Farmacogn.* **2016**, *26*, 787–796. [CrossRef]
27. Gupta, S.; Lakshmi, A.J.; Manjunath, M.N.; Prakash, J. Analysis of nutrient and antinutrient content of underutilized green leafy vegetables. *LWT Food Sci. Technol.* **2005**, *38*, 339–345. [CrossRef]
28. Fári, M.G.; Domokos-Szabolcsy, É. Method for Producing Plant Protein Coagulum. Hun Patent WO/2019/150144, 8 August 2019.
29. Gibbs, P.A. Novel uses for lactic acid fermentation in food preservation. *J. Appl. Bacteriol.* **1987**, *63*, 51S–58S. [CrossRef]
30. Kovács, B.; Győri, Z.; Prokisch, J.; Loch, J.; Dániel, P. A study of plant sample preparation and inductively coupled plasma emission spectrometry parameters. *Commun. Soil Sci. Plant Anal.* **1996**, *27*, 1177–1198. [CrossRef]
31. Boór, A.; Bélafiné Bakó, K. Determination of antioxidant content in sloe (*Prunus spinosa* L.) and dogwood (*Cornus mas* L.) fruits. In Proceedings of the Conference of Chemical Engineering '2010, Singapore, 26–28 February 2010; pp. 55–58. (in Hungarian).
32. Sparks, D.L.; Page, A.L.; Helmke, P.A.; Loppert, R.H.; Soltanpour, P.N.; Tabatabai, M.A.; Johnston, C.T.; Summner, M.E. *Methods of Soil Analysis: Chemical Methods*; Part 3; ASA and SSSA: Madison, WI, USA, 1996.
33. Bradford, M. A rapid and sensitive method for the quantitation of microgram quantities of protein utilizing the principle of protein-dye binding. *Anal. Biochem.* **1979**, *72*, 248–254. [CrossRef]

34. Roxas, V.P.; Smith, R.K.; Allen, E.R.; Allen, R.D. Overexpression of glutathione S-transferase/glutathione peroxidase enhances the growth of transgenic tobacco seedlings during stress. *Nat. Biotechnol.* **1997**, *15*, 988–991. [CrossRef] [PubMed]
35. Woodbury, W.; Spencer, A.K.; Stahmann, M.A. An improved procedure using ferricyanide for detecting catalase isozymes. *Anal. Biochem.* **1971**, *44*, 301–305. [CrossRef]
36. Zhang, Z.; Huang, R. Analysis of Malondialdehyde, Chlorophyll Proline, Soluble Sugar, and Glutathione Content in Arabidopsis seedling. *Bio-Protocol* **2013**, *3*, 1–8. [CrossRef]
37. Porra, R.J.; Thompson, W.A.; Kriedemann, P.E. Determination of accurate extinction coefficients and simultaneous equations for assaying chlorophylls a and b extracted with four different solvents: Verification of the concentration of chlorophyll standards by atomic absorption spectroscopy. *Biochim. Biophys. Acta* **1989**, *975*, 384–394. [CrossRef]
38. Duncan, D.B. Multiple range and multiple F-tests. *Biometric* **1955**, *11*, 1–42. [CrossRef]
39. Balfour, E. Anomalous secondary thickening in Chenopodiaceae, Nyctaginaceae and Amaranthaceae. *Phytomorphology* **1965**, *15*, 111–122.
40. Myśkow, E.; Gola, E.; Tulik, M. Continuity of procambium and anomalous cambium during formation of successive cambia in Celosia argentea. *J. Plant. Growth Regul.* **2019**, *38*, 1458–1466. [CrossRef]
41. Carlquist, S. Successive cambia revisited: Ontogeny, histology, diversity, and functional significance. *J. Torrey Bot. Soc.* **2007**, *134*, 301–332. [CrossRef]
42. Reddy, G.U.; Deshmukh, V.R.; Joshi, R.N.; Kayama, R. Utilization of alfalfa (*Mediacago sativa* L.) whey as a fertilizer in irrigation. *J. Jpn. Grassl. Sci.* **1987**, *33*, 32–37.
43. Monteros, J.M.; Bouton, H.J. The future of alfalfa and forage crops. In Proceedings of the Western Alfalfa & Forage Conference, Reno, Nevada, 2–4 December 2009.
44. Martin, N.P. Alfalfa: Forage crop of the future. In *Proceedings of the 28th Kentucky Alfalfa Conference*; Lacefield, G., Forsythe, C., Eds.; CaveCity Convention Center: Cave, KY, USA, 2008; pp. 17–26.
45. Ream, H.W.; Smith, D.; Walgenbach, R. Effects of deproteinized alfalfa juice applied to alfalfa-bromegrass, bromegrass, and corn. *Agron. J.* **1977**, *69*, 685–689. [CrossRef]
46. Novik, G.; Olga Meerovskaya, O.; Savich, V. Waste Degradation and Utilization by Lactic Acid Bacteria: Use of Lactic Acid Bacteria in Production of Food Additives, Bioenergy and Biogas. In *Food Additives*; Karunaratne, D.N., Pamunuwa, G., Eds.; IntechOpen: Rijeka, Croatia, 2017. [CrossRef]
47. Zubaidah, E.; Akhadiana, W. Comparative study of inulin extracts from dahlia, yam, and gembili tubers as prebiotic. *Food Nutr. Sci.* **2013**, *4*, 8–12. [CrossRef]
48. Bautista-Trujillo, G.U.; Cobos, M.A.; Ventura-Canseco, L.M.C.; Ayora-Talavera, T.; Abud-Archila, M.; Oliva-Llaven, M.A.; Dendooven, L.; Gutierrez-Miceli, F.A. Effect of Sugarcane Molasses and Whey on Silage Quality of Maize. *Asian J. Crop Sci.* **2009**, *1*, 34–39.
49. Cheng, X.; Dong, Y.; Su, P.; Xiao, X. Improvement of the fermentative activity of lactic acid bacteria starter culture by the addition of Mn. *Appl. Biochem. Biotechnol.* **2014**, *174*, 1752–1760. [CrossRef]
50. Dimitrovski, D.; Velickova, E.; Dimitrovska, M.; Langerholc, T.; Winkelhausen, E. Synbiotic functional drink from Jerusalem artichoke juice fermented by probiotic Lactobacillus plantarum PCS26. *J. Food Sci. Technol.* **2016**, *53*, 766–774. [CrossRef]
51. Kwaw, E.; Ma, Y.; Tchabo, W.; Apaliya, M.T.; Wu, M.; Sackey, A.S.; Xiao, L.; Tahir, H.E. Effect of lactobacillus strains on phenolic profile, color attributes and antioxidant activities of lactic-acid fermented mulberry juice. *Food Chem.* **2018**, *1*, 148–154. [CrossRef]
52. Valero-Cases, E.; Nuncio-Jáuregui, N.; Frutos, M.J. Influence of fermentation with different lactic acid bacteria and in vitro digestion on the biotransformation of phenolic compounds in fermented pomegranate Juices. *J. Agric. Food Chem.* **2017**, *65*, 6488–6496. [CrossRef]
53. Kim, S.Y. Production of fermented kale juices with Lactobacillus strains and nutritional composition. *Prev. Nutr. Food Sci.* **2017**, *22*, 231–236. [PubMed]
54. Pirie, N.W. *Leaf Protein and Other Aspects of Fodder Fractionation*; Cambridge University Press: London, UK, 1978; pp. 118–123.
55. Anugoolprasert, O.; Kinoshita, S.; Naito, H.; Shimizu, M.; Ehara, H. Effect of low pH on the growth, physiological characteristics and nutrient absorption of sago palm in a hydroponic system. *Plant Prod. Sci.* **2012**, *15*, 125–131. [CrossRef]

56. Long, A.; Zhang, J.; Yang, L.T.; Ye, X.; Lai, N.W.; Tan, L.L.; Lin, D.; Chen, L.S. Effects of low pH on photosynthesis, related physiological parameters, and nutrient profiles of citrus. *Front. Plant Sci.* **2017**, *8*, 185. [CrossRef] [PubMed]
57. Anjum, N.A. Book review: Oxidative damage to plants-antioxidant networks and signaling. *Front. Plant. Sci.* **2015**, *6*, 452. [CrossRef]
58. Szollosi, R. Superoxide dismutase (SOD) and abiotic stress tolerance in plants: An overview. In *Oxidative Damage to Plants*; Academic Press: New York, NY, USA, 2014; pp. 89–129.
59. Santamaría-Fernández, M.; Karkov Ytting, N.; Lübeck, M. Influence of the development stage of perennial forage crops for the recovery yields of extractable proteins using lactic acid fermentation. *J. Clean. Prod.* **2019**, *218*, 1055–1064. [CrossRef]
60. Solati, Z.; Jørgensen, U.; Eriksen, J.; Søegaard, K. Dry matter yield, chemical composition and estimated extractable protein of legume and grass species during the spring growth. *J. Sci. Food Agric.* **2017**, *97*, 3958–3966. [CrossRef] [PubMed]
61. Kovács, Z.; Fári, M.G.; Kaszás, L.; Koroknai, J.Á.; Domokosné-Szabolcsy, É. Obtention of functional leaf protein concentrate (F-LPC) based on Ereky-process: Historical survey and preliminary results. *New Biotechnol.* **2018**, *44*, s100–s101. [CrossRef]
62. Kamm, B.; Hille, C.; Schönicke, P. Green biorefinery demonstration plant in Havelland (Germany). *Biofuels Bioprod. Biorefin.* **2010**, *6*, 253–262. [CrossRef]

14

Biostimulant Seed Coating Treatments to Improve Cover Crop Germination and Seedling Growth

Yi Qiu [1,2,†], Masoume Amirkhani [1,*,†], Hilary Mayton [1], Zhi Chen [2] and Alan G. Taylor [1]

1 Cornell AgriTech, Horticulture Section, Cornell University, SIPS, Geneva, New York, NY 14850, USA; qyi0508@emails.imau.edu.cn (Y.Q.); hsm1@cornell.edu (H.M.); agt1@cornell.edu (A.G.T.)

2 College of Mechanical and Electrical Engineering, Inner Mongolia Agricultural University, Hohhot, Inner Mongolia 010018, China; sgchenzhi@imau.edu.cn

* Correspondence: ma862@cornell.edu

† Co-first authors.

Abstract: Biostimulant seed coating formulations were investigated in laboratory experiments for their potential to increase maximum germination, germination rate, germination uniformity, and seedling growth of red clover (*Trifolium pratense* L.) and perennial ryegrass (*Lolium perenne* L.) seeds. Red clover and perennial ryegrass seeds were coated with different combinations of soy flour, diatomaceous earth, micronized vermicompost, and concentrated vermicompost extract. Coated and non-coated seeds of red clover and perennial ryegrass were evaluated for germination and growth after 7 and 10 days, respectively. Red clover seed was maintained at a constant 20 °C with a 16/8 h photoperiod, whereas for perennial ryegrass seed, the germinator was maintained at 15/25 °C, with the same photoperiod as red clover. Coated treatments significantly improved germination rate and uniformity with no reduction in total germination, compared to the non-treated controls in red clover. In contrast, for perennial ryegrass, the total germination percentage of all coated seeds was reduced and displayed a delayed germination rate, compared with the non-treated controls. Shoot length, seedling vigor index, and dry weight of seedlings of coated seed treatments of both crops were significantly higher when compared to controls for both species. In addition to growth metrics, specific surface mechanical properties related to seed coating quality of seeds of both species were evaluated. Increasing the proportion of soy flour as a seed treatment binder in the coating blend increased the integrity and compressive strength of coated seeds, and the time for coatings to disintegrate. These data show that seed coating technologies incorporating nutritional materials and biostimulants can enhance seedling growth and have the potential to facilitate the establishment of cover crops in agriculture and land reclamation.

Keywords: seed coating; cover crop; vermicompost; biostimulant; growth enhancement

---

1. Introduction

Exponential growth in human global population, from 1.7 billion in 1900 to approximately 7.6 billion in 2019, has led to the over use and degradation of agricultural landscapes, including grasslands used for grazing, forage, and food production [1,2]. The rapid growth of populations in pastoral areas, including Inner Mongolia, China, have caused intensive overutilization of grasslands. Approximately, 40% of land area in China is classified as grassland and accounts for 13% of the world's grassland [2,3]. Overgrazing and conversion of grassland to cropland has led to declines in overall agricultural productivity due to increased soil erosion, degraded soil structure, and reduced soil fertility. Recently, China implemented vegetation restoration programs to improve biodiversity in agriculture environments, soil health, and productivity, and to reduce erosion and desertification [2]. Legumes and ryegrasses are widely used as cover crops to reduce desertification and restore productivity on degraded grasslands [4] and are commonly used for land reclamation and restoration of abandoned mine land [5,6]. Perennial ryegrass (*Lolium perenne* L.) is a cool season grass native to southern Europe, the Middle East, Central Asia, and North Africa [7]. Ryegrass is often used to stabilize soils for erosion control and is frequently seeded with red (*Trifolium pratense* L.) or white (*Trifolium repens* L.) clovers for increased productivity in grazing and to provide nitrogen and aid in weed suppression [8]. Red clover is particularly tolerant to drought conditions and helps to improve soil structure due to its large, fast-growing (more than 60 cm/year) tap root [8]. The benefits of cover cropping in both organic and conventionally managed systems are well documented [9].

Cover crops increase soil organic matter, soil structure, nutrient retention and cycling, and reduce soil erosion [8]. However, under drought conditions, and in areas with poor soils such as arid degraded grasslands, germination and subsequent growth of cover crops are inadequate, and sowing is often unsuccessful. Seed enhancements, which can include seed priming, coating, and conditioning are frequently used to improve seed delivery during planting, and to increase seed germination, stand uniformity, seedling growth, and suppress disease [10]. Seed priming increased germination rate and overall seedling emergence in a study investigating perennial ryegrass for fall seeding under cool temperatures and improved wheat stand establishment under marginal soil conditions [11,12]. Seed treatments with fungicides and fertilizers enhanced stand establishment of perennial ryegrass in field experiments in New Zealand [13,14]. Seed enhancements via seed coating can also provide micro and macronutrients or biostimulant materials to increase germination, seedling vigor, and stand establishment [15]. Seed coating technology has been used as a promising and effective approach for enhancing establishment and yield of different grass and forage species (*Lolium perenne, Trifolium pratense, Elymus dahuricus*) in semi-arid areas of China such as Inner Mongolia [16–18].

Biostimulants are materials that can augment plant growth when applied to plants and seeds, but are not classified as fertilizers, pesticides, or soil amendments [19]. Commonly applied biostimulants include microbial inoculants, beneficial bacteria and fungi, nitrogen containing compounds, biopolymers, and plant extracts [19]. Research and use of biostimulants in agriculture has increased in recent years in an effort to reduce reliance on less sustainable conventional pesticides and fertilizers, which are often overused, in agricultural cropping systems [15,19–22]. Seed treatments require even smaller amounts of active ingredients per hectare than foliar applied treatments, primarily due to the reduced surface area treated, and increase germination and plant growth when compared to non-treated seed [23].

Modern seed coating technology utilizes different approaches depending on the shape and size of the seed and the type and amount of materials added to seeds [10,24–27]. Currently, seed pelleting, film coating, and seed encrusting are the most common coating/treatment procedures used in the seed industry to enhance plant and seedling performance. While seed pelleting, often employed to develop more uniform seeds for mechanical planting, can increase seed weight from 200 to > 5000%, film coating or encrusting utilizes much smaller quantities of materials resulting in a build-up in seed weight of between 0.5–10% and 20–200%, respectively [24]. The physical properties and thickness of the seed treatment/coating are critical factors that influence seed germination and seedling vigor. A thick hard seed coating can reduce, delay, or cause abnormal germination or may even be toxic, while a minimal, fragile seed coating can break or disintegrate before planting or not have a high enough dosage of an active ingredient to be effective. Therefore, specialized seed coating formulations must be developed and evaluated in order to be utilized effectively for any given plant species and agronomic purpose.

The specific objectives of this research were to explore plant-derived bio-based biostimulant seed coatings to enhance germination and growth of two cover crop species, red clover and perennial ryegrass, as an approach for seeding cover crops for grassland restoration. Previous research on seed coatings of broccoli and tomato with soy flour and compost materials showed promising results related to maximum germination, germination uniformity, and seedling vigor [15,25–27].

2. Materials and Methods

2.1. Seed and Coating Materials

Two species of cover crops were selected to evaluate biostimulant seed coatings in this study. Red clover 'VSN-variety not stated' seed was obtained from King's AgriSeeds, Inc., Lancaster, PA, USA, and 'Tetraprime' perennial ryegrass seed was provided by SeedWay, LLC, Penn Yan, NY, USA. The red clover and perennial ryegrass seeds were coated with different combinations of soy flour (SF), diatomaceous earth (DE), micronized vermicompost (MVC-2 and 3), and concentrated vermicompost extract (CVE) to identify the most stable and effective coating formulations (Table 1). Specific treatments and ratios of coating materials evaluated were (SF:DE 30:70, 40:60, 50:50 and 60:40, SF:MVC-2 (30:70), SF:MVC-3 (30:70), SF:DE:CVE (30:70)). A mechanical Ro-Tap shaker (Ro-Tap Testing Sieve Shaker No. 1506; The W.S. Tyler Co., Cleveland, OH) was utilized to sieve the SF to obtain a particle size smaller than (<75 μ), as previous studies have shown that smaller particle size results in more even distribution of the SF coating on seeds [15]. Seed coating biostimulant materials used in this research were previously analyzed by the Cornell Soil Health Nutrient Analysis Laboratories and recently reported in [27].

Table 1. Materials used as seed coating biostimulant treatment formulations in this study.

Coating Materials	Abbreviation	Source
Soy Flour	SF	Archer Daniels Midland Co., Decatur, IL, USA
Diatomaceous Earth	DE	Perma-Guard, Inc., Albuquerque, NM, USA
Concentrated Vermicompost Extract (liquid)	CVE	Worm Power, Avon, NY, USA (concentrated by Caloris Engineering, Easton, MD, USA)
Micronized Vermicompost	MVC-2	Worm Power, Avon, NY, USA
Micronized Vermicompost	MVC-3	TerraVesco, Sonoma Valley Worm Farm, Sonoma, CA, USA

2.2. Seed Coating

A 15-cm diameter, R-6 (Universal Coating Systems, Independence, OR, USA) laboratory-scale rotary pan coater was used to coat seeds in all experiments (Figure 1). Each seed coating treatment consisted of two components: dry powder and liquid. SF and DE, SF and MVC-2, and SF and MVC-3 were applied as dry powder to the seed surface with distilled water. For the SF, DE, and CVE treatments, water was replaced with liquid compost extract. For each treatment, the powder and liquid were applied to the surface of the seeds in incremental amounts as they rotated in the R6 to achieve uniform results. To clean the residual dust of each coating batch and avoid cross contamination of treatments, the R-6 pan was cleaned with a sponge and hot water-liquid soap solution. This was followed by cleaning with disinfectant wipes (three times), and R-6 with high pressure air flow was applied around the pan and cylinder to ensure completely drying. In this study, coating combinations of SF:DE and SF:MVC for both crops were applied on separate days.

Twenty-five grams of seed were used for red clover (1000 seed weight = 2.5 g) treatments, and 15 g of seed were used for the perennial ryegrass (1000 seed weight = 1.5 g). Therefore, we had an equal amount of treated seeds (~10,000 seeds) for each treatment of each crop. The total build up percentage was approximately 30% for clover and 70% for perennial ryegrass (Figure 2). Variation in the percent build up needed to achieve uniform coverage reflects the need for specifically developed seed coatings for each seed type and species. The size and shape of seeds influences uniformity of treatments on the seeds. After coating, the seeds were dried at room temperature for 24 h (h) until completely dry [15,28]. To improve observation of the seed coating uniformity of the SF:DE and SF:DE:CVE blend, a red dye (Pro-Ized red colorant, Bayer Corp, Research Triangle Park, NC, USA) was added to the binder (1.0 mL dye per 10 mL binder). Due to the dark brown color of the MVC materials, no dye was used for SF and MVC combinations; the natural color showed coating quality and uniformity of application.

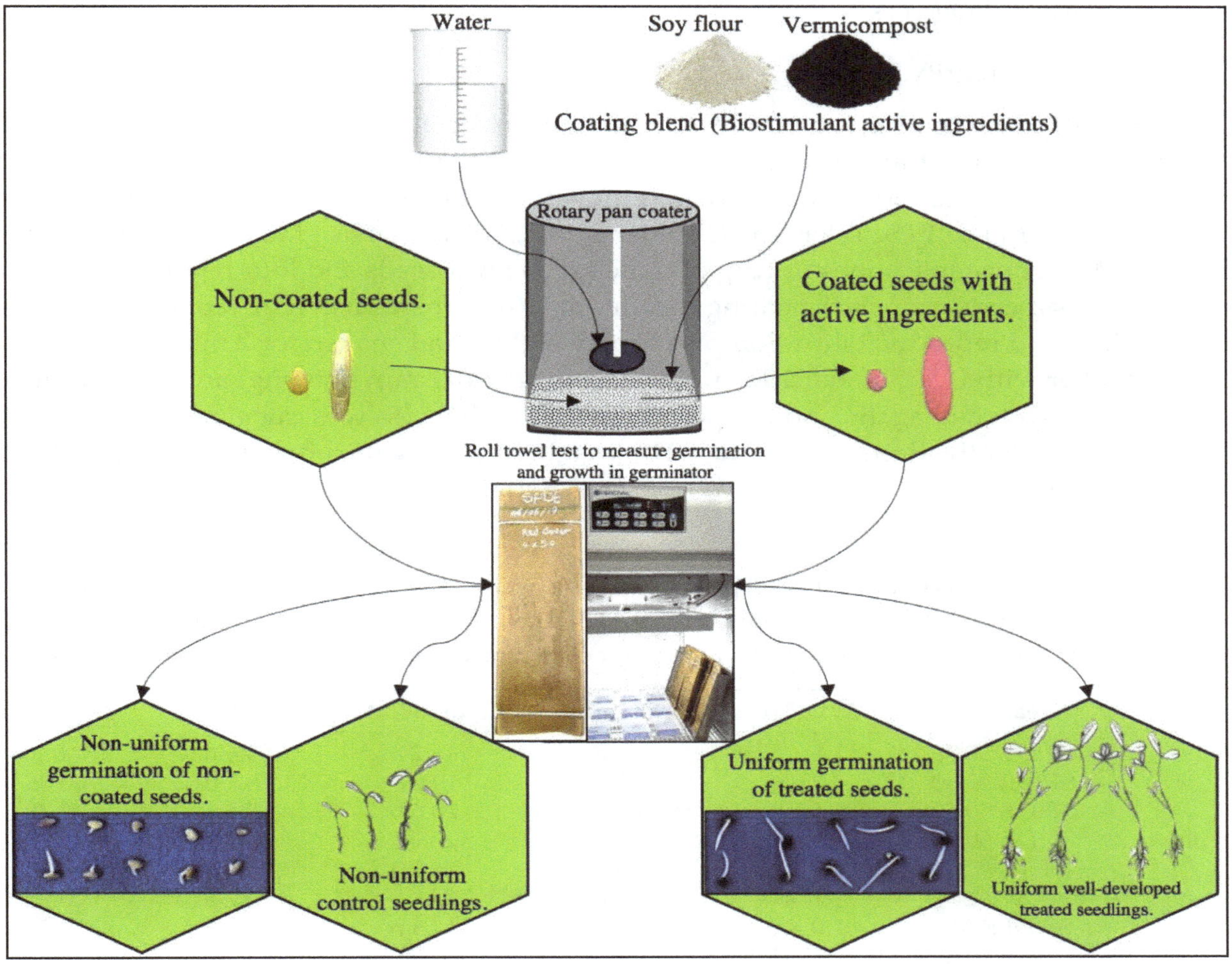

Figure 1. Figure of seed coating methodology used as an approach for application of biostimulant compounds for sustainable agriculture.

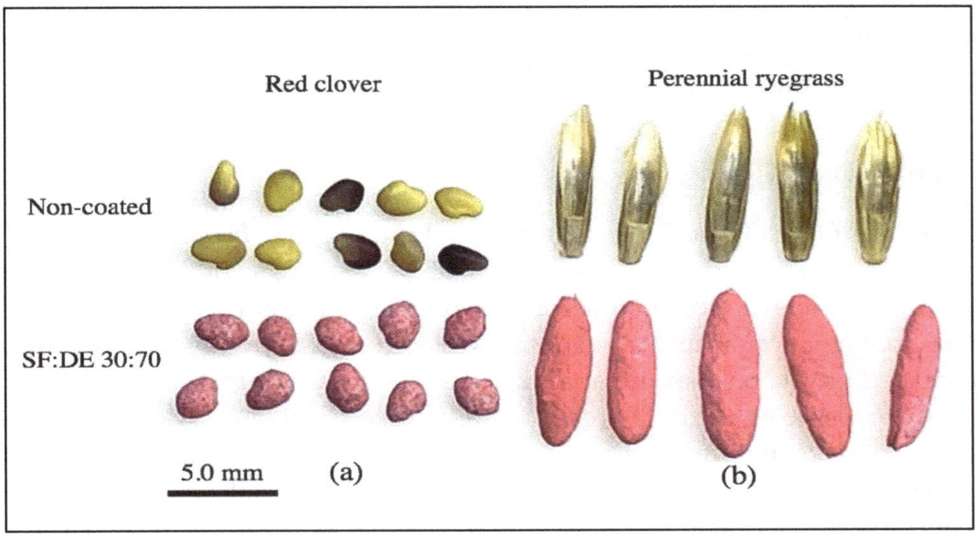

Figure 2. Non-coated and coated (SF:DE 30:70) red clover seeds with 30% build up (**a**) and non-coated and coated (SF:DE 30:70) perennial ryegrass with 70% build up (**b**). To improve observation of seed coating uniformity, a red dye was added to the binder.

2.3. Seed Coat Physical Properties

2.3.1. Seed Coating Integrity Test

The strength of the coating is an important quality as it relates to germination and potential for damage during handling, transportation, and planting. The surface material of coated seeds must have good mechanical properties to ensure that they do not crack or disintegrate before sowing. A Ro-Tap sieve shaker (The W.S. Tyler Co., Cleveland, OH, USA) was used to test the integrity of the coated seeds [15]. Four replications of 1.5 g of coated red clover seeds and four replications of 1.5 g of perennial ryegrass seeds from each coating formulation (treatments listed in Table 2) were tested to assure reliability and reproducibility. Samples were weighed and shaken for 2 min using a standard Ro-tap test shaker with U.S. Standard Testing Sieve No. 25 (0.71 mm opening) and a solid catch pan. Each sample was weighed again, and the percentage of coating loss was calculated according to the weight before and after the Ro-tap procedure. The weight of coating material, which passed through a No. 25 sieve was reported as weight loss (WL %) of material.

Table 2. Results of seed coating physical property testing, weight loss (WL, %), disintegration time (DT, min), compressive strength (Force N), time to break (TB) seed coating measured in s, relaxation time (RT) after the seed coating was fractured measured in s for coated seeds of red clover and perennial ryegrass.

Crop	Treatment	WL (%)	DT (min)	Force (N)	TB (s)	RT (s)
Red clover	SF:DE 30:70	1.5 a *	58 a	16.2 a	4.4 a	0.3 a
	SF:DE 40:60	1.2 ab	75 b	19.2 b	5.1 b	0.36 b
	SF:DE 50:50	0.6 bc	100 c	20.6 b	5.6 c	0.48 c
	SF:DE 60:40	0.4 c	103 c	23.9 c	5.5 c	0.5 c
	SF:MVC-2	1.2 ab	78 b	16.8 a	4.4 a	0.38 b
	SF:MVC-3	1.1 ab	83 b	16.5 a	4.7 a	0.32 a
	SF:DE:CVE	1.0 ab	80 b	17.5 b	4.9 a	0.31 a
Perennial ryegrass	SF:DE 30:70	1.4 A *	40 A	15.3 A	5.7 A	0.2 A
	SF:DE 40:60	1.2 AB	58 B	17.9 B	6.0 B	0.38 B
	SF:DE 50:50	0.5 B	90 C	20.9 C	6.4 C	0.51 C
	SF:MVC-2	1.3 B	60 B	15.6 A	5.4 A	0.24 A
	SF:MVC-3	0.9 BC	55 B	16.1 A	5.5 A	0.22 A
	SF:DE:CVE	0.9 BC	60 B	15.8 A	5.8 A	0.31 B

* Different letters within each column for each crop indicate significant differences using a Least Significant Difference (LSD) test at a significance level of $p < 0.05$. Lower case letters represent significant differences in red clover treatments and upper case letters denote perennial ryegrass treatment differences.

2.3.2. Mechanical Property Test

A texture analyzer (TA-XTplusC, Texture Technologies Corp., Hamilton, MA) was used to test the compressive strength of coated seeds. The TA-XTplusC is a precision instrument used to measure the surface mechanical properties of coated seeds and the compressive strength of a single seed. The arm of the texture analyzer containing a weighing sensor moves in a downward motion to compress the coated seed placed on the base of the analyzer and then returns to its original position. Data are assessed as the compressive strength (Force N) and time to breakage (TB, measured in seconds (s)) required to fracture the seed coating. The relaxation time (RT), which is the time required to completely rupture the seed coating was also measured [29,30]. After the seed coating was completely broken, the force (N) increases until the seed embryo was crushed (Figure 3). Texture analyzer software (Exponent Connect, version 7.0.2.0, S. Hamilton, MA, USA, 2018) was used to record force for TB and RT [26]. Ten coated seeds were randomly selected from batches of different formulations (SF:DE = 30:70, 40:60, 50:50 and 60:40, SF:MVC-2 (30:70), SF:MVC-3 (30:70), SF:DE:CVE (30:70)) to test their surface compressive strength for both red clover and perennial ryegrass coated seeds.

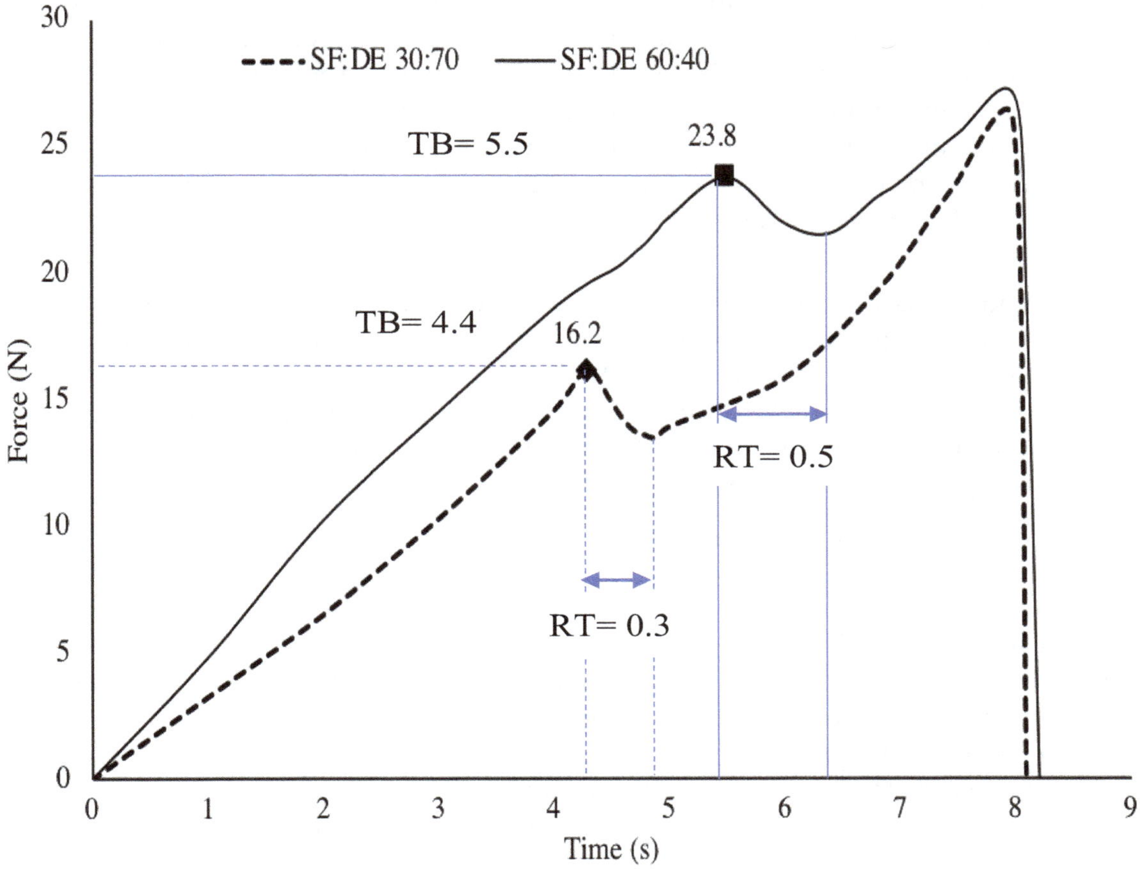

Figure 3. The values of peak load force required to break the seed coat of a single seed from two different seed coating blends of soy flour (SF) and diatomaceous earth (DE) tested at room temperature (SF:DE 30:70 and 60:40) for red clover are 16.2 and 23.8 N, respectively. The maximum force value (N) is a measure of coating strength and shows the maximum force needed to break the seed coat. Time to break seed coating (TB) and Relaxation Time (RT) after seed coat fracture until the coat is completely broken, both measured in s, are shown for a single seed. Force, TB, and RT data shown in Table 2 are the means of 10 seeds (Equipment: TA-XTplusC, Texture Technologies Corp., Hamilton, MA, USA, Software: Exponent Connect, version 7.0.2.0).

2.3.3. Seed Coating Hydration Test

The wet strength of a seed coating is largely dependent on the adhesion of the components after immersion in water. In theory, the slower the decomposition rate of coated seeds in water is, the more likely it is to delay germination. The hydration test was used to investigate the integrity of the coating materials when immersed in water. Hydration tests were conducted to determine the potential for seed coatings to prevent or delay germination. Four replicates of 100 coated seeds with different proportions of soy flour and diatomaceous earth, SF:DE (30:70, 40:60, 50:50 and 60:40), and the soy flour micronized vermicompost treatments SF: MVC-2, SF: MVC-3 were placed in 5 mL of distilled water to determine disintegration time. Disintegration time was measured in minutes.

2.4. Seed Germination and Seedling Growth Measurements

Four replicates of 50 treated and non-coated control seeds were placed on two 30 cm × 45 cm moistened germination paper towels (Anchor Paper Company, St. Paul, MN, USA); then, an additional moistened standard germination paper towel was placed on top of the seeds. The towels were rolled and positioned in a germinator (Percival germinator, Model I-36LL, Perry, IA, USA). For perennial ryegrass seed, the germinator was maintained at 15/25 °C, with a 16/8 h photoperiod [31]; red clover, was maintained at a constant 20 °C with the same photoperiod [31]. Radical emergence (>2 mm) was used to determine successful seed germination. The number of newly germinated seeds for both red clover and perennial ryegrass was recorded every 24 h. For perennial ryegrass, the total germination percentage (Gmax %) was recorded after 10 days. The Gmax % for red clover was recorded after seven days. Gmax %, the number of germinated seeds and germination uniformity (GU), (GU = time required for 90% germination subtracted by time required for 10% germination) were calculated [32] for each treatment. In addition, germination rate (T50, the time in h to reach 50% total germination) was calculated according to the equation developed by Coolbear et al. [33].

Root and shoot measurements (cm) were conducted in separate roll towel experiments (under the same growing conditions described above for seed germination) for each treatment using four replicates of 50 seeds. The seed vigor index (SVI) was equal to Gmax % multiplied by seedling length (combined root and shoot lengths) divided by 100 [34]. Seedlings were measured a week after full emergence for both crops. After measuring shoot and root lengths, all seedlings from each treatment were dried in an oven at 80 °C for 48 h to obtain the dry weight data.

2.5. Statistical Analysis

In all experiments, normality tests were conducted prior to ANOVA and all data passed the normal distribution test at a significance level of 0.05. Analysis of variance (ANOVA) ($\alpha = 0.05$) and Fisher's least significant difference test for seed coating physical property and Dunnett test for germination and seedling growth data were performed on each of the significant variables measured by Minitab Express [35]. All Gmax % (Tables 2–4) and WL % (Table 2) data were arcsine transformed for analysis. Data for Gmax % and WL% are presented as non-transformed means (Tables 2–4). Pearson correlation was conducted for coating physical properties data using Minitab Express (Table 5).

Table 3. Germination and growth metrics of soy flour formulations as measured by total germination (Gmax %), germination rate (T50) measured in hours (h), germination uniformity (GU) measured in hours (h), shoot and root length (cm), and Seedling Vigor Index (SVI = Gmax % × Seedling length) of different coating formulas of SF:DE for red clover and perennial ryegrass.

Crop	Treatment	Gmax (%)	T50 (h)	GU (h)	Shoot (cm)	Root (cm)	SVI
Red clover	Control	95 b *	35 a	37 a	3.6 b	2.4 b	5.7 b
	SF:DE 30:70	98 a	27 b	27 b	4.1 a	3.0 a	7.0 a
	SF:DE 40:60	99 a	29 b	25 b	4.3 a	2.9 a	7.1 a
	SF:DE 50:50	96 b	30 b	27 b	4.1 a	2.9 a	6.7 a
	SF:DE 60:40	96 b	34 a	28 b	3.9 a	2.7 a	6.3 b
Perennial ryegrass	Control	85 A *	75 B	39 B	6.5 B	5.5 B	10.2 B
	SF:DE 30:70	83 A	79 A	40 B	7.6 A	6.2 A	11.5 A
	SF:DE 40:60	83 A	80 A	42 B	7.4 A	6.3 A	11.4 A
	SF:DE 50:50	80 B	83 A	47 A	7.2 A	5.9 A	10.5 A

* Different letters within each column for each crop indicate significant differences using a Dunnett test at a significance level of $p < 0.05$. Lower case letters represent significant differences between each red clover seed coating treatment compared with the control and upper case letters denote each of perennial ryegrass treatment differences compared with the control.

Table 4. Germination and growth metrics of soy flour/vermicompost formulations as measured by total germination (Gmax %), germination rate (T50) measured in hours (h), germination uniformity (GU) measured in hours (h), shoot and root length (cm), seedling dry weight (DW) recorded in grams (g), and Seedling Vigor Index (SVI = Gmax % × Seedling length) from evaluation of different coating materials for red clover and perennial ryegrass. The proportion of all coating materials is 30:70 (30% SF and 70% of DE or MVC).

Crop	Treatment	Gmax (%)	T50 (h)	GU (h)	Shoot (cm)	Root (cm)	DW (g)	SVI
Red clover	Control	94 b *	36 a	35 a	3.7 b	2.5 b	0.05 b	6.0 b
	SF:DE	98 a	26 b	27 b	4.2 a	2.8 a	0.07 a	6.9 a
	SF:MVC-2	99 a	26 b	24 b	4.4 a	2.9 a	0.07 a	7.3 a
	SF:MVC-3	99 a	25 b	25 b	4.7 a	3.0 a	0.08 a	7.6 a
	SF:DE:CVE	98 a	27 b	25 b	4.5 a	3.2 a	0.07 a	7.5 a
Perennial ryegrass	Control	87 A *	76 A	41 B	6.5 B	5.9 B	0.10 B	10.8 B
	SF:DE	85 A	78 A	43 B	7.9 A	6.5 A	0.13 A	12.3 A
	SF:MVC-2	86 A	77 A	40 B	8.1 A	6.8 A	0.14 A	12.8 A
	SF:MVC-3	82 B	81 A	48 A	8.3 A	6.6 A	0.13 A	12.2 A
	SF:DE:CVE	85 A	77 A	43 B	8.4 A	6.6 A	0.15 A	12.8 A

* Different letters within each column for each crop indicate significant differences using a Dunnett test at a significance level of $p < 0.05$. Lower case letters represent significant differences between each red clover seed coating treatment compared with the control and upper case letters denote each of perennial ryegrass treatment differences compared with the control.

Table 5. Correlation coefficients between disintegration time (DT, min), weight loss (WL, %), and compressive strength (Force, N) from seed coating applications of soy flour and diatomaceous earth on red clover and perennial ryegrass seeds.

Crop		WL (%)	Force (N)
Red clover	DT (min)	−0.99 ***	+0.92 **
	WL (%)	-	−0.94 **
Perennial ryegrass	DT (min)	−0.99 ***	+0.99 ***
	WL (%)	-	−0.96 ***

, * Significant at $p < 0.001$, 0.0001, respectively.

3. Results and Discussion

3.1. Seed Coating Physical Properties

The integrity and physical properties of coated and pelleted seeds are critical for overall performance. The production of dust can lead to health and environmental risks; therefore, it is important to quantitatively analyze the potential for breakage and weight loss that may occur during transportation and handling. In contrast, a seed coating that is too hard or impermeable to water may hinder germination. In this study, physical properties of the various seed coating formulations were tested by employing three different tests, including mechanical, texture, and hydration analyses.

Experimental results from the seed coating mechanical Ro-tap and Texture analysis (TA-XTplusC) tests are presented in Table 2 and Figure 3. For both crops, increasing the proportion of soy flour in the coating blend increased the compressive strength (Force N) of coated seeds. The time required to break the seed coating (TB), measured in s and relaxation time (RT) after breaking seed coating (Table 2) increased as soy flour proportion increased (Table 2). For example, for red clover, the average force (N) increased from 16.2 to 23.9 N as the soy flour content increased from 30% to 60%, which is an increase of approximately 48%. As soy flour content increased, the TB of coated seeds gradually increased from 4.4 to 5.5 s for red clover and 5.7 to 6.4 s for perennial ryegrass (Table 2). Although the same seed coating blend were used for both crops, the TB ranges were different most likely due to the difference in build-up percentage. There was a 1.1 s delay in breakage time for red clover when the content of soy flour increased from 30% to 60%, and 0.7 s delay for perennial ryegrass when the content of soy flour increased from 30% to 50%. The force value to break down the coating for perennial ryegrass significantly increased by approximately 37% as soy flour increased from 30% (15.3) to 50% (20.9). Additionally, the weight loss of coated seeds from the Ro-tap test gradually decreased from 1.5% to 0.4% for red clover and 1.4% to 0.5% for perennial ryegrass (Table 2) as the proportion of soy flour in the seed coating increased. Interestingly, the mechanical properties of each seed formulation were not significantly different, even though the shape and surface properties of the seeds of both species differed.

There was no significant difference in both crops in terms of weight loss and disintegration time of coated seeds in water (Table 2) of soy flour and vermicompost. For red clover, when seed coatings with soy flour and vermicompost (30:70) (SF:MVC-2, SF:MVC-3 and SF:DE:CVE) were compared with SF:DE 30:70, the force (N) to break the seed coating increased by 0.6, 0.4, and 1.3 N, respectively and time to break (TB) the seed coating were non-significant, indicating that soy flour could serve as a binder for both types of materials used in this study (DE and vermicompost). There was no significant difference observed on force and TB data among treatments for perennial ryegrass (Table 2). Relaxation time after breaking red clover seed coating (RT) of SF:MVC-2, SF:MVC-3, and SF:DE:CVE increased slightly by 0.08, 0.02, and 0.01s, and the RT of perennial ryegrass increased by 0.04, 0.02, and 0.11s compared with that of SF:DE 30:70. In conclusion, the surface mechanical strength of different seed coating formulations with soy flour and vermicompost blends were non-significant but slightly higher than that of soy flour and diatomaceous earth.

The hydration test measures the time required to dissolve the coating materials. The proportion of soy flour in the seed coating blends had a significant effect on disintegration time (DT) (Table 2). A higher proportion of soy flour in the coating blend for red clover increased the DT from 58 to 103 min. This pattern was also observed for perennial ryegrass, as the proportion of soy flour ratio increased from 30% to 50%, the DT significantly increased from 40 to 90 min (Table 2). There were no significant differences in WL or DT for micronized vermicompost and compost extract (SF:MVC-2, SF:MVC-3 and SF:DE:CVE) seed treatments on either crop species (Table 2).

According to the American Seed Trade Association (ASTA), the key to a successful seed treatment is high physical integrity with low dust production. Determination of mechanical integrity of coated seeds is an important step in order to meet environmental safety standards [36]. Amirkhani et al. [15,27] tested the mechanical properties of several different broccoli coated seeds with Ro-tap and Texture Analyzer methods. The weight loss percentage that Amirkhani et al. [27] reported were slightly higher than the data collected from the red clover and perennial ryegrass seed treatments evaluated in this study for the same seed coating formulations. Total peak load force to break the seed coats for broccoli seeds were slightly lower than the value that was observed in this study for the SF:DE combination. Overall, the mechanical integrity of the coated red clover and perennial ryegrass seeds were more stable for the same seed coating formulation treatments, compared to broccoli. This difference might be because of the size and shape of clover and perennial ryegrass seeds contrasted to the broccoli seeds. Accinelli et al. [37] also attributed differences in seed dust emission for seed coating treatments of maize (*Zea mays* L.) and canola (*Brassica napus* L.) to seed physical characteristics. The mechanical integrity data observed for the different formulations in this study are in accordance with European Standards (Italy and France) and meet the benchmarks for safety of dust production of coated seeds [38].

3.2. Germination and Seedling Growth of Soy Flour and Diatomaceous Earth Seed Coating

In seed coatings with diatomaceous earth, soy flour served as the biostimulant component of the formulations. The results presented in Table 3 show that all coated treatments of red clover seeds significantly improved T50 and GU with no reduction in Gmax % compared to the non-treated control

(Table 3), except for the SF:DE 60:40 treatment. Although soy flour proportions higher than 40% resulted in stronger and more durable seed coating mechanical properties, it had a negative effect on germination parameters (Gmax % and T50). For example, seeds treated with 30% and 40% soy flour had 98 and 99% Gmax % and T50 of 27 and 29 h, respectively, However, increasing the soy flour to 60% resulted in maximum germination of 96% and delayed the T50 to 34 h (Table 3). The negative effect in germination was attributed to the hard mechanical barrier of the seed coating with high soy flour content.

In contrast, for perennial ryegrass seeds, the Gmax % of all coated seeds was slightly reduced and showed delayed germination rates compared with the non-treated control. Control seeds of perennial ryegrass had the greatest Gmax % (85%) and significantly faster T50 (75 h) compared to all coating formulations. Application of 50% soy flour to the seed coating (SF:DE 50:50) reduced the Gmax % to 80% and T50 by 8 h and GU, compared with the non-treated control seeds. Due to the high percentage of coating build up (70%), the delay and a slight reduction in Gmax % was not unexpected. Several studies have indicated that the seed coating can act as a mechanical barrier for water absorbance and radical emergence [10,15].

Shoot and root length and seedling vigor index (SVI) are important indicators that determine whether the treated seeds promote seedling growth. Shoot and root length and seedling vigor index of treated seeds were significantly higher than those of non-treated control seeds for both crops (Table 3). The lowest application of soy flour (SF:DE 30:70) to the red clover seeds resulted in 4.1 cm shoot length, 3.0 cm root length, and 7.0 SVI, respectively, which were 14, 25, and 23 % higher than those of non-treated seeds. The same application rate of soy flour (SF:DE 30:70) to perennial ryegrass seeds improved the shoot growth by 17% and increased both root length and SVI values by approximately 13% compared to the control seeds. Amirkhani et al. [15,25–27] reported similar results for seed coatings that combined soy flour with diatomaceous earth. In their research, the seed coating blends had significant and positive effects on the above and below ground growth parameters of broccoli, tomato, radish, and hemp. They hypothesized that since soy flour is a plant-based protein and a rich source of several amino acids, it may have led to the increase in plant shoot and root growth and dry matter content and influenced uptake of nitrogen.

3.3. Germination and Seedling Growth of Soy Flour and Vermicompost Seed Coating

In addition to soy flour and diatomaceous earth, co-application of soy flour and vermicompost as rich sources of nutritional materials were tested as seed coatings and their effect on germination and seedling growth were recorded for red clover and perennial ryegrass. Shoot and root length, dry weight, and seedling vigor index of seedlings of all coated seed treatments were significantly higher compared to the non-treated controls (Table 4).

All treated red clover seeds germinated significantly faster (approximately 10 h) and had higher Gmax % (Table 4 and Figure 4) than non-coated seeds. They also germinated more uniformly than the non-treated control seeds. Gmax % was ≥ 98% for all treated seeds, which was significantly higher than control with 94% Gmax %. Red clover data showed that the shoot and root length and seedling vigor index of treated seeds were significantly higher than the non-treated control seeds. For example, compared with non-treated seeds control, the shoot length of SF:DE, SF:MVC-2, SF:MVC-3, and SF:DE:CVE increased by 14%, 19%, 27%, and 22%, respectively. Moreover, the root length of treated seeds increased by 12%, 16%, 20%, and 28%, respectively. All treatments showed a 40 to 60% increase in seedling dry weight (DW) compared with the control. The seedling vigor indexes (SVI) were 15%, 22%, 27%, and 25% higher than control, respectively (Table 4).

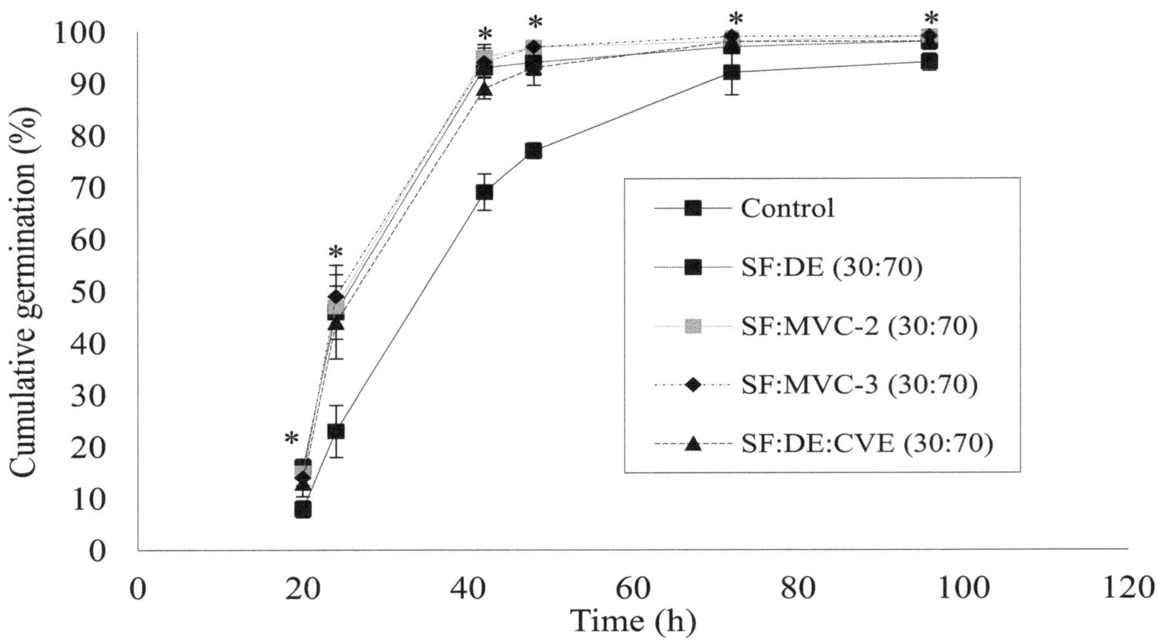

Figure 4. Cumulative germination percentage of red clover non-treated control seeds versus biostimulant coated seeds. * Significant at $p \leq 0.05$.

For perennial ryegrass, application of soy flour (SF:DE) and co-application with vermicomposts (SF:MVC-2 and 3, and SF:DE:CVE) increased shoot length by 22%, 25%, 28%, and 29% and root length by 10%, 15%, 12%, and 12%, respectively, compared to the non-treated control. The highest root length was observed in the SF:MVC-2 treatment. All treatments had higher DW than the control. Additionally, the highest SVI was observed in the SF:DE:CVE treatment, which was approximately 40% higher than the SVI of non-treated control (Table 4).

Statistical analysis (Pearson's correlation) of seed coating formulations and germination showed significant negative correlations between seed coating WL (%) and DT (min) (r = −0.99 ***). There was also a significant positive correlation between DT (min) and force (N) (r = +0.92 **). A significant negative correlation between WL (%) and force (N) (r = −0.94 **) from SF:DE coating formulations of red clover evaluated was observed (Table 5). For perennial ryegrass, the significant correlation coefficient between WL (%) and DT (min) was r = −0.99 *** and the correlation between WL (%) and force (N) was r = −0.96 ***. Lastly, for perennial ryegrass, the correlation between DT (min) and force (N) was r = +0.99 *** (Table 5). These data indicate that a higher proportion of SF in the seed coating formulation resulted in harder coatings but had only a slight impact on the Gmax %. For red clover, increasing the soy flour from 30% to 60% in seed coating formula reduced the Gmax % by 2%; however, T50 was significantly delayed by 7 h (Table 3). Similarly, for perennial ryegrass, increasing soy flour from 30% to 50% in the seed coating resulted in a 3% reduction in Gmax % and a minor delay on T50 (4 h).

In the present study, the seedling growth data for both cover crops evaluated indicate that seed coating can be an efficient and effective delivery method for application of nutritional biostimulant materials at the time of sowing for rangeland and grassland restoration. Several previous studies showed that applications of plant-based proteins and vermicompost can improve biometric growth parameters, related to production and yield of horticultural, field, and cover crops. Karlsons et al. [39] showed that a 10% addition of vermicompost in pure sand significantly increased fresh and dry weight of winter rye shoots by 578% and 265%, respectively. Tognetti et al. [40] found that application of vermicompost to degraded volcanic soil (to the extent of 20 and 40 g/kg soil) sown with ryegrass (*L. perene*) significantly increased ryegrass yields compared to the control due to the large nutrient concentrations and high microbial populations, when mixed with the soil. The positive effect of vermicompost on plant growth in this study agrees with the results of Alwaneen [41] on alfalfa and Amirkhani et al. [27] on broccoli. Amirkhani et al. [27] found that dairy manure-based vermicompost can supply essential nutrients to plants to enhance growth as well as increase the organic matter contents of soil for higher crop production. Moreover, in the recent decade, several researchers have been working on treating plants with biostimulants to stimulate crop productivity and increase stress tolerance under dynamic abiotic stresses [42–44]. The cover crop seeds treated with biostimulants in combination with other bio-effectors, such as superabsorbent polymers to investigate the response of these plants to drought, can be an area of future studies.

4. Conclusions

Seed coating technology can be an effective strategy to maximize early stand establishment of cover crops. Biostimulants applied as seed coatings have the potential to effectively promote seedling growth, and early stand establishment of red clover and perennial ryegrass. In this study, biostimulant seed coatings promoted the seedling growth of red clover and perennial ryegrass seeds and accelerated the germination of red clover compared to the non-treated control seeds. More rapid germination could aid in establishment under arid conditions and in areas with poor soils. However, further studies are needed to determine if vermicompost and plant-based proteins can be developed for economical commercial applications as seed treatments. The seed industry commonly includes fertilizers and *Rhizobium*, nitrogen fixing bacterial inoculants for red clover seeding. Additional research will be needed to determine if the biostimulant materials used in these experiments are compatible with seed inoculants. The use of biostimulants in combination with vermicompost and other biofertilizers as seed coatings may offer a great opportunity to increase plant production and long-term sustainability in agricultural landscapes.

Author Contributions: Y.Q., M.A., and H.M. contributed equally in writing the manuscript. Z.C. was involved in reviewing the manuscript. A.G.T. assisted with conceptualization of the overall experiments, funding, writing, and reviewing the manuscript. All authors have read and agreed to the published version of the manuscript.

Acknowledgments: The authors appreciate the technical assistance of Michael Loos in laboratory experiments and Zhen Wang for assistance with this project.

References

1. Roser, M.; Ritchie, H.; Ortiz-Ospina, E. *World Population Growth*; Our World in Data; Oxford University-Global Change Data Lab (GCDL), 2019; Available online: https://ourworldindata.org/world-population-growth (accessed on 20 January 2020).
2. Mu, S.; Zhou, S.; Chen, Y.; Li, J.; Ju, W.; Odeh, I.O.A. Assessing the impact of restoration-induced land conversion and management alternatives on net primary productivity in Inner Mongolian grassland, China. *Glob. Planet. Chang.* **2013**, *108*, 29–41. [CrossRef]
3. Ni, J. Carbon storage in grasslands of China. *J. Arid Environ.* **2002**, *50*, 205–218. [CrossRef]
4. Qiu, Y.; Chen, Z.; Hou, Z.F.; Song, T.; Mi, L.K.; Shao, Z.W. Numerical simulation and experiment on improving pelleted coating of forage grass seeds by vibration force field. *Trans. CSAE* **2017**, *33*, 86–93.
5. Sheoran, V.; Sheoran, A.S.; Poonia, P. Soil reclamation of abandoned mine land by revegetation: A review. *Int. J. Soil Sediment Water* **2010**, *3*, 13.
6. Skousen, J.; Zipper, C.E. Reclamation guidelines for surface mined land in South-Western Virginia. Chapter: Revegetation species and practices. *Va. Coop. Ext. Publ.* **1996**, *460*, 120.
7. Balfourier, F.; Imbert, C.; Charmet, G. Evidence for phylogeographic structure in lolium species related to the spread of agriculture in Europe. A cpDNA study. *Theor. Appl. Genet.* **2000**, *101*, 131–138. [CrossRef]

8. McKenna, P.; Cannon, N.; Conway, J.; Dooley, J.; Davies, W.P. Red clover (*Trifolium pratense*) in conservation agriculture: A compelling case for increased adoption. *Int. J. Agric. Sustain.* **2018**, *16*, 342–366. [CrossRef]
9. Daryanto, S.; Fu, B.; Wang, L.; Jacinthe, P.A.; Zhao, W. Quantitative synthesis on the ecosystem services of cover crops. *Earth-Sci. Rev.* **2018**, *185*, 357–373. [CrossRef]
10. Taylor, A.G.; Allen, P.S.; Bennett, M.A.; Bradford, K.J.; Burris, J.S.; Misra, M.K. Seed enhancements. *Seed Sci. Res.* **1998**, *8*, 245–256. [CrossRef]
11. Snapp, S.; Price, R.; Morton, M. Seed priming of winter annual cover crops improves germination and emergence. *Agron. J.* **2008**, *100*, 1506–1510. [CrossRef]
12. Harris, D.; Raghuwanshi, B.S.; Gangwar, J.S.; Singh, S.C.; Joshi, K.; Rashid, A.; Hollington, P.A. Evaluation by farmers of on-farm seed priming in wheat in India, Nepal and Pakistan. *Exp. Agric.* **2001**, *37*, 403–415. [CrossRef]
13. Vartha, E.W.; Clifford, P.T.P. Effects of seed coating on establishment and survival of grasses, surface-sown on tussock grasslands. *N. Z. J. Exp. Agric.* **1973**, *1*, 39–43. [CrossRef]
14. Falloon, R.E.; Fletcher, R.H. Increased herbage production from perennial ryegrass following fungicide seed treatment. *N. Z. J. Agric. Res.* **1983**, *26*, 1–5. [CrossRef]
15. Amirkhani, M.; Netravali, A.N.; Huang, W.; Taylor, A.G. Investigation of soy protein-based biostimulant seed coating for broccoli seedling and plant growth enhancement. *HortScience* **2016**, *51*, 1121–1126. [CrossRef]
16. Ou, C.M.; Mao, P.S. Progress in research and application of coating technology for grasses. *Seed* **2019**, *11*, 63–67. (In Chinese)
17. Li, C.Y.; Zhang, F.; Liu, C.H.; Yu, F.S.; Li, Y.C. Screening of coating materials of the forage seeds. *J. Northeast Agric. Univ.* **2013**, *4*, 94–100, (In Chinese with English abstract).
18. He, Z.X.; Mao, P.S.; Sun, Y.; Li, M. Review of grass seed coating technology. *Acta Agrestia Sin.* **2016**, *2*, 270–277, (In Chinese with English abstract).
19. Rouphael, Y.; Colla, G. Synergistic biostimulatory action: Designing the next generation of plant biostimulants for sustainable agriculture. *Front. Plant Sci.* **2018**, *9*, 1655. [CrossRef]
20. Colla, G.; Hoagland, L.; Ruzzi, M.; Cardarelli, M.; Bonini, P.; Canaguier, R.; Rouphael, Y. Biostimulant action of protein hydrolysates: Unraveling their effects on plant physiology and microbiome. *Front. Plant Sci.* **2017**, *8*, 2202. [CrossRef]
21. Wilson, H.T.; Amirkhani, M.; Taylor, A.G. Evaluation of gelatin as a biostimulant seed treatment to improve plant performance. *Front. Plant Sci.* **2018**, *9*, 1006. [CrossRef]
22. Calvo, P.; Nelson, L.; Kloepper, J.W. Agricultural uses of plant biostimulants. *Plant Soil* **2014**, *383*, 3–41. [CrossRef]
23. Schmitt, A.; Koch, E.; Stephan, D.; Kromphardt, C.; Jahn, M.; Krauthausen, H.J. Evaluation of non-chemical seed treatment methods for the control of Phoma valerianellae on lamb's lettuce seeds. *J. Plant Dis. Prot.* **2009**, *116*, 200–207. [CrossRef]
24. Taylor, A.G. Seed treatments. In *Encyclopedia of Applied Plant Sciences*; Thomas, B., Murphy, D.J., Murray, B.G., Eds.; Elsevier Acad. Press: Oxford, UK, 2003; pp. 1291–1298.
25. Amirkhani, M.; Netravali, A.N.; Taylor, A.G. Improving seedling growth uniformity and seed vigor index by using plant-based protein seed coating in tomato and broccoli NYSAES research symposium. In Proceedings of the NYSAES Research Symposium 2017, Geneva, NY, USA, 23 June 2017; p. 6.
26. Amirkhani, M.; Mayton, H.S.; Loos, M.T.; Netravali, A.N.; Taylor, A.G. Biostimulant effect of a plant-based protein applied as a seed coating on selected crops. In Proceedings of the 3rd Annual Cornell AgriTech Research Symposium, Geneva, NY, USA, 29 June 2018; p. 4.
27. Amirkhani, M.; Mayton, H.S.; Netravali, A.N.; Taylor, A.G. A seed coating delivery system for bio-based biostimulants to enhance plant growth. *Sustainability* **2019**, *11*, 5304. [CrossRef]
28. Talley, B. *Status and New Developments in Seed Coatings*; Summit Seed Coating: Caldwell, ID, USA, 2011; Available online: https://uknowledge.uky.edu/ky_alfalfa/2012/Session/6/ (accessed on 28 August 2019).
29. Cespi, M.; Bonacucina, G.; Misici-Falzi, M.; Golzi, R.; Boltri, L.; Palmieri, G.F. Stress relaxation test for the characterization of the viscoelasticity of pellets. *Eur. J. Pharm. Biopharm.* **2007**, *67*, 476–484. [CrossRef] [PubMed]
30. Texture Analysis. Texture: Measure and Analyze Properties. Disintegration. Available online: https://www.stablemicrosystems.com/MeasureDisintegration.html (accessed on 15 August 2019).
31. AOSA. Rules for testing seeds. Principles and procedures. *Assoc. Off. Seed Anal.* **2014**, *1*, 6–25.

32. Egli, D.B.; Hamman, B.; Rucker, M. Seed vigor and uniformity of seedling emergence in soybean. *Seed Technol.* **2010**, *32*, 87–95.
33. Coolbear, P.; Francis, A.; Grierson, D. The effect of low temperature pre-sowing treatment on the germination performance and membrane integrity of artificially aged tomato seeds. *J. Exp. Bot.* **1984**, *35*, 1609–1617. [CrossRef]
34. Abdul-Baki, A.A.; Anderson, J.D. Vigor determination in soybean seed by multiple criteria. *Crop Sci.* **1973**, *13*, 630–633. [CrossRef]
35. Minitab (Version Express for PC). 2019. Available online: http://www.minitab.com/en-us/products/express (accessed on 28 August 2019).
36. American Seed Trade Association (ASTA). The Guide to Seed Treatment Stewardship. For Applicators. Available online: www.seed-treatment-guide.com (accessed on 5 December 2019).
37. Accinelli, C.; Abbas, H.K.; Little, N.S.; Kotowicz, J.K.; Mencarelli, M.; Shier, W.T. A liquid bioplastic formulation for film coating of agronomic seeds. *Crop Prot.* **2016**, *89*, 123–128. [CrossRef]
38. Foqué, D.; Devarrewaere, W.; Verboven, P.; Nuyttens, D. Physical and chemical characteristics of abraded seed coating particles. *Asp. Appl. Biol.* **2014**, *122*, 85–94.
39. Karlsons, A.; Osvalde, A.; Una Andersone-Ozola, U.; Ievinsh, G. Vermicompost from municipal sewage sludge affects growth and mineral nutrition of winter rye (*Secale cereale*) plants. *J. Plant Nutr.* **2015**, *39*, 765–780. [CrossRef]
40. Tognetti, C.; Laos, F.; Mazzarino, M.J.; Hernández, M.T. Composting vs. vermicomposting: A comparison of end product quality. *Compost Sci. Util.* **2005**, *13*, 6–13. [CrossRef]
41. Alwaneen, W.S. Effect of cow manure vermicompost on some growth parameters of alfalfa and *Vinca rosa* plants. *Asian J. Plant Sci.* **2016**, *15*, 81–85. [CrossRef]
42. Fleming, T.R.; Fleming, C.C.; Levy, C.C.B.; Carlos, R.; Franck, H.; Nolasco, J.B.; Liu, F.Q. Biostimulants enhance growth and drought tolerance in Arabidopsis Thaliana and exhibit chemical priming action. *Ann. Appl. Biol.* **2019**, *174*, 153–165. [CrossRef]
43. Rouphael, Y.; Cardarelli, M.; Bonini, P.; Colla, G. Synergistic action of a microbial-based biostimulant and a plant derived-protein hydrolysate enhances lettuce tolerance to alkalinity and salinity. *Front. Plant Sci.* **2017**, *8*, 131. [CrossRef]
44. Van Oosten, M.J.; Pepe, O.; De Pascale, S.; Silletti, S.; Maggio, A. The role of biostimulants and bioeffectors as alleviators of abiotic stress in crop plants. *Chem. Biol. Technol. Agric.* **2017**, *4*, 5. [CrossRef]

Permissions

All chapters in this book were first published in MDPI; hereby published with permission under the Creative Commons Attribution License or equivalent. Every chapter published in this book has been scrutinized by our experts. Their significance has been extensively debated. The topics covered herein carry significant findings which will fuel the growth of the discipline. They may even be implemented as practical applications or may be referred to as a beginning point for another development.

The contributors of this book come from diverse backgrounds, making this book a truly international effort. This book will bring forth new frontiers with its revolutionizing research information and detailed analysis of the nascent developments around the world.

We would like to thank all the contributing authors for lending their expertise to make the book truly unique. They have played a crucial role in the development of this book. Without their invaluable contributions this book wouldn't have been possible. They have made vital efforts to compile up to date information on the varied aspects of this subject to make this book a valuable addition to the collection of many professionals and students.

This book was conceptualized with the vision of imparting up-to-date information and advanced data in this field. To ensure the same, a matchless editorial board was set up. Every individual on the board went through rigorous rounds of assessment to prove their worth. After which they invested a large part of their time researching and compiling the most relevant data for our readers.

The editorial board has been involved in producing this book since its inception. They have spent rigorous hours researching and exploring the diverse topics which have resulted in the successful publishing of this book. They have passed on their knowledge of decades through this book. To expedite this challenging task, the publisher supported the team at every step. A small team of assistant editors was also appointed to further simplify the editing procedure and attain best results for the readers.

Apart from the editorial board, the designing team has also invested a significant amount of their time in understanding the subject and creating the most relevant covers. They scrutinized every image to scout for the most suitable representation of the subject and create an appropriate cover for the book.

The publishing team has been an ardent support to the editorial, designing and production team. Their endless efforts to recruit the best for this project, has resulted in the accomplishment of this book. They are a veteran in the field of academics and their pool of knowledge is as vast as their experience in printing. Their expertise and guidance has proved useful at every step. Their uncompromising quality standards have made this book an exceptional effort. Their encouragement from time to time has been an inspiration for everyone.

The publisher and the editorial board hope that this book will prove to be a valuable piece of knowledge for researchers, students, practitioners and scholars across the globe.

List of Contributors

Giuseppe Mannino and Cinzia M. Bertea
Department of Life Sciences and Systems Biology, Innovation Centre, Plant Physiology Unit, University of Turin, 10135 Turin, Italy

Ivano Viglianten and Valeria Contartese
Green Has Italia S.p.A, 12043 Canale (CN), Italy

Cristina Campobenedetto
Department of Life Sciences and Systems Biology, Innovation Centre, Plant Physiology Unit, University of Turin, 10135 Turin, Italy
Green Has Italia S.p.A, 12043 Canale (CN), Italy

Carla Gentile
Department of Biological, Chemical and Pharmaceutical Sciences and Technologies (STEBICEF), University of Palermo, 90128 Palermo, Italy

Shumaila Khan, Hongjun Yu, Qiang Li, Basheer Noman Sallam, Heng Wang, Peng Liu and Weijie Jiang
Institute of Vegetables and Flowers, Chinese Academy of Agricultural Sciences (CAAS), No.12 Zhongguancun South Street, Haidian District, Beijing 100081, China

Yinan Gao
College of Horticulture and Landscape, Tianjin Agricultural University, Tianjin 300000, China

Carlo Andreotti
Faculty of Science and Technology, Free University of Bozen-Bolzano, Piazza Università, 5-39100 Bozen-Bolzano, Italy

Markus Kelderer, Claudio Casera, Michele Bassi, Peter Robatscher and Aldo Matteazzi
Laimburg Research Centre, Laimburg 6, 39051 Pfatten-Vadena, Italy

Sebastian Soppelsa
Faculty of Science and Technology, Free University of Bozen-Bolzano, Piazza Università, 5-39100 Bozen-Bolzano, Italy
Laimburg Research Centre, Laimburg 6, 39051 Pfatten-Vadena, Italy

Hosam O. Elansary
Plant Production Department, College of Food and Agriculture Sciences, King Saud University, Riyadh 11451, Saudi Arabia
Floriculture, Ornamental Horticulture, and Garden Design Department, Faculty of Agriculture (El-Shatby), Alexandria University, Alexandria 21545, Egypt

Eman A. Mahmoud
Department of Food Industries, Damietta University, Damietta 3417, Egypt

Diaa O. El-Ansary
Precision Agriculture Laboratory, Department of Pomology, Faculty of Agriculture (El-Shatby), Alexandria University, Alexandria 21545, Egypt

Mohamed A. Mattar
Department of Agricultural Engineering, College of Food and Agriculture Sciences, King Saud University, Riyadh 11451, Saudi Arabia
Agricultural Engineering Research Institute (AEnRI), Agricultural Research Centre, Giza 12618, Egypt

Anna Kocira, Joanna Lamorska, Rafał Kornas, Natalia Nowosad and Marzena Tomaszewska
Institute of Agricultural Sciences, State School of Higher Education in Chełm, Pocztowa 54, 22-100 Chełm, Poland

Danuta Leszczyńska
Institute of Soil Science and Plant Cultivation, State Research Institute, Czartoryskich 8, 24-100 Puławy, Poland

Katarzyna Kozłowicz
Department of Biological Bases of Food and Feed Technologies, University of Life Sciences in Lublin, Głęboka 28, 20-612 Lublin, Poland

Sylwester Tabor
Department of Production Engineering, Logistics and Applied Computer Science, Agricultural University Kraków, Balicka 116B, 30-149 Kraków, Poland

Eugenio Cozzolino
Council for Agricultural Research and Economics (CREA)—Research Center for Cereal and Industrial Crops, 81100 Caserta, Italy

Maria Giordano, Nunzio Fiorentino, Christophe El-Nakhel, Antonio Pannico, Ida Di Mola, Mauro Mori and Youssef Rouphael
Department of Agricultural Sciences, University of Naples Federico II, 80055 Portici, Italy

Marios C. Kyriacou
Department of Vegetable Crops, Agricultural Research Institute, 1516 Nicosia, Cyprus

Giuseppe Colla
Department of Agriculture and Forest Sciences, University of Tuscia, 01100 Viterbo, Italy

Matteo Caser, Sonia Demasi, Dario Donno and Valentina Scariot
Department of Agricultural, Forest and Food Sciences, University of Torino, Largo Paolo Braccini 2, 10095 Grugliasco, Italy

Andrea Berruti, Erica Lumini and Valeria Bianciotto
Institute for Sustainable Plant Protection, National Research Council (CNR), Viale Mattioli 25, 10125 Torino, Italy

Íris Marisa Maxaieie Victorino
Institute for Sustainable Plant Protection, National Research Council (CNR), Viale Mattioli 25, 10125 Torino, Italy
Department of Life Sciences and Systems Biology, University of Turin, Viale Mattioli 25, 10125 Torino, Italy
Biological Science Department, Science Faculty, Eduardo Mondlane University (UEM), Av. Julius Nyerere-Campus Universitário, Maputo 3453, Mozambique

Laura Barsanti and Paolo Gualtieri
Istituto di Biofisica, Consiglio Nazionale delle Ricerche (CNR), Via Moruzzi 1, 56124 Pisa, Italy

Primo Coltelli
Istituto Scienza e Tecnologie dell'Informazione, CNR, Via Moruzzi 1, 56124 Pisa, Italy

Marek Gugała and Krystyna Zarzecka
Faculty of Agrobioengineering and Animal Husbandry, University of Natural Sciences and Humanities in Siedlce, ul. Prusa 14, 08-110 Siedlce, Poland

Anna Sikorska
Department of Agriculture, Vocational State School of Ignacy Mo´scicki in Ciechanów, ul. Narutowicza 9, 06-400 Ciechanów, Poland

Pavol Findura
Department of Biosystems Engineering; Slovak University of Agriculture Nitra, Tr. A. Hlinku 2, 949-76 Nitra, Slovak Republic
Faculty of Production and Power Engineering; University of Agriculture in Krakow, ul. Balicka 116B, 30-149 Krakow, Poland

Urszula Malaga-Toboła
Faculty of Production and Power Engineering; University of Agriculture in Krakow, ul. Balicka 116B, 30-149 Krakow, Poland

Silvana Francesca, Amalia Barone and Maria Manuela Rigano
Department of Agricultural Sciences, University of Naples Federico II, Portici, 80055 Naples, Italy

Carmen Arena
Department of Biology, University of Naples "Federico II", via Cinthia 4, 80126 Napoli, Italy

Bruno Hay Mele
Stazione Zoologica Anton Dohrn, Villa Comunale, 80121 Napoli, Italy

Carlo Schettini
ALMA SEGES Soc. Coop., Eboli, 84025 Salerno, Italy

Patrizia Ambrosino
Agriges S.r.l., San Salvatore Telesino, 82035 Benevento, Italy

Lucia Ottaiano
Department of Agricultural Sciences, University of Naples Federico II, Portici 80055, Italy

Sławomir Kocira
Department of Machinery Exploitation and Management of Production Processes, University of Life Sciences in Lublin, Akademicka 13, 20-950 Lublin, Poland

Agnieszka Szparaga and Tomasz Piskier
Department of Agrobiotechnology, Koszalin University of Technology, Racławicka 15–17, 75-620 Koszalin, Poland

Maciej Kuboń
Institute of Agricultural Engineering and Informatics, Agricultural University Krakow, Balicka 116B, 30-149 Krakow, Poland

Ewa Czerwińska
Department of Biomedical Engineering, Koszalin University of Technology, Śniadeckich 2, 75-453 Koszalin, Poland

Nóra Bákonyi, Döme Barna, Ibolya O. Tóth, Szilvia Kovács and Miklós Gábor Fári
Department of Agricultural Botany, Plant Physiology and Biotechnology (MEK), Debrecen University, Böszörményi Street 138, 4032 Debrecen, Hungary

Szilvia Kisvarga
Research Institute for Fruitgrowing and Ornamentals, National Agricultural Research and Innovation Center (NARIC) Szent-Györgyi Street 4., 2100 Gödöllő, Hungary

Hassan El-Ramady and Tarek Alshaal
Department of Agricultural Botany, Plant Physiology and Biotechnology (MEK), Debrecen University, Böszörményi Street 138, 4032 Debrecen, Hungary
Soil and Water Department, Faculty of Agriculture, Kafrelsheikh University, Kafr El-Sheikh 33516, Egypt

Neama Abdalla
Department of Agricultural Botany, Plant Physiology and Biotechnology (MEK), Debrecen University, Böszörményi Street 138, 4032 Debrecen, Hungary
Plant Biotechnology Department, Genetic Engineering Division, National Research Center, 33-El-Behouth Street, Cairo 12622, Egypt

Margaréta Rozbach and Csaba Fehér
Department of Applied Biotechnology and Food Science (BME-ABÉT), Szent Gellért Square 4, 1111 Budapest, Hungary

Nevien Elhawat
Department of Agricultural Botany, Plant Physiology and Biotechnology (MEK), Debrecen University, Böszörményi Street 138, 4032 Debrecen, Hungary
Department of Biological and Environmental Sciences, Faculty of Home Economic, Al-Azhar University, Sperpay Number 4 Rd, Tanta, 31732, Egypt

Masoume Amirkhani, Hilary Mayton and Alan G. Taylor
Cornell AgriTech, Horticulture Section, Cornell University, SIPS, Geneva, New York, NY 14850, USA

Zhi Chen
College of Mechanical and Electrical Engineering, Inner Mongolia Agricultural University, Hohhot, Inner Mongolia 010018, China

Yi Qiu
Cornell AgriTech, Horticulture Section, Cornell University, SIPS, Geneva, New York, NY 14850, USA
College of Mechanical and Electrical Engineering, Inner Mongolia Agricultural University, Hohhot, Inner Mongolia 010018, China

Index

www.ingramcontent.com/pod-product-compliance
Lightning Source LLC
LaVergne TN
LVHW082134200926
848484LV00008B/2052